Studies in Computational Intelligence

Volume 954

Series Editor

Janusz Kacprzyk, Polish Academy of Sciences, Warsaw, Poland

The series "Studies in Computational Intelligence" (SCI) publishes new developments and advances in the various areas of computational intelligence—quickly and with a high quality. The intent is to cover the theory, applications, and design methods of computational intelligence, as embedded in the fields of engineering, computer science, physics and life sciences, as well as the methodologies behind them. The series contains monographs, lecture notes and edited volumes in computational intelligence spanning the areas of neural networks, connectionist systems, genetic algorithms, evolutionary computation, artificial intelligence, cellular automata, self-organizing systems, soft computing, fuzzy systems, and hybrid intelligent systems. Of particular value to both the contributors and the readership are the short publication timeframe and the world-wide distribution, which enable both wide and rapid dissemination of research output.

Indexed by SCOPUS, DBLP, WTI Frankfurt eG, zbMATH, SCImago.

All books published in the series are submitted for consideration in Web of Science.

More information about this series at http://www.springer.com/series/7092

Siddharth Swarup Rautaray · Phani Pemmaraju ·
Hrushikesha Mohanty
Editors

Trends of Data Science and Applications

Theory and Practices

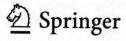
Springer

Editors
Siddharth Swarup Rautaray ⓘ
School of Computer Engineering
Kalinga Institute of Industrial Technology
Deemed to be University
Bhubaneswar, Odisha, India

Phani Pemmaraju
Global Delivery Services
Infor India
Hyderabad, Telangana, India

Hrushikesha Mohanty
School of Computer Engineering
Kalinga Institute of Industrial Technology
Deemed to be University
Bhubaneswar, Odisha, India

School of Computer and Information
Sciences
University of Hyderabad
Hyderabad, Telangana, India

ISSN 1860-949X   ISSN 1860-9503   (electronic)
Studies in Computational Intelligence
ISBN 978-981-33-6817-0   ISBN 978-981-33-6815-6   (eBook)
https://doi.org/10.1007/978-981-33-6815-6

This Springer imprint is published by the registered company Springer Nature Singapore Pte Ltd.
The registered company address is: 152 Beach Road, #21-01/04 Gateway East, Singapore 189721,
Singapore

Preface

Data science is an interdisciplinary field that uses scientific methods, processes, algorithms, and systems to extract knowledge and insights from many structural and unstructured data. The goal of data science is to "turn data into data products." Data science has been established as an important emergent scientific field and paradigm driving research evolution in disciplines such as statistics, computing science and intelligence science, and practical transformation in domains such as science, engineering, the public sector, business, social science, and lifestyle. The rising importance of scientific data, both big and small, brings with it a wealth of challenges to combine structured but often siloed data with messy, incomplete, and unstructured data from text, audio, and visual content such as sensors and weblog data. New trends to extract, transport, pool, refine, store, analyze, and visualize data are needed to unleash their power while simultaneously making tools and workflows easy to use by the public at large.

Becoming a successful practitioner of data science is a real challenge. The knowledge base incorporates demanding topics from statistics, computer science, and mathematics. On top of that, domain-specific knowledge, if not critical, is very helpful. Along with domain-specific knowledge, the real-time applications of data science are necessary. This needs to be brought together as a coherent package of data science course. The book covers the theoretical and practical applications, which are being developed and implemented within the industry.

The book includes contributions ranging from theoretical and foundational research, platforms, methods, applications, and tools in the area of data science. The book will provide theory and practices in the area of data science that adds a social, geographical, and temporal dimension to data science research, as well as application-oriented chapters that prepare and use data in discovery research. Furthermore, it has real-world examples drawn from a broad range of application areas, including health care, bioinformatics, social analytics, and natural language processing. The text also covers, typically from an application-oriented perspective, advances in data science in areas such as big data collection, searching, analysis, and knowledge discovery. This book will be helpful for the students, practitioners, and researchers working in the area of data science and analytics.

Bhubaneswar, India   Siddharth Swarup Rautaray
Hyderabad, India   Phani Pemmaraju
Bhubaneswar/Hyderabad, India   Hrushikesha Mohanty

Acknowledgements

The genesis of the book is the 11th Industry Symposium 2021 on "Trends of Data Science and Applications: Theory and Practices," held as a satellite event of the 17th International Conference on Distributed Computing and Internet Technology (ICDCIT, Kalinga Institute of Industrial Technology (KIIT), Deemed to be University, Bhubaneswar, Odisha, India) during January 7–10, 2021. The full chapters have been submitted by the invited speakers and the authors of the symposium, which have been peer-reviewed to be included in the book. Based on the reviewer comments, the authors have made necessary changes in the chapter. We extend our sincere thanks to all these invited speakers and authors. We also extend our sincere thanks to Springer for support in publishing this book.

We are thankful to the ICDCIT team for their confidence in us in shaping up this industry symposium. Shri Darpendra Narayan Dwivedy has been a great mover to the ICDCIT series since its beginning. We are grateful to the founder of KIIT and KISS, Prof. Achuyta Samanta, the patron of the ICDCIT series from its inception, for his endless support toward ICDCIT and its satellite events.

Siddharth Swarup Rautaray
Phani Pemmaraju
Hrushikesha Mohanty

About This Book

The book includes extended version of selected papers presented at 11th Industry Symposium 2021 held during January 7–10, 2021. The book covers contributions ranging from theoretical and foundation research, platforms, methods, applications, and tools in all areas. It provides theory and practices in the area of data science, which add a social, geographical, and temporal dimension to data science research. It also includes application-oriented papers that prepare and use data in discovery research. This volume contains chapters from academia as well as practitioners on big data technologies, artificial intelligence, machine learning, deep learning, data representation and visualization, business analytics, healthcare analytics, bioinformatics, etc. This book will be helpful for the students, practitioners, researchers as well as industry professionals.

Contents

About the Editors

Siddharth Swarup Rautaray is an Associate Professor at the School of Computer Engineering, KIIT, deemed to be University, Bhubaneswar, India. He received his Ph.D. in Information Technology from the Indian Institute of Information Technology, Allahabad, U.P., India. He has over 9 years of teaching and research experience. His research interest areas are big data analytics, image processing, intelligent systems, human-computer interaction, and similar innovative areas. His research contribution includes 05 co-edited proceedings/books, over 75 research publications in reputed conferences, book chapters, and journals. As an organizing chair, he has organized 05 International conferences and has been part of different core committees of other conferences and workshops. He has delivered invited talks in different workshops and conferences.

Phani Pemmaraju is the Director of Consulting Services at Infor India. He received his MCA degree from the School of Computer and Information Sciences, University of Hyderabad, Hyderabad. He has over 22 years of experience in the field of software development. He has got certifications in APICS—CPIM Certified in basics of supply chain management and Microsoft certified professional. He has been a member of advisory board, e-governance implementation at University of Hyderabad. He has published several research articles and delivered invited lectures in different institutes.

Hrushikesha Mohanty is currently a Vice Chancellor of KIIT, deemed to be University, Bhubaneswar, India. Prior to this appointment, Dr. Mohanty was associated as a Professor at the School of Computer and Information Sciences, University of Hyderabad, Hyderabad. He received his Ph.D. from IIT Kharagpur. His research interests include Distributed Computing, Software Engineering and Computational Social Science. Before joining University of Hyderabad, he worked at Electronics Corporation of India Limited for developing strategic real-time systems. Other than computer science research publications, he has penned three anthologies of Odia poems and several Odia short stories.

NLP for Sentiment Computation

Hrushikesha Mohanty

Abstract Sentiment is natural to human beings. Sentiments are expressed in different forms ranging from written, spoken to exhibiting. Our ancient scriptures record classifications of sentiments and propose Rasa Theory as the earliest study on human mind and its expressions. At the advent of digital era, sentiments are being poured into social media. The study on computational linguistics has become encouraging in developing computational models for automatic detection of sentiments in a given text be it a document or a social media posting though both the varieties project different class of problems for finding sentiments with them. This article discusses on some linguistic approaches known in the paradigm of natural language processing and their uses in sentiment detection. The approaches discussed are lexical analysis, corpora based, aspect based, social semantics and the trends of research in this field. It also discusses on exploring sentiments in a text of multiple domains.

1 Introduction

The *Theory of Rasa* as proposed by Bharat Muni is the ancient treatise that lists the sentiments usually invoked by actors while enacting a play. In *Natya Sastra* as the saint says, a play is only complete when it has *naab rasha* i.e. the nine types of sentiments viz. Hashya (laugh), Raudra (fury), Veera (fighting),Srungar (love), Shantam (peace), bibhatsa (disgust), Karunya (kindness), Bhaya (frightening) and abdhutam (wow, amazing). The work of the saint is dated somewhere in 500 BCE to 500 CE. Indian classical dance forms are based upon this theory of Bharat Muni. Further, the works of the great poet Kalidasa pens poetic treatises like Meghaduta, Kumarasambhab and Raghubansha in Sanskrit narrating the sentiments of the characters in his works.

H. Mohanty (✉)
School of Computer & Information Sciences, University of Hyderabad, Hyderabad, India
e-mail: hmcs@uohyd.ac.in; h.mohanty@kiit.ac.in

KIIT Deemed University, Bhubaneswar, India

© The Author(s), under exclusive license to Springer Nature Singapore Pte Ltd. 2021
S. Rautaray et al. (eds.), *Trends of Data Science and Applications*,
Studies in Computational Intelligence 954,
https://doi.org/10.1007/978-981-33-6815-6_1

Modern times linguists extend the classic idea of *rasa* in dance to languages and view those words that express cognitive states of speakers as emotive or sentiments words. Linguists label a word indicating the sentiment(s) the word conveys. Psychologists go further in studying emotions to define its affective sentiments. But, the sociologists claim that sentiments are not only affective due to cognitive state of a person but also depends on its socio-cultural ambience. While identifying sentiment characteristics, the representation of emotions are researched to visualise the semantic continuity of emotions of a person. Psychologists no more view emotions as discreet events. While emotion is viewed as a point in a dimensional space that models human cognitive space, then there can be plausible paths in the defined space so that dynamic change of emotions can be explained. The first such dimensional approach is presented in [6]. It labels one hundred plus emotions in a space with dimensions showing arousal and valence. Another such two dimensional space to represent emotions is reported in [7]. Latest, it is shown that emotion can be effectively expressed in four dimensional space viz. valence, potency, arousal and unpredictability [8]. Thus research on emotions has been from different domains including psychologists, sociologists and linguists. In recent times, computational linguists and computing scientists have taken up the lead to understand the sentiments and emotions expressed in digital contents like text, image and video etc.

On receiving a message, one reacts either as an active or a passive responder. A passive responder i.e. ignoring a message to react, conveys its unacceptability to the message. Whereas an active responder may react to the message by conveying its understanding of the message. This understanding includes recognising the issues in a communication. Other than this, there is a cognitive expression of an active responder. An active responder, conveys an emotion like happy, jubilant, unhappy, perturbed, confused and etc. aroused due to the message. Thus, a communication makes an affect at its receiver. While the research in Natural Language Processing has been engaged in understanding a text by syntactic and semantic analysis; the recent trend is to understand the emotions conveyed in a message. Because, emotions are the effects, messages make at the receivers. Now, in the renewed quest for artificial intelligence, there is also an interest in study on emotion as one can't deny its role in making of intelligence. Because, emotions resulting to sentiments in a person, play decisive role in decision makings. This is giving rise to a new discipline of study called *Affective Computing* [1]. As sentiment analysis is foundational base of this new computing paradigm and it has many useful applications in different domains ranging from governance to business, the study on sentiment analysis has assumed immense interest among researchers.

Sentiment analysis, at early stages of research, has been with text understanding and summarisation resulting to categorisation of a text as either positive or negative in its sense. Thus the sentiment here is bipolar. Further, based on some measures like the numbers of positive and negative words present in a text, order of a bipolar sentiment can be ascertained. But, some more researchers, use multiple labels to express sentiments; that is the extension of bipolar sentiment label scheme to multipolar one. So, there are many emoticons for communication on cyber spaces. The research interest is in auto classification of texts to the sentiment labels. This has been one of the research interests for the researchers engaged in sentiment analysis. Further for growing importance in social media mining, the sentiment analysis of the posts in Twitter, facebook, blogs and other online platforms is a recent research trend. The techniques used for sentiment analysis can be categorised to statistical approach [2], knowledge-based approach [4] and hybridised approach [5] that combines both statistical and knowledge based approaches. Statistical techniques in analysing texts for sentiments are void of consideration of text semantics. This gives a blunt view on emotions, at times out of context. Again, these techniques are sensitive to the quality as well as bulk of data collected. Some researchers report the use of soft computing techniques for labelling a text. This classification also depends on quality of samples collected for supervised classification. An unsupervised technique for the purpose also depends on quality and amount of data collected. On the other hand, knowledge based approach considers semantic of the text in classifying and labelling a text. Being associated with semantics,sentiments assigned by these techniques remain contextual. But the difficulty with these techniques is with the coverage as well as quality of the knowledge collected for solving a problem. This chapter wishes to explore the roles of natural language processing techniques in sentiment analysis. It doesn't claim an experimental analysis of the techniques mentioned but plans to trace some of the landmark points in the course of the research in this field. In the next section, a general concept on natural language processing in sentiment analysis has been explained. The research works on three dimensions viz. lexical, corpora and aspect are introduced in the following three sections in sequence. Each section presents some of the important works carried out following a research dimension. The chapter ends with a brief concluding remark.

2 Natural Language and Sentiments

Semantics indeed flow from language. Computational linguistics have shown interest in developing techniques to understand sentiments of a text. Basically, they study the polarity of a text in its sense either positive or negative. The positive polarity of a text is computed from the count of positive words the text has. For example let's take the sentences:

The lockdown announced by the government was appreciated.
The lockdown brought devastating effect on the national economy.

The first sentence is positive for having the word *appreciated* that gives a positive connotation to the sentence. For the second sentence, the word *devastating* presents a negative mark. A text is positive when its count of positive words outnumbers that of negative words. In order to characterise the sentiment of a word, conventional natural language processing techniques follow the ideas of linguistics. Some of the techniques include corpus based technique, lexical analysis, statistical measures of word associations and semantic analysis. All these methods of course use a prior defined polarity of a word. Lexical analysis technique is used to identify the words of interest (adjectives and adverbs) in a text. For sentiment detection, usually adjectives in a text are traced. And the Corpus based techniques use constraints on co-occurrence of words with the similar or dissimilar polarity. Statistical measures on co-occurrence of words with prior priority are computed to ascertain the distribution of such words and their formulations based on which the sentiment of a text is derived.

The above methods assume, words are assigned with prior priority. Some researchers ascertain prior priority of a word by referring to glosses available with WordNet. Computational linguists look for strategies to label a word with a sentiment choosing it from a given set of sentiments. One of such work is reported in [3]. The work reports a novel technique for sentiment recognition and polarity detection of a given text. First the author lists twenty four types of emotions under four categories viz. pleasantness, aptitude, attention and sensitivity. And each category has six types of emotions. The observations from fMRI experiments show that different parts of a brain gets activated at different types of emotions. The transitions of intensities of brain activations is modelled as an invertible bell function that shows positive and negative emotions along the range of values the function assumes i.e. $(-1, 1)$. The spatial arrangements of twenty four emotions is seen as an hour glass so the method is named as Hour Glass of Emotions. The paper while identifying twenty four types of emotions presents a scale for assigning priority to each. This idea can be translated to assign polarity to the words of emotions. A scale can be designed to map sentiments to polarity numbers from the range $(-1, 1)$. Then determining polarity of a text is a problem of summarisation. However, identifying a sentiment of a word with context reference is a linguistic problem because identifying sentiment of a word along with its neighbouring words becomes illusive. For example:

He was not a great runner for the day.
Anita ruthlessly smashed all the records.

In the first case the sentiment of the sentence is negative, though the word *great* is positive. But, the second sentence has positive priority though it has the word *ruthless* of negative connotation. Thus, understanding sentiment has to take context into account as the work [2] proposes. In order to find contextual polarity of a word, first the words are annotated as positive or negative polarity and in the second step disambiguity resolution is carried out. A disambiguity in word polarity occurs when the word is situated in the midst of conflicting polarities. Annotation of words with respect to its context is carried out by using a corpus that is developed by multi-perspective question answering scheme applied on a given set of demonstrative sentences. Then the second step is executed to disambiguate conflicting contextual priorities in a sentence. For the purpose machine learning techniques are applied. Features like positive, negative and neutral words, sentences and documents are considered by learning algorithms in designing of classifiers that learn from training data. The four types of learning algorithms viz. boosting, memory-based learning, rule learning, and support vector learning, are used for the recognition of contextual polarity. Finding sentiments of a document uses a lexicon-based model or a corpus-based model. A lexicon based model is primarily an unsupervised one. It uses an available dictionary with sentiment words. But, the technique is limited by the size of the dictionary. It is found at times comparatively inefficient than statistical methods for processing large number of words. But, corpus-based approach follows a supervised technique to train SVM like classifiers. Thus, again, its performance is dependent on training data sizes. Considering the drawbacks as well as strength of both the methods, [9] proposes a hybrid technique that uses both the models in designing of sentiment classifiers. In a given text, some words are tagged with their sentiments using a dictionary of words with annotated sentiments discovered manually. At the first stage, lexicon analysis of a text is performed. It divides a sentence into different zones based on structural analysis of a statement. Then for each zone, sentiment words are recognised and their impacts on zones are computed; this computation returns sentiment of a sentence by aggregating impacts of words in zones. The process carried on for different texts so the dictionary grows. From a given text, there is a set with sentiment classified words and there is another set with unclassified words. Now the second step of the proposed method starts taking the two sets of words obtained from the first stage as inputs. The features of the classified words are found. Similarly features of the words in unclassified set are found. Now classifiers like SVM are used for learning so unclassified words are subjected to classification. The authors by experiment have found that hybrid technique performs better because both the techniques individually are biased. While lexicon based is positive priority biased, corpus based model is better to identify negative priority. With experimentation results the authors vouch for a hybrid of both the techniques for classifications of sentiments in text documents.

Another work uses NLP techniques for sentiment findings in microblogs. The authors in [10] reports an algorithm that performs two functionalities viz. realtime sentiment detection and microblog user sentiment profiling. The techniques applied for sentiment detection on microblogs, are classified into two categories viz. subject dependent and subject independent. For the prior case, hashtags indicate the domain of discussion so the related words are searched to identify priority of words. Whereas the later is domain independent. Sentiment of a microblog is determined by processing reactions of people to a microblog. While the previous case analyses the contents the later analyses the reactions to a content. Analysing sentiments in reactions to a microblog provides a sentiment profile on an issue. Similarly sentiment profile of a user can be carried out. The paper follows a general approach in computing priority of a microblog sentence. The technique it follows for sentence priority calculation has three steps. First, it finds segments i.e. sub sentences; and then words in a segment are annotated with the part of speech. In second step, the polarity of a segment is computed. For it, key words of the segment is matched with a dictionary and patterns of words are matched with repository of mined patterns. On matching, polarity of a segment is computed based on the polarity assigned to patterns. Then the third and the final step computes the sentence polarity as an aggregation of polarity of its segments.

$$\lambda(s_i) = \frac{1}{\min(1, N - i + 1)}, 1 \leq i \leq N \tag{1}$$

$$P(S) = \sum_{i}^{N} [\lambda(s_i) \times p(s_i)] \tag{2}$$

An idea of sentence segments and their contributions towards polarity of sentence is modelled in the formula above. Linguistics have observed that position of a segment in a sentence has its impacts in computation of a sentence sentiment. Usually, segments at the begin and towards end have more importance then the middle ones. Let's define variables used in the formula as: s_i is the ith segment and its polarity is $p(s_i)$ and $\lambda(s_i)$ is its positional value, N the number of segments and S is the sentence. The summation of products of segments' positional and polarity values results to $P(S)$ the polarity of a sentence S. For fastness of computation the authors used binary search and parallelize computation so the demand of online sentiment assessments of microblogs is met.

The works multilingual sentiment analysis is reviewed in [11]. The use of multilingual sentiment analysis is obvious as social media population is gradually turning to a heterogeneous society with people of diversified culture and languages. Multilingual sentiment analysis though largely follows three approaches viz. lexicon based, corpora based and hybrid models for sentiment analysis still needs special attention to deal with a challenge that is language specific. One approach is to translate statements from other languages to English and then use the readily available tools for the purpose. But, the approach is vulnerable to the loss of information caused by the translation. The process of translation including tokenisation and segment detection is tricky and so also is semantic interpretation. Then the second approach is to deal with each language separately. But, it requires a semantic mapping for words of one language to the other. The mapping at times could be ill defined. However, a good number of researchers engaged in multilingual sentiment analysis is developing multi-language corpora that help in summarisation of sentiments expressed in multiple languages. The paper gives a comparison of works in this area. The interested readers may refer to the paper. The variety of approaches the authors follow includes lexicon based, seed word selection, SentiWordNet, supervised machine learning and some heuristics. For semantic inadequacy in lexicon based strategies and knowledge incompleteness in corpora, researchers have seen efficacies in hybrid techniques in both monolingual and multilingual sentiment analysis. Having a good research progress in the field, users of various domains have shown interests in sentiment driven applications like marketing, pedagogy, counselling, community development and governance. Each kind of users wishes to analyse comments and reviews on a particular aspects of its interests. This has given rise to research on aspect based sentiment analysis.

Aspects are the dimensions on which people view in its world of communication. For example, on a product, customers view on its cost, functionality, maintenance support, durability and issues like these in interest of users. So, a review can be one of these aspects. In order to assess sentiments of a product, user sentiment on each aspect of a product is required to be assessed and then an aggregation of these sentiments is to be done to compute the product sentiment as a sum total of its aspect sentiments. The process of aspect based sentiment analysis is divided into aspect identification, classification and aggregation. In analysis of a review, the first work is to identify the words conveying aspects. This is called sentiment-target pairing. Next, the classification of sentiment-target pairs is done based on the polarity of a pair i.e. either positive or negative pair. However, for some target pairs there may be already defined priority based on the paired sentiment words. Finally, the sentiment values of each aspect is aggregated to provide a comprehensive view. The paper [13] gives a review of works carried out in this area. It provides a taxonomy in categorising techniques applied for aspect based sentiment analysis. Reported works are put into three divisions viz. aspect identification, sentiment detection and the third category includes the both. The algorithms used for aspect detection are further divided into frequency based, syntax based, supervised machine learning, unsupervised machine learning and hybrid approaches. While the sentiment analysis follows techniques those are dictionary referencing, supervised machine learning or unsupervised machine learning types. The third type is jointly finding aspects and sentiments following one of the techniques among syntax-based and supervised, unsupervised learning and hybrid machine learning.

On having an over all idea on natural language based sentiment analysis we will look into each approach to understand basic nuance of it.

3 Lexical Based

Lexical based approach deals with vocabulary used in a text for sentiment analysis. Ideally we would like to understand the subjects and the sentiments a given text has. In that sense, it's natural to look for the nouns and their adjectives. A noun,associated verb, adverb and associated words may also throw some light on sentiment(s) a sentence has. For example lets take sentences as:

- He was happy at the news.
- She sang not so sweet.

While the first sentence gives a positive polarity having the word 'happy', the second sentence have negative polarity for the phrase 'not so sweet'. In the later case though there exists a positive word 'sweet' still accompanying words 'not so' brings in just opposite. This shows, the first step is to identify words conveying sentiments in a text. Then the polarity due to these words are identified. Polarity of a text, is due to the aggregation of sentiments of opinion words found in a text. The aggregation indicates the number of positive and negative sentiments the text has. But, in case of microblogs, the sentiment analysis is done at sentence level. Usually, first microblog sentences like Tweets are put into a cleaning process that removes noises like unusual abbreviations. Then the part-of-speech (pos)of the words in the sentences are found. A sentence with pos information is put into sentiment analysis study. A sentiment analyser consults dictionary and SentiNet like repository to identify describing words and their sentiments. Though, in aggregation sentiment is bipolar i.e. positive and negative, but in reality there could be subdivisions to positive emotions viz. happy, praising, cheerful, smart and etc. In such case, a graded emotion quantification scheme can be worked out. For polarity aggregation, an average score of polarity of sentiment words is usually carried out.

The paper [14] presents a detail work on lexicon-based sentiment analysis emphasizing the need for semantic based approach for having some applications that need analysis at sentence level for not having corpora as such available. As the most of the corpora based system needs a large size of corpora , availability of such large amount of data becomes a limitation. The paper makes a distinction between semantic orientation and semantic analysis. Semantic orientation is about judging polarity of words in a sentence or a text. Whereas semantic analysis is about general methods to find subjectivity and its polarity in a text. The paper discusses semantic orientation that's appropriate for computing sentiment of a microblog. The calculation goes with metrics to quantify the impacts of features viz. i. *adjectives* ii. *noun,verb* and *adverb* iii. *intensification* iv. *negation* v. *irrealis moods* and vi. text level *repetitions* present in a sentence . For a given word and its semantic orientation feature, a measure of the sentiment the word may invoke at the reader is determined. For each feature, a scale has been defined considering two extreme emotions e.g. love and hate with scale (+5, -5) while in between eight more associated emotions can be placed. The authors have suggested for auto generation of dictionary by association of words to a given set of seed words. For the purpose a set of six features named text is self explanatory except few; those we will put here. The words like *slightly, somewhat, pretty, really, beautiful* and *excellent* are the intensifiers for they express a degree of a emotion on goodness. One can define a range of values associated to these words in a range (1, 6) where the intensifier 'slightly' assumes score (1) and (6) for 'excellent'. Similarly, negation in a sentence affects polarity of a statement. A negation to a positive feature may be assigned a negative of its positive value e.g. 'not excellent' may be assigned sentiment score as −6. However, such convention is open to criticism as 'not excellent' is open to interpretation. In a language there are some negators like *never, noway, none, nobody* and *nothing* can be considered negative sentiments. Verbs and prepositions like *lack* and *without* also convey negative semantics e.g. *lacks an order., it was without discipline.*

While considering lexicons contributing to sentiments, there are some lexicons which should be ignored as argued by the authors as these in reality are not contributing types. They have identified them of two types viz. *irrealis* and *repetitions*. In linguistic domain the former refers to the features those are irrelevant in the sense of their reliability e.g. *But for middle aged people, a visit to the place could have been enjoyable.* The sentiment expressed in this sentence is obfuscated. Similarly, a sentence with repeating emotions may not add to the sentiments e.g. *The place has nice roads and nice gardens.* This repetition of the word *nice* doesn't add to sentiment score.

Devising strategy for sentiment detection as seen in the paper [14], is language specific, and needs expertise of linguists. Having done with the identification of language specific sentiment orientation then sentiment score is calculated. The authors propose to compute semantic orientation scoring for a word. It refers to SentiNet dictionary for getting positive and negative score of a word with respect to a sense. It may be noted here that a word may have use in different senses and an order is defined over these senses. Semantic score computation may choose the positive and negative scores in the best sense which usually is the first in order. However, another approach could be the average of the scores due to all the senses. A score value assumes a value within (0, 1). The computation of semantic orientation is carried out as:

$$SO(w) = (Pos(w_f) - Neg(w_f))$$ (3)

i.e. $SO(w)$ semantic orientation of a word w is the difference between its the first sense positive and negative scores (w_f is the first sense semantic orientation of the word w). With respect to all the senses the semantic orientation for the word w is computed as:

$$SO(w) = \frac{1}{|\text{ senses }|} \sum_{x \in \text{senses}} (Pos(w_x) - Neg(w_x))$$ (4)

where $Pos(w_x)$ and $Neg(w_x)$ are respectively positive and negative scores of the word w with respect to a sense x where senses is a set indicating the different aspects of sentiments a word may invoke. This way the paper proposes a novel approach of semantic orientation for sentiment computation that uses sentiment associated with a word as well as of words in its neighbourhood. That way the semantic expressed with a word is considered for sentiment computation. In case of microblogs, because of limitation on text length as it is for Twitter, though the scope to compute semantics and sentiments is limited still some researchers vouch for NLP techniques for sentiment computation. Extending the role of language in assessing sentiments, some researchers have considered corpora that stores a genre of writings giving much scope for semantics in sentiment computation. The next section discusses on the corpora based sentiment computation.

4 Corpora Based

A corpora is a database of linguistic information of a genre that shows distinguishable features identifying a genre uniquely. A corpora can be used for understanding a text by matching words of the text to its genre and scooping out the meanings in the context of the text. Words exhibiting a unique aspect of a genre are identified and classified to different aspects. These words are weighted based on their degrees of participations in distinguishing a genre. Then, a group of words useful to describe an aspect is associated to a function as its parameters. These functions are considered as discriminating function used for classifications. This general concept of corpora based NLP is also used for sentiment analysis. With a given sentiment corpora, a text can be classified to a sentiment in applying corpora defined discriminating functions. Understanding the subjectivity of a text by using corpora is reported in [15]. Subjectivity of a text refers to aspects of a language used to express opinions. The factual information a text presents, give objectivity of the text. A language mainly has features to express positive, negative and speculative subjectivity. The words like *beautiful* and *terrible* reflect positive and negative subjectivity whereas a word like *possible* is speculative one. In order to identify such polarity in expression, corpora is used. Further, there could be an associated degree to a polarity e.g. 'pretty, attractive, beautiful' like words exhibit degrees of positive polarity in describing something beautiful. Corpora may generate a grading of such words associating each with a weight according to its grade. This process of grading can be automated by bootstraping with a given seeding positive word; at times the process could be manual tagging too. An experiment on corpora generation for sentiment analysis is reported in [16]. In two steps corpora has been created. First, a text is taken from customer opinions on different web services. Then opinion polarity is identified and annotated manually by a set of annotators. The project built corpora for detecting irony and sarcasms. Further [17] reports another experiment to find the words that express sentiments. They use two methods primarily based on LSA: Latent Semantic Analysis technique that's a statistical method for extracting a semantic space from a large corpora. This work tells, the corpora performs better when it remains domain specific. The two methods viz. SO-LSA (Semantic Orientation Latent Semantic Analysis) and DI-LSA (Dictionary based Latent Semantic Analysis). In SO method, affective aspect i.e. sentiment due to a word is estimated from its frequency of occurrences with other surrounding words. These words with a given seed word form a vector. This word vector is matched against given set of aspect vectors for similarity detection. For the purpose a cosine similarity of an aspect vector with word vectors having known aspects is computed. The degree of similarity provides the degree polarity to a known aspect. But, the technique may not be useful for aspect words not having patterns of frequent co-occurrences. In that case, lexicon based LSA method is used. The technique instead of looking for frequent word patterns, it finds the occurrence of similar lexicons in a dictionary. A dictionary is initiated with known seed words and its range of words from negative to positive polarities. Based on word matching in dictionary, polarity value of a word is ascertained. The paper in view that for specific domain with limited corpora size, the performance in sentiment analysis is appreciable. The observation though sounds trivial still the method proposed in the paper is useful for many business applications looking for automation of sentiment detection in customer comments.

Translation of a text from one language to another has enormous applications. Among the difficulties in language translation, maintaining emotion in one language to another adds to its complexity. The problem is addressed in [18] suggesting creation of parallel corpora that for a word projects word(s) in other languages, conveying the same sentiment so that consistency of an emotion in a text is maintained in the translated text. Translating to parallel languages need each language specific tools. The researchers have proposed a simple mechanism that only uses language specific NER: Named Entity Recognition software and emotion dictionary. The strategy for translation, does not follow method of summarisation because in that local sentiment may get lost. In translation one has to look for sentence level translation to maintain consistency of emotion in translation. For the purpose, the paper follows a window approach. In a sentence, emotion is viewed around a subject. Around a subject word a word-window of six words width, is created to find two levels of positive and negative words. It's found that this simple heuristic works well for text translations in seven European languages. Then emotion dictionary of the text language is referred to find the polarity of words in the window. Then, the sum total polarity of words in a window is computed. Further, the emotion intensifiers may add or reduce polarity based on their positive or negative connotations. This simple approach is appreciable for it is not computation heavy. Next we present a work following the trend.

The simultaneous usage of multi language corpora for sentiment analysis is particularly tried for sentiment analysis in tweets. In [19], researchers process tweets in multiple languages using multi-lingual Twitter corpora. These investigations are classified into two categories; one when tweet language is known and the other unknown. For unknown case, they follow multilingual and monolingual approaches to identify sentiment. In former case, they use a multilingual trained corpora to identify sentiment expressed in one among the trained languages. In later case, the processing is in two stages i.e. first language identification and then sentiment finding. The case is ofcourse easy when the language of the tweets is apriori known. The experiment the paper reports uses tweets in English and Spanish. Following a multilingual approach, they trained the system with English and Spanish sentiment words and mixed two training sets for uses in sentiment analysis without recognising the language of a given tweet. They introduced a concept called code-switching corpus with sentiment labels and have exhibited the robustness of the multilingual sentiment analysis. The process for multilingual method has tow phases viz. one does feature extraction and another on context detection. For, multilingual tweets processing [19], after tokenisation of a tweet,tokens are matched with multilingual corpora. A tweet can be of a language among the many a corpora has. Once a token is matched with a word in multilingual corpora then its language is known. After that parsing and other steps are carried with respect to the known language. This process, is called code-switching. The features it extracts are words, lemmas, parts of speech tags and psychometric properties. Words of interests are selected on the basis of their number of occurrences. Root words are identified in lemmatisation so the processing of different forms of the same word is avoided. Then the words with psychometric properties are identified. These words basically indicate the sentiments of a tweet. The part-of-speech tags for words are identified with respect to the grammar and the corpora of the language. Now for Indian languages there are many parsers, syntax analysers and corpora available. Using these available software the above steps can be carried out. Now from a tweet we get a set of words and with each word tagged with part-of-speech and psychometric properties i.e. $T = (t_1...t_i...t_n)$. Then the associations among words are explored. We may get $((t_j)_{j=1}^p, r, (t_k^q{}_{k=1}))$ where r

is the relation that binds tweet words $(t_j...t_k)$ telling about the sentiments the tweet expresses. Computational linguistics use *n-gram* technique to extract meaning of a sentence. The same technique can be followed to find sentiments in a sentence. An algorithmic description of the said method is given below.

Algorithm 1 Multilingual_Tweet_Sentiment(t_w: tweet, K: Corpora)

---

t_w: a tweet;
$T = Pick Freq Word(t_w)$
$T_{post} = Part Of SpeechT(T)$
$T_o = Get Origin Word(T_{post})$
$T_o^s = Chk Psycho Prop(T_o)$
$T_{ngf} = Find N gram Feature(T_o^s)$

The function PickFreqWord with a tweet t_w as input finds the words those are not frequent and tweet slang. On finding the set of words T of the tweet t_w, it recognises the language of the words on using a given multilingual corpora K. It is to be noted that, for processing a tweet must be in one of the language of the corpora K. Then in sequence the part of speech tagging of the words, lemmatisation and psychological aspects of the words are carried out respectively by the functions PartOfSpeechT(T), GetOriginWord(T_{post}); the function FindNgramFeature finds the dominant features that reflect some relationships of entities with words and their psychological features. The found dominating ones or identified with many n-gram feature(s) reflect sentiments and accompanying other less dominant sentiments present in a tweet. These can be corroborated on further analysis of the tweets emerging from the same tweet-handler. A person in consequence to an event, may exhibit several types of emotions that we call aspects of emotions. For example, the dominating features can be seen as different emotional aspects. In the next section we will discuss on some work on aspects of sentiments.

5   Aspect Based

An event can be viewed in more than one ways and so may induce different types of emotions. For each type there is a context associated. This is termed as aspect based emotion. For example, a pandemic like Covid-19 may induce several emotions like fear, concern, disgust and helplessness. Now while detecting sentiments the aspects of emotions are identified. A work [20] reports different emotions people at large express on suffering migraine. They have collected tweets (in English) for a time period and processed to find out different aspects of emotions people have expressed using their tweeter handlers. At the beginning cleaning of tweets is carried out by transforming tweet-centric words to formal English words.Tweet-centric words, we mean the words in distorted forms used in tweets. Then lemmatisation of words is carried out and the root words are identified. Aspect based sentiment detection follows qualitative approach while quantitative approach resorts to statistical methods in finding frequencies of occurrences of polarised words. The former approach being qualitative finds out different signatures of emotions by finding n-gram features in words of tweets. In a collection of tweets on migraine, the n-gram features (combination of tweet words) are recognised and then a feature is assigned to a sentiment category following Latent Dirichlet Allocation method for topic modelling. The researchers have found that their method is able to find out relations between different aspects like treatment and cure, suffering and profanity etc. While finding aspects of sentiments in general with the disease, they have also profiled sentiments of frequent tweeterites on determining their activeness on social media and performing tweeter level sentiment analysis.

Aspect based segment detection not only goes beyond the recognition of bipolarity sentiments i.e. positive and negative but also identifies sentiments people may have on different contexts associated with an entity. The problem of context identification turns difficult for many reasons like detection of isolated features and overlapping features. Overlapping features may mean the same or similar contexts. The approach to meet the challenge is by understanding word semantics instead of resorting only to statistical approach. Having words embedded with meanings, it is possible to get semantic associations the sentiment words make. Aspect category detection problem is a part of aspect based semantic detection. The problem is studied in [21]. The authors report a hybrid model that blends association rules with semantic rules where the prior is statistics based and the later follows semantic association. The usual approach is to identify aspect words and corresponding opinion words and then association mining is carried out following statistical approach. These established associations find rules among aspect words and views. This association rules further augmented by word embedded meanings. Thus a new set of class based semantic rules is generated and these rules are used to detect aspect categories. According to the researchers, the work is the first attempt to find semantic rules for aspect category detection.

While syntactic analysis is an established method for aspect based sentiment detection, it has limitation to analyse natural impulse based sentiments as such expression often doesn't follow a grammar in strict sense so there exists chances of missing some words of sentiments and aspects. A work reported in [22] proposes a new framework that's novel sentiment detection and claims as better than a syntax based approach for the purpose. The authors have analysed Chinese language for aspect based sentiment detection. Even though the proposed approach has some language specific processing still the approach is generic. The first step is to compress a sentence. This compression is meant to reduce a sentence simpler form for syntactic analysis. This transformation is done in two ways viz. extractive and abstractive. In case of the former, the words of aspects and verbs are extracted. Whereas in case of abstraction, an abstraction of a given sentence is carried out by generalisation of word semantics. These approaches work well when a given sentence is grammatically correct. But, for natural expression like *My God atleast camera works good.* is put for sentence compression to generate a simple sentence preserving its sentiment as *Camera works good.*. Again a simple transformation based on aspect-sentiment words may not be enough as seen in this case: *Great atleast camera works good*. This case just picking up aspect-sentiment word associations as *Camera works good* is not enough as it misses sentiment polarity words like *great* and *atleast*. In order to get over the challenge the researchers have propose Conditional Random Fields (**CRF**) model to label words and their associations with probability based on which the words are selected to form a compressed sentence for sentiment analysis. The work proposes a framework that has three steps viz. (1) Sentence Compression (2) Aspect based Sentiment Analysis (3) Aspect and Polarity word collocation extraction and then recognition of Aspect Polarity . The first step i.e. sentence compression, is important for not allowing the loss of sentiment as well as filtering out the words not required for the purpose. The compression process has two main steps viz. the first to pick up basic features that common sentence compression technique achieves. The features include words, Part of Speech (PoS) tags and semantic features of the words. The second step is the recognition of sentiment features in a sentence. Sentiment features include perception features e.g. *red* and polarity feature *deep* in the phrase *deep red*. And the third step is to collect potential semantic features. These are the prefix or suffix to a polarity words. The idea here is to capture some semantic features that are language specific. In some language, words prefix or suffix to a feature word bear some hidden sentiments. Further, it considers brown clustering of words that collects words that look different but have the same semantic. Similarly, it also considers soft clustering of words that collects words with the similar meanings to one class. This kind of classification helps to compress words without loss of meaning. In addition another two techniques viz. word embeddings and dependency search can be taken up in finding features. Word embeddings is another technique [23] to find similarity in words when each word is judged from different dimensions i.e. mapped to a multidimensional space. The closed colocating words in multidimensional space represent similarly among the words. Exploring dependency among features requires knowledge on syntactic relation that forbids deletion of a word as the words related to it may loose meanings without the word. Such dependencies are to be recognised by experience as studied in [24]. Having these feature words identified, then a structure, like parse tree is generated showing the dependency structure exists among the feature words. A generated structure is labelled by conditional probability using CRF Modelling. And then a bi-directional traversal of the labelled tree is performed to pick up aspect words and corresponding priority words. A sentiment word is taken as a root and then at both the sides aspect and priority words are picked up. These colocated words give an aspect-sentiment pair. Usually aspect based semantic search considers aspect semantic pairs of words. This work is special for considering priority of semantic so it is capable of identifying major and minor aspects of a sentiment. The to-and-fro traversal continues till the parse tree is traversed completely.

Aspect based sentiment detection looks at granular level exploring all the possible dimensions responsible for evoking sentiments in an observation. This technique is useful in studying market reaction of a product. Customers' comments can be analysed to find the aspects of a product they like and the aspect they don't. Having a good deal of business potential, there has been growing interest in sentiment study. It's true the sentiment is an imprint of mental status of a speaker. In the next section we will discuss a seminal work that explores the relation between the study of psychology and that of language in expressing sentiments. The next section first referring to early works with psychology perspective then charts the trends of research in this field.

6 Trends

A Communication carries cognitive status of a speaker and so reversely from communication it should be possible to identify speakers' cognitive status. Language is a medium of a communication. Language features present cognitive status of a speaker. The researchers of computational linguistics supported by computing technology are in search of algorithms so they can churn the messages and identify sentiments present in. In this context we find it's fitting to discuss an early seminal work reported in [25]. This paper provides a glimpse of several text analysis techniques and presents a method called Linguistic Inquiry and Word Count (LIWC) that detects cognitive meanings to words in experimental settings. Historically, psychologists, first have tried to understand human cognition from its communication. This follows the general understanding that the mind hidden is visible behind the words spoken. Probably, that's why news analysts on television panel discussions try to read aloud many things in the words spoken by spokespersons. Initially, the methods psychologists follow for mind study are indirect in nature. For example a technique is *stimulus-based*. A subject is put to an experiment setting e.g. answering to a set of questions on looking at a picture. It is not transparent to the person under observation for cognitive mapping. Again, the interpretation of an answer and mapping it to a cognitive state is a business of a psychologists. Later, the researchers got interest in content analysis which is a *direct* means not stimulus based. At the advent of the computing technology and computational linguistic study, the possibility of algorithm based generic text analysis became a reality [26]. The program *General Inquirer* the authors developed, was a complex one with algorithms to detect several psychological disorders through the analysis of word usages. However, the algorithms used the expertises of psychologists while manipulating word weights and deriving possibility of mental disorders based on word associations and their weights. Thus the analysis is hidden to the people whose text is being analysed. In a sense, the first generic transparent method for cognitive study through text analysis is reported in [27]. The researcher opines that the words used in day to day life are powerful enough to paint cognitive states of people. With this premise a computer software was developed and used for many applications. However, LIWC is a transparent and popular computational method that has looked into the same problem. This being a foundational work on language processing and sentiment recognition, the next para presents a brief on the method.

LIWC, the first computer based method for sentiment recognition has two components, one is the software that reads text word by word and the second is a dictionary. A read word is matched with the dictionary. On matching a word, it's required to measure the psychological impact the word makes. Psychometrics of words depend upon the usage of words. The paper identifies those usages and counts the number of such usages to provide an aggregate sentiments a text projects. Some of the aspects include: (1) the use of pronouns (2) the types of words associated with those pronouns. e.g. 'he is excellent' (3) association of verb with a pronoun e.g. 'I did this'—an assertive expression. (4) Social relationship e.g. 'under me'—a social hierarchy for dominating attitude (vertical relationship), 'you and me'—A cordial relationship (horizontal relationship) (5) deceptive identification e.g. 'heard from some one'— referring to an unidentified speaker to assert on an issue. In order to understand sentiment of a text LIWC proposes to keep count of such linguistic occurrences and assign their weights statistically according to the occurrence distribution. This early work is simple and seminal. On the advent of microblogging the focus of sentiment study has shifted to sensing sentiments in short texts like tweets, facebook postings, images and videos. Analysing short texts like tweets is challenging for its peculiar writing styles. Various attempts including NLP techniques are being applied for the purpose of sentiment detection. Till now we have discussed on lexical, corpora, aspect based techniques. In order to chart the *trend of research*, we present two issues in the following subsections.

6.1 Social Semantic

Not only the grammar of a language plays a role in analysing sentiments in a text but also the context of a communication and its participants have roles to play. For example, in the context of a common interest a person may have a certain view and his tweets reflect the view. As the birds of the same feather flock together, a person's social contacts may have the same view as the person has. These aspects in communication may add in devising strategy for sentiment analysis, particularly in case of tweets. Tweets are noisy and limited in size. These make tweet processing challenging. A work reported in [28] proposes a strategy that uses social as well as contextual relationships of postings in sentiment analysis. The factors considered include: *topic context*, *user context* and *friend context*. Topic context is a matrix that shows the topics a message has. Let $M \in R^{N \times N_T}$ be message-topic matrix where N and N_T are the number of messages and topics respectively and $M_{i,j} = 1$ means the the ith message has jth topic otherwise $M_{i,j} = 0$. Connection matrix T provides relations among microblog messages as $T = M \times M^T \in R^{N \times N}$ such that $T_{i,j} > 0$ the value indicates the number of topics the messages i and j share between. Next message-user matrix $P \in R^{N \times N_U}$ where N and N_U are the number of messages and users respectively. Now $P_{i,j} = 1$ means the message i has been posted by the user j otherwise 0. Then the message-user-topic relation is given by $U = (P \times P^T) \circ T \in R^{N \times N}$ where $\circ$ is Hadamard product as $(A \circ B)_{i,j} = A_{ij} \times B_{i,j}$; $(P \times P^T)$ matrix indicates two messages are posted by the same user. $U_{i,j} > 0$ means the message i and the message j are posted by the same user on the same topic, otherwise $U_{i,j} = 0$. Retweets and shares of microblog posts show a homophily tendency of people i.e. a kind of herd behaviour that usually seen in social media. Such similar postings not only shows topic-user context but also shows *Friend context* that gives an idea on polarity of a message. A connection matrix defined by friends, the authors define as $F = (P \times S \times P^T) \circ T \in R^{N \times N}$ where $S \in R^{N_U \times N_U}$, the matrix $(P \times S \times P^T)$ indicates whether two messages posted by two users have social relationship. $F_{i,j} > 0$ indicates the messages i and j are posted by two users with social relationships and the messages are of the same topics. Now for sentiment computation these two factors like talking on the same topics and talking by many on the same topic, contribute to content based sentiments.

$$s_{m_i} = Cs(m_i) + \alpha \sum_{i \neq j, j=1}^{N} U_{ij} \times (| (Cs(m_i) - Cs(m_j)) |)$$

$$+ \beta \sum_{i \neq j, j=1}^{N} F_{ij} \times (| (Cs(m_i) - Cs(m_j)) |)$$

where for the message m_i, $Cs(m_i)$ is the content based sentiment computed on the basis of language features like lexicons and aspects or by any other technique that considers message context but not the social context for sentiment computation. The above formula for sentiment computation has two additives viz. user and friend contexts. The values of the multiplicative factors α and β can be assigned considering the domain specific requirements. For example in some cases sentiments on social contexts may be treated negative e.g. panic sentiments during pandemic, so α and β may assume a value from a range of values $(-1, 1)$. However, the paper [28], instead of computing sentiment directly as it is shown above, has considered the problem as an optimisation problem so that contributions of social contexts towards sentiment computation could be minimal and normalised so these factors don't overwhelm the sentiment computation. For details the readers may refer to the paper [28].

6.2 Multi Domain

Another challenging area is the multi-domain sentiment assessment that is necessary when one needs to analyse opinions on objects of different domains also when opinions are expressed in more than one language. In such cases, a simple method may use labelled corpora of each domain and annotated mappings to corpora of other domains. But, this is an expensive approach for creating corpora of different domains. In general sentiment computation problem is considered as a binary classification problem. A classifier trained for a domain may not work well for other domains for their corpora mismatches. Thus the problem is challenging, at the same time attracts researchers for its practical implications. A work reported in [29] presents a method for multi-domain sentiment assessment. First, it creates a thesaurus considering texts of different domains. Each text has a topic vector. Then a distribution of topic vectors is generated. The topic vectors of the texts having known sentiments are labelled accordingly. This gives a sentiment sensitised thesaurus having word distributional similarity. This thesaurus is used as the base for further expansion during sentiment classification of new texts.

In plain speaking, the premise of the concept lies with co-occurrence of the words in a text, expressing the same and similar sentiments. Based on this premise a method is proposed that first builds a sentiment-sensitive thesaurus giving distribution of sentiment words in a text. Then, for a given lexical element or sentiment element u in a text, let there be a co-occurring lexical or sentiment element w that adds to the feature vector of the element u. $f(u, w)$ returns a value for the feature w in vector u. The vector u can be seen as a compact representation of the distribution of u over a range of words those co-occur with u in a given text. The distribution hypothesis says the words that have similar distributions have the similar sentiment. A quantification of $f(u, w)$ following the principle of point wise mutual information between a lexical element u and feature w is made by

$$f(u, w) = \log \left(\frac{\frac{c(u,w)}{N}}{\frac{\sum_{i=1}^{n} c(i,w)}{N} \times \frac{\sum_{j=1}^{m} c(u,j)}{N}} \right) \tag{5}$$

where $c(u, w)$ is the number of sentences in which element u and w cooccur. $N = \sum_{i=1}^{n} \sum_{j=1}^{m} c(i, j)$, n is the total number of lexical elements and m is the total number of features exist in the given sentences.

Basically $f(u, w)$ is a measure of cooccurrence of u and w with respect to cooccurrence of w with all other elements multiplied by the number of cooccurrence of u with each feature elements. All the cooccurrences are seen in proportion to total number of lexical elements and features. Then, The relatedness between two lexical or sentiment elements say **u** and **v** is computed as

$$\gamma(v, u) = \frac{\sum_{w \in (x | (f(v,x) > 0)} f(u, w)}{\sum_{w \in (x | (f(u,x) > 0)} f(u, w)} \tag{6}$$

The above gives a measure of proportion of pointwise mutual information features of u is shared with element v. It's to be noted that, relatedness of two elements is not symmetric i.e. $\gamma(u, v) \neq \gamma(v, u)$. This metric helps to define similarity among elements in a thesaurus. It's to be noted that the authors emphasise on the uniqueness of their thesaurus for its sentiment-sensitivity i.e. associating sentiment to a semantic.

$$Score(u_i, \mathbf{d}) = \frac{\sum_{j=1}^{N} d_j \gamma(w_j, u_i)}{\sum_{l=1}^{N} d_l} \tag{7}$$

For a text d there is a vector $\mathbf{d}$ having features $w_1, \ldots w_i \ldots w_N$ where N is the total number of feature elements. Let u_i be a base entry of the text d. d_j is the number of uni-grams or bi-grams the w_j has in d. Score of an entry word u_i in a text d is a normalised value of summarisation of relatedness of u_i with all the other features of d. The metrics relatedness and score are used to expand multi-domain thesaurus. Expansion of it is required as it enables to store a hitherto unknown basic feature. A lexical or a sentiment element of a domain may not have a direct mapping element in other domain. In that case for an unknown basic element b in a domain, there is a need of finding an equivalent lexical or sentiment word in other domain. Now, the relatedness of b with each feature in thesaurus is to be computed and then score value for each is also computed. A feature having high score with b is considered as its equivalent and the sentiment score of the matched feature is also ascribed to b and included as a new entry to the thesaurus; thus the thesaurus is expanded. For details of the proposed scheme the readers may refer to [29]. The problem of multi-language and multi-domain sentiment translation is further complex by one more level for finding a lexical or sentiment term in a language to its equivalent term in another language. The problem is challenging but important for having many applications e.g. analysis of the reviews of a product(s) expressed in different languages.

7 Conclusion

Emotions are not always expressed but it sits in words in between. Psychologists see the real personality that is hidden, distorted and occluded, in the words the person speaks. A higher degree of sentiment study lies with a higher degree understanding of natural language. Again understanding emotion also depends on the similarity of the realms both speaker and listener reside in. It's not only content of a communication but the style of communication matters in conveying sentiments. In the context of social media conversation, it is important to note the style of communication and lexical used in communication. The socio-psychological basis has a definite role in expressing self through languages. If both the communicators are in different language frames then the mapping in sentiment metrics of word usages is a tricky issue.

References

1. Cambria, Erik. 2016. Affective computing and sentiment analysis. *IEEE Intelligent Systems* 102–107.
2. Wilson, Theresa, Janyce Weibe, and Paul Hoffmann. Recognizing contextual polarity: an exploration of features for phrase-level sentiment analysis. *Computational Linguistics 35* (3): 393–433.
3. Cambria, Erik, Andrew Livingstone, and Amir Hussain. 2012. The Hourglass of emotions. *LNCS 7403*: 144–157.

4. Cambria, E., D. Olsher, and D. Rajagopal. 2014. SenticNet 3: A common and common-sense knowledge base for cognition-driven sentiment analysis. In *Proceedings of 28th AAAI Conference Artificial Intelligence*, 1515–1521.
5. Bravo-Marquez, F., M. Mendoza, and B. Poblete. 2014. Meta-level sentiment models for big social data analysis. *Knowledge-Based Systems* 69 (2014): 86–99.
6. Russell, J. 1979. Aective space is bipolar. *Journal of Personality and Social Psychology* 37: 345–356.
7. Whissell, C. 1989. The dictionary of affect in language. *Emotion: Theory Research, and Experience* 4: 113–131.
8. Fontaine, J., K. Scherer, E. Roesch, and P. Ellsworth. 2007. The world of emotions is not two-dimensional. *Psychological Science* 18 (12): 1050–1057.
9. Qiu, Likun, Weishi Zhang, Changjian Hu, and Kai Zhao. 2009. SELC: A self-supervised model for sentiment classification, CIKM'09, 929–936.
10. Zhao, Yiming, Kai Niu, Zhiqiang He, Jiaru Lin, and Xinyu Wang. 2013. Text sentiment analysis algorithm optimization & platform development in social. In *Sixth International Symposium on Computational Intelligence and Design*, 410–413. https://doi.org/10.1109/ISCID.2013.108.
11. Dashtipour, Kia, Soujanya Poria, Amir Hussain, Erik Cambria, Ahmad Y. A. Hawalah, Alexander Gelbukh, and Qiang Zhou. 2016. Multilingual sentiment analysis: state of the art and independent comparison of techniques. *Cognitive Computing* 8: 757–771 https://doi.org/10.1007/s12559-016-9415-7.
12. Kumar, Ashish, Mayank Saini, and Aditi Sharan. 2020. Aspect category detection using statistical and semantic association. *Computational Intelligence 2020*: 1–22. https://doi.org/10.1111/coin.12327.
13. Schouten, K., and F. Frasincar. 2016. Survey on aspect-level sentiment analysis. *IEEE Transactions on Knowledge and Data Engineering* 28 (3): 813–830.
14. Taboada, Maite, Julian Brooke, ilan Toloski, Kimberly Voll, and Manfred Stede. 2011. Lexicon-based methods for sentiment analysis. *Computational Linguistics 37* 2: 267–307.
15. Janyce, M. 2000. Wiebe, learning subjective adjectives from corpora. In *Proceedings of AAAI*.
16. Bosco, Cristina, Viviana Patti, and Andrea Bolioli. Developing corpora for sentiment Aanalysis: The case of Irony and Senti-TUT. *Proceedings of the Twenty-Fourth International Joint Conference on Artificial Intelligence (IJCAI 2015)*, 4158–4162.
17. Bestgen, Yves. 2008. Building affective Lexicons from specific corpora for automatic sentiment analysis. *Proceedings of the International Conference on Language Resources and Evaluation, LREC*, 496–500.
18. Steinberger, Josef, Polina Lenkova, Mijail Kabadjov, Ralf Steinberger, and Erik van der Goot. 2011. Multilingual entity-centered sentiment analysis evaluated by parallel corpora. In *Proceedings of Recent Advances in Natural Language Processing*, 770–775.
19. Vilares, David, Miguel A. Alonso, and Carlos Gomez-Rodrguez. Sentiment analysis on monolingual, multilingual and code-switching twitter corpora. In *Proceedings of Workshop on Computational Approaches to Subjectivity, Sentiment and Social Media Analysis*, 2–8.
20. Deng, Hao, Qiushi Wang, Dana P. Turner, Katherine E. Sexton, Sara M. Burns, Matthias Eikermann, Dianbo Liu, Dan Cheng, and Timothy T. Houle. Sentiment analysis of real-world migraine tweets for population research. Cephalalgia Reports, vol. 3, 1–9. https://doi.org/10.1177/2515816319898867.
21. Kumar, Ashish, Mayank Saini, and Aditi Sharan. 2020. Aspect category detection using statistical and semantic association. *Computational Intelligence* 1–22. https://doi.org/10.1111/coin.12327.
22. Che, Wanxiang, Yanyan Zhao, Honglei Guo, Zhong Su, and Ting Liu. 2015. Sentence compression for aspect-based sentiment analysis. *IEEE/ACM Transaction on Audio, Speech and Language Processing, 23* (12): 2111–2124.
23. Turian, J., L. Ratinov, and Y. Bengio. 2010. Word representations: A simple and general method for semi-supervised learning. In *Proceedings 48th Annual Meeting Association for Computing Linguistics*, Stroudsburg, PA, USA, 2010, 384–394. Series ACL'10.

24. McDonald, R. 2006. Discriminative sentence compression with soft syntactic evidence. In *Proceedings of EACL*.
25. Tausczik Yla R., and James W. Pennebaker 1. 2010. The psychological meaning of words: LIWC and computerized text analysis methods. *Journal of Language and Social Psychology*, *29* (1): 24–54.
26. Stone, P.J., D.C. Dunphy, M.S. Smith, and D.M. Ogilvie. 1966. *The General Inquirer: A Computer Approach to Content Analysis*. Cambridge: MIT Press.
27. Weintraub, W. 1989. *Verbal Behaviour in Everyday Life*. New York.
28. Fangzhao, Wu, Yongfeng Huang, and Yangqiu Song. 2016. Structured microblog sentiment classification via social context regularization. *Neurocomputing* 175: 599–609.
29. Bollegala, Danushka, David Weir, and John Carroll. 2013. Cross-domain sentiment classification using a sentiment sensitive thesaurus. *IEEE Transation on Knowledge on & Date Engineering*, *25* (8): 1719–1731.

Productizing an Artificial Intelligence Solution for Intelligent Detail Extraction—Synergy of Symbolic and Sub-Symbolic Artificial Intelligence Techniques

Arunkumar Balakrishnan

Abstract Businesses are today moving from a data-driven focus for becoming 'Insight-driven' businesses. This paradigm shift hinges on the use of artificial intelligence technology. This shift needs accurate and smooth input of all pertinent details. This is facilitated by using artificial intelligence to provide intelligent detail extraction from documents and other sources. The techniques behind building an intelligent detail extraction product are discussed. The path from an initial solution through identifying and developing the characteristics that mark a product are described. A powerful synergy is created by appropriately using symbolic machine learning as well as deep learning techniques.

1 Introduction

Intelligent detail extraction (or intelligent document processing) is a keystone in the shift that is happening across business ventures, where organizations are moving from a data-centric approach to an insight-driven one. Computers and software were until now used to store data and then retrieve required information. The data and the details within it were not fully analyzed or processed to find hidden patterns that reflect trends and anomalies. In a world today where change is happening at a much faster rate, it has become a necessity to be prepared for change. This need is coupled with the proliferation of scanned images of documents caused by easy availability of cameras and scanners on mobile devices. While the availability of these images provides the required data, the vast quantity of such a media of data makes it a daunting task to manually upload all data to an analytics system. Expanding on this point, the first level of preparedness for an organization in this fast-changing world is to envision what could be the probable changes. Finding the patterns that lie in the data is done by using Artificial Intelligence (AI) techniques like long short-term memories and generative adversarial networks on the history of data to arrive at

A. Balakrishnan (✉)
Ikval Softwares LLP, Coimbatore, India
e-mail: arun@ikval.com

© The Author(s), under exclusive license to Springer Nature Singapore Pte Ltd. 2021
S. Rautaray et al. (eds.), *Trends of Data Science and Applications*,
Studies in Computational Intelligence 954,
https://doi.org/10.1007/978-981-33-6815-6_2

Fig. 1 Transformation to an
insight-driven business [1]

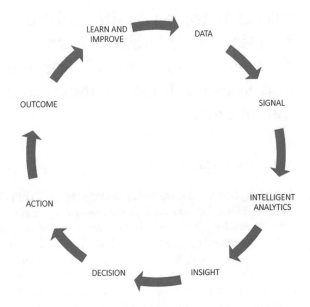

predictions of what is to happen. This could include price change as well as customer preference changes and other influencers of strategy and decisions. To use these AI techniques, there has to be sufficient, up-to-date, and accurate data. A large number of processes are still tied to details on paper. Until recent times, the data extraction and entry into information technology systems from these paper-based documents have been manual. This manual process has resulted in errors in entry as well as a significant latency in up-to-date details. The recent change in technology has to been to use AI techniques to speed up as well as to ensure correctness of the extraction of details from various sources. AI technology is required as the documents and other sources of data are not all in a structured format, and the variations in these documents are many. Documents could be in a semi-structured format or could even be in a loose informal manner. Each of the prime components in a document that need to be extracted can possibly be specified in a variety of ways (various types of documents are detailed in the following section).

2 Problem Description of Intelligent Detail Extraction

Prior to the development and application of recent advances in AI technology to the extraction of data from documents, optical character recognition systems were used. The accuracy of optical character recognition systems has not been good enough to be employed in a standalone manner. With the advent of deep learning techniques for vision applications, this has now become achievable. The capability to learn modifications as well as new entities of the document to be processed by using machine learning techniques has been a major contributor to the increased usage of "Intelligent Document Extraction (IDE)" within may industries. It has also been noted that these new techniques have to be used along with a suite of other methods engineered to handle complexities seen in processing a document. In one dimension, the documents have a large variation in the entities within it: tables, text-boxes, charts, blocks, and arrows. These variations form a major challenge. In another perspective, the documents are often provided to the IDE system after being scanned and converted into a PDF/Image format. The process of scanning creates "noise:" in the result, the angle of the document can be tilted, sometimes the whole page is turned 90°, often there are smudges created on the image, and the paper is sometimes folded in parts.

There are also cases, where a document is bi-lingual, especially in vernacular languages, where a few English characters are accepted today as part of the colloquial language.

Another challenge occurs due to the font and spacing used in documents. Often the characters have a very small spacing between them; this is especially so for the important numbers relating to a company's (tax) identification. These numbers (identifications) are important and need to be retrieved exactly as it is. It is even more challenging as it does not match common words.

Usage of IDE applications also includes retrieval of proper nouns, like names of products and of people. As these are not commonly recognized, there has to be a specific process/module developed to recognize these.

Overall, the challenges for an IDE to overcome are;

1. Low image quality
2. High character density
3. Multiple variations in document details
4. Multi-lingual documents
5. Combination of hand-written and printed text
6. The presence of uncommon words (like proper nouns).

These challenges make IDE a significant task. In the rest of this section, more detail is provided on the variations in documents. Documents are classified into three types:

1. Structured documents
2. Semi-structured documents
3. Unstructured documents.

Structured documents follow a fixed set of layouts with very few variations. Once the application learns these limited set of layouts, it can then identify the layout of a new document and process it accordingly. Semi-structured documents have a large number of variations with data appearing in different places of the document. This type of documents has to be approached differently. In these documents, the domain knowledge can be used to narrow down the variations. This approach along with the use of NLP allows data to be extracted in an efficient manner. The "Case study" section will discuss the approach to handle such semi-structured documents. Unstructured documents do not follow any repeatable layout. To extract data from such documents, a complete understanding of the domain is required. An ontology of the domain and its associated words can then be utilized by the NLP to understand the meaning of sentences and thereby extract relevant data from the document.

3 Components of an IDE

Document Input: IDE commences with the document that contains data significant to the business. This can be physical documents or digitized documents or email. The IDE application has to be capable of processing all these formats: PDFs of the scanned physical documents, GIF, BMP, PNG, JPEG images, HTML formats are all included.

Document Pre-process: Stage two involves pre-processing of the document; more specifically this involves ten operations:

1. Removal of smudges and other dirt from the document
2. Aligning the document
3. Removing background images
4. Removing seals
5. Removing watermarks
6. Changing inverse printing
7. Recognizing logos
8. Recognizing signatures
9. Recognizing bar codes
10. Identifying and handling tables in the document.

In addition to implementing these processes with their variations, it is also a challenge to identify when and where these processes are required. Sometimes, the change is required across the document while at other times the change is only for certain parts of the document.

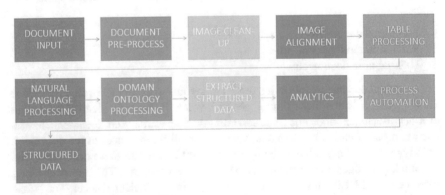

Fig. 2 Overall composition of an IDE [1]

Natural Language Processing, Domain Ontology, Extract Structured data: In stage three, the document is read using a software function/tool that can disambiguate words and sentences. This capability is primarily based on using a Long Short-Term Memory (LSTM) deep learning architecture. Even though state-of-the-art LSTM architecture is used, still proximity of words and combinations of alphabets and numbers leads to errors in disambiguation. To overcome, these various techniques are used to normalize the appearance of each character and symbol. Once the words are retrieved natural language processing is applied in order to develop a meaning and intent for the contents of the document. Another challenge related to Natural Language Processing (NLP), applies to documents that have words that are misspelt. It is laudable that we humans have a capability to identify words appropriately given just the first letter of a word. AI applications are faced with a daunting task when the words are misspelt. The wrong spelling leads to possibilities, where a word can be formed by changes to any of the alphabets in the word. Trying out all the possibilities is an NP problem. A need arises to bring in domain-specific knowledge to narrow down the possibilities to a comparable set.

A major NLP task is to

- disambiguate,
- find co-references,
- identify topics,
- identify entities,
- identify concepts,
- notice negations.

The result of the NLP processing has to generate relationships between entities that are represented in the paragraphs of the document. This is the beginning of creating a structure in the information. This result provides classification, categorization, and a summarization of the document. This is what allows the next set of functions to (1) generate reports, analytical results, graphs, (2) allow easy access to relevant data, and (3) generate insights to the management.

4   Survey of Work on Extraction of Characters

Viswanathan et al. [3] provide a survey of work on extracting text from image. Li C. and Wand M.'s approach matches documents with pre-stored models. The extracted text is matched against a set of words along with their variants in spelling. Cesarini F., Gori M. Marinai S Soda G's work required the user to input a conceptual model of the document which was used to match with the new documents. This depended on the user's input and expertise in classifying documents.

DeepReader [3] uses low-level vision functions to identify text-based items and then store them in a relational database. It then utilizes a sequence-to-sequence model to generate answers to user queries.

5   Case Study: Invoice Processing

5.1   *Details*

Invoice processing is a prime activity in maintaining accounts within an organization. The digitization and automation of activities within an organization depend on the related digitization of invoices received. This directly translates to the accounts payable maintained by the organization. While the invoices generated by the organization can be from its internal information technology department, the invoices that it receives from its vendors may be in either a paper form or in a PDF form. In these cases, there is a significant effort involved in reading each invoice and extracting important details from it that are required for the accounting software and other financial processes.

A specific requirement was to process hotel invoices and extract details from close to twenty fields in each invoice. Over three thousand invoices were to be processed daily. The fields in the invoices did not have any specific pre-determined locations; the legends for a field could be many, various fonts, and sizes could be used for the characters, and there were particular forms of overwriting/dirt that could appear on the invoices.

Printed invoices would be scanned and converted to a PDF/JPEG format. This would be input to the application. Data had to be extracted from these PDF/JPEG documents. The data required were, primarily from the fields of:

1. Hotel name
2. Invoice number
3. Invoice date
4. GST number(s)
5. SGST number(s)
6. Tariff
7. Other charges: Food and beverages
8. Other charges: Laundry
9. Other charges: Transportation
10. Discount
11. Advance paid
12. Net amount
13. Total taxes.

Invoices could be in different formats;

- Single customer,
- Group of customers (over different dates),
- Summary sheets.

Overall, there were many possibilities in which the details could appear in an invoice.

5.2   Architecture

Our approach is to combine knowledge base AI (domain ontologies and linguistic rules) and ML-based AI (supervised or unsupervised) to create a synergy between each individual approach. Image/character extraction is done through a combination of Natural Language Processing (NLP) and deep learning along with computer vision to understand shapes/objects on the document, form, or email. This way additional details got from the context is used to improve accuracy.

5.2.1   Natural Language Processing

At the core of our application is a homegrown natural language processing software. Prior to actually developing the NLP capability in house, other NLP software were evaluated, but there were specific limitations in them. A deep learning approach to building the NLP software was also considered. Given the context of application and the time constraints, it was decided that a knowledge-based approach to NLP was better. This NLP software helps significantly in disambiguating and in providing valuable narrowing down of the options at many points of the overall IDE.

5.2.2   Learning Components of the Invoice

Learning structures from examples [4] provide the base for the architecture used to learn the structure of different invoices. This idea was combined with the machine learning strategy of using "background knowledge" in inductive learning [2]. A knowledge base was built which identified the components of an invoice. For each of these components, the many varieties in which they can be expressed were stored in the knowledge base.

Examples of components (for a hotel invoice) are:

- Hotel name,
- Invoice details,
- Tariff details,
- Tax details,
- Other charges,
- Summary.

5.2.3   Ontology of Image De-noising

An ontology was built of image de-noising and enhancement techniques. These were used as part of the pre-processing of the invoice in order to help in extracting the details correctly. Over twelve techniques were stored in this ontology. The application would use appropriate techniques based on (1) input characteristics of the invoice and (2) status of the extraction process. A failure driven cascaded approach was also tried by the application when direct application did not produce expected results. By this, different image enhancement techniques would be tried if a previous technique failed. Different combinations of these techniques were also tried as in many cases, a single technique would not be sufficient. For example, a "dilate" process would need to be followed by a width increase of the characters in order for characters like Z to be extracted correctly ("Z" would otherwise be mistaken for "2").

5.2.4   Domain Specific Constraints

Identifying characters correctly also required usage of constraints that were particular to the application. For example, identifying characters of a GST number was aided by the rules by which a GST number has to be formed: State code, PAN number, 14th character is a "Z" and other constraints which were useful in disambiguating the characters. A knowledge base of these rules was built and used appropriately.

5.2.5   Retrieving Data

The knowledge about the possible structures of each component were used to identify the data elements and retrieve the relevant data. The NLP software worked in tandem with the knowledge base during this retrieval phase, to disambiguate the words and ensure that the correct data was extracted.

5.2.6   Python Library for Character Extraction

After pre-processing the characters/words, the result is given to the Python library for character extraction. This result is passed through the domain constraints module to resolve any remaining ambiguities.

Even after all these techniques are used, there are still characters (especially upper-case letters) that prove elusive to identify correctly. In such cases, the python library for character extraction has to be trained with images of both the contradicting characters. This resolves the issue.

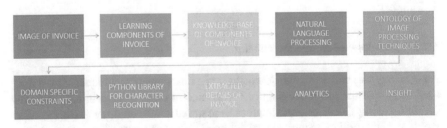

Fig. 3 Architecture of solution for processing invoices

5.3 Challenges

One of the major challenges in processing invoices lies in the hotel name. Often the hotel name is a GIF image or is a logo of the hotel. Extracting the name of the hotel from this was very difficult. Multiple options were considered including finding alternative means to find the hotel name. The email of the hotel as well as the URL of the hotel were used to find the hotel name, when required.

Another challenge was in extracting the GST number as the characters are tightly packed in the GST number. In addition, this field is alphanumeric. When the python character recognition library fails to provide the correct character, then the recognition library has to be trained with instances of the contradicting characters so that it learns to differentiate between them. Python's character recognition library is based on a long short-term memory deep learning architecture.

5.4 Insight

The extracted data are used to do trend analysis with previous years data (1) Number of occupants (2) Geographical distribution of occupants (3) Room types (4) Cancellation frequency (5) Add On like sightseeing travels.

5.5 Discovery and Productizing

This software is to be used for over twelve thousand customers. First, the primary set of customers were identified. This amounted to eight hundred. The application's knowledge base was populated with the attributes for these eight hundred customers. This was then deployed, and the results were confirmed for these eight hundred customers.

For the balance eleven thousand two hundred customers, there was only a few invoices per customer. Therefore, a discovery feature was added to the software. Discovery of the attributes of a new customer worked by searching for the possible attributes of each component in the knowledge base. Once the attributes are discovered, then a new entry is created in the knowledge base. The new entry could be a link to existing component structures or could be an instance of a new component structure depending upon the results of the discovery phase.

The discovery phase involved the use of the ontology of image processing techniques to try different approaches until the details are extracted correctly. Along with the image processing techniques, the knowledge about the domain of application as well as the constraints applicable in the region of hotel invoices were used to constantly filter out the values that would not be plausible. In this manner, the application uses the knowledge base along with a set of discovery techniques to identify the component structure of new customer invoices. The application was successfully deployed to work with all twelve thousand customers.

6   Results and Conclusion

The objective of extracting details from hotel invoices is met through the architecture and program that was developed. The usage of LSTM-based python libraries, knowledge base of components, machine learning of structural properties, discovery to address new customer invoices, and an overall synergy between symbolic artificial intelligence and deep learning has been detailed.

Productization required that the application go beyond specific hotel invoice formats and be able to handle a wider set of hotel formats. This was facilitated by starting with a knowledge base of component formats, followed by a discovery process that searched and identified the location of data elements. Discovery involved (1) learning new terminology for the required fields and (2) handling image noise through different composition of techniques.

The application successfully demonstrated the capability to learn new formats of the invoices and the capability to create new compositions of techniques to handle noise in the image. A continuous learning was demonstrated of techniques to handle noise in the image and learning new structures (formats) of invoices.

This application differs from other approaches to text extraction by first identifying components of a document that is seen often across categories. It then builds a knowledge base of these components to arrive at the overall structure of the document. This approach reduces the complexity significantly, without losing any important detail. This makes the approach feasible and viable.

Successful deployment and usage of the application to handle over three thousand hotel invoices per day along with a continuous acceptance of new customer invoices are demonstrated. Capability to identify components of an invoice as well as to add new component structures allows the software to be productized and applied with no limitations.

References

1. Forrester. 2020. Reimagine enterprise document comprehension- do more with AI. Webinar.
2. Muggleton, S. 1994. Inductive logic programming: Theory and methods. *The Journal of Logic Programming* 19–20, Supplement 1:629–679, May–July 1994.
3. Viswanathan, et al. 2018. Deep Reader, information extraction from document images via relation extraction and Natural Language. arXiv 1812.04377v2 14.
4. Winston, P.H. Learning structural descriptions from examples. http://hdl.handle.net/1721.1/6884 DSpace@MIT 1970-09-01 AITR-231.

Digital Consumption Pattern and Impacts of Social Media: Descriptive Statistical Analysis

Rabi N. Subudhi

Abstract Dependence on digital media has increased manifold during the COVID—pandemic lockdowns, across the globe. Most of the offices and academic institutions started operating, on 'work-from-home' mode, on digital platform. Even senior citizens, non-working homemakers and kids spent more time in social (networking and communication) media. The 'difficult times' of COVID—pandemic has also shown the world the 'different times' and the difference in our preferred way of functioning. The 'digital consumption pattern' changed substantially, both by its scale and by diversity. The present paper discusses the issue of 'effectiveness and impact' of 'digital medium' on 'digital life' of its users (/digital consumers), particularly during this extended lockdown. The objective is to discuss issues relating to the 'effect of extended/longer use of net, during this pandemic, for academic and professional activities from home, continuously'. The study is about the life in virtual world, particularly during this extended lockdown. The paper adopts a method of descriptive statistical analysis, covering different categories of users of Eastern India. Using a structured questionnaire method, descriptive primary data, relating to digital consumption, were collected from around 1350 respondents, from Odisha and its neighbouring states. Result shows how different categories prefer and use their preferred social media. Study finds a significant contribution of Internet-based social media (SM), countering 'social isolation' of senior citizens.

Keywords Social impacts · Statistical analysis · Digital consumption · Social media · Internet consumption pattern

1 Introduction

Gone are the days when benefit of computers and Internet was limited to qualified technocrats and professionals. As computing devices have become affordable and user-friendly, over the years, the digital divide (of those who are capable of using

R. N. Subudhi (✉)
School of Management, KIIT Bhubaneswar, Bhubaneswar, India
e-mail: rabisubudhi@gmail.com

© The Author(s), under exclusive license to Springer Nature Singapore Pte Ltd. 2021
S. Rautaray et al. (eds.), *Trends of Data Science and Applications*,
Studies in Computational Intelligence 954,
https://doi.org/10.1007/978-981-33-6815-6_3

33

computers and the rest, those who are not using it) is being abridged very fast. Availability of most of the services through mobile applications, available in smart phones, in a very user-friendly manner, has taken millions of non-users to users of Internet. We shall call such 'users', for the purpose of this research article, as 'consumers' of Internet, or the digital consumers.

It is of academic interest to study the digital consumption pattern of diverse categories of users, particularly in Eastern India. It is more interesting to study, if and how 'digital diversity' and 'digital consumption pattern' has changed during this unprecedented COVID lockdown.

This paper, after review of existing literature, analyses a primary data set of over 1300 Internet users, from different categories, from Odisha and neighbouring states (of Eastern India). It adopts a descriptive statistical analysis method.

2   Review of Literature

Kulesz [15], in his study on 'digital technology on cultural diversity', opines that 'digital technology acts as an opportunity in chain of consumption, production and distribution'. The author identified contrast of culture across countries, where every country focused on 'socio-economic engagement'. Serrano and Brusco [17] focussed on social, demographic and economic aspects related to digital divide. The social networking through Internet brings digital inclusion and enhances the scope of e-commerce, e-government services. The digital awareness and training along with effective policies and regulations are suggested by authors to minimize the barrier of exclusion. Nayak and Krishnamurthy focus on digital technology used in inclusive social health models of BPO customers. The result of the qualitative exploratory research is that the digital technology is helpful in strategic management of customer satisfaction, dealing with risk and benefits of both insurer and policy maker. The authors choose participants from different medias through virtual interviews and find that the 'culture and subculture also influence digital music'. Gvili and Levy [9] study on e-Word Of Mouth (WOM) and attitudinal differences. The study is based on examining 864 participants of different social networks and ANOVA is used to test the differences in attitudes and structural equation model to test e-Word Of Mouth (WOM) attitude model. The conclusion is on the basis of differences in channels, the attitudes towards WOM are different. Goldsmith and Lafferty [16] have studied on the response of consumers regarding the effectiveness of advertisement through Internet. They have also focused on advantages and disadvantages associated with online ads.

Internet is the affordable means of communication and transactions along with entertainment and work-related information. 'It creates digital inclusion as in organizations older adults learn technical skill from young volunteers and build a sense of community feeling'. Illingworth et al. [13] have made cost–benefit analysis of Internet. As a marketing and communication tool, how Internet influences the perceptions and beliefs of non-profit organizations. They have concluded that irrespective of size and level, Internet is very useful, cost-effective and time-effective marketing weapon for organizations. The virtual experience influences the perception of consumers and their pattern of consumption of products and services are changed. Dwivedi, Bobek and Zabukovsek have studied on online teaching and learning. They have viewed that the teacher-centric environment is shifted to student centric. By analysing online learning behaviour of 152 postgraduate students, the author concluded that if the online teaching progresses according to the syllabus, there would be enhancement in the scope of engagement of students. Internet has enough power to run and control the entire organization by focusing on storing, transmitting and retrieving data. It influences the perception of members of organization and helps to enhance knowledge with updated information. Internet has converted the old organizational system with new organizational setup and new technology.

3 Access of Internet Across Generations

According to Koch, Internet provides a social platform of connectivity of people based on usages of Internet, preferences and concern of privacy. Every generation has its own wants, which can be easily fulfilled with the help of different apps or smart phones or other webs of digital device or Internet. Vrakl states that Internet is used by Gen-X, Gen-Y and Boomers to make a congruency across them. Boomers stay connected with their millennial children and their friends trough Facebook and email with mobile phones and computers. Though Xers do not get enough time to spend with their friends because of their profession, they use Facebook, YouTube, mail, etc. to be connected and they prefer flexible schedule of work and work-from-home. Both Gen-X and Gen-Y adopt Internet as a smooth medium of communication and social interaction, and at the same time, they prefer online buying by using Internet to visit the shop. Viens views that according to Global Web Index, nearly 114,000 of diversified generations use Internet to be connected socially. 4.4 billion of people of the world are using Internet through multiple devices. Kamber states about the adults' digital world with digital culture and immigration of digital technology. Their digital life makes them more ambitious, responsible and updated with the world as compared to other generations who embrace digital life. According to Harris, Baby Boomers use Internet not just like other adult generations. They utilize Internet for checking mails, political news, economic situation, online-banking transaction and weather-related information, etc. rather adult generations (Gen-X, Gen-y and Gen-Z) spend enough time with Internet for instant messaging, videos, charting, etc. Davenport states about the use of Internet by different generations in organizations. Traditionalists or Veterans prefer face-to-face and oral or written communication to Internet. Boomers use email for communication, but more use phone and fax. Gen-X quickly adapt technology and regular user of Internet. According to Harris et al. (2017), customers of different cohort prefer different patterns of grocery shopping. E-grocery is both time and cost effective for technically sound and busy customers, whereas traditional super market is preferred by many customers who meet their friends and relatives and get varieties of products while making shopping as traditional market creates crowd and a broad based. Normalini focuses on Internet banking in Malaysia through cross-sectional study by the respondents using online banking. The author also states that Millennials are born and brought up in the era of online transactions and very adaptive with technology. Internet banking converts the traditional banking system to millennium banking trend with the help of which customers can access their accounts through online wherever in the world. They usually bridge the communication gap between Boomers and Millennials. Millennials use Internet from very early stage of life. They use short and sweet communication medium like WhatsApp, Twitter, Facebook, etc. and get immediate and easy response. As generations are different, their characteristics and ideologies are different and digital technology provides a common link to make them together.

4   Impact of Internet on Business-Management

In organizations for smooth communication, strategy formulation, implementation and evaluation, to achieve the goals and objectives, Internet plays very vital role. The apps like: Skype, Zoom, GoToMeeting, etc. are helpful to share information with customers, venders, staffs and other stake holders connected any corner of the world. Business to business and business (B2B) to customer (B2C) all transactions can run with online and within a fraction of time. The manual works and time taking transactions have become quicker with the help of software like: QuickBooks and Wave Accounting, etc. These help companies to handle timely transactions, storing and retrieving valuable data, error detection and automatic routine tasks. The setup of online stores are possible with e-commerce which can easily upload information related to product, services, prices and features for customers through advertisement and marketing around the world. The business research and analysis work and focusing on movements of competitors are possible for business intelligence platform and online updating of information. The document-signing services for clients and employees, Internet enabled camera for knowing happenings of organizations, sensory security alarm, etc. are the helping tools from Internet to business. The online payments like ATM, Paytm, credit card, debit card, etc. create a broad scope for business and customers enjoy branded products near to their door bells through online delivery from amazon, flip cart, snap deal, etc. The online food delivery to online groceries, medicines and other consumer products is easily come to us with a simple order. At the same time, many organizations now opt for e-recruitment, online interview through video conference, assign online tasks and measuring online performance. To monitor performance of employees, the applications like SAP, Workday, etc. are very useful and the organizations arrange training and development programs after identifying the weaknesses of employees. The modern versions of Microsoft Office helps in preparing presentation, writing business letters, sending mail, etc. through Internet makes the administrative work effective. Similarly, cloud storage plays very significant role to store and protect data of companies. By using Internet Relay Chat (IRT), the organizations arrange virtual meetings, new ideas are shared across members, new products are developed and services are carried out. It is no matter from which region or geographic location they belong, but they are easily connected with each other. Similarly, the research and development works can run smoothly with the help of online libraries. The service sectors enjoy wide range of e-services in different businesses like, IT firms, banking, insurance, trading, travelling, tourism, education, entertainment, restaurant, consultancy, real estate, recruitment, career, research and development, online publications, etc. There are virtual markets of buyers and sellers and online order and delivery of products. The primary purpose is to save time, money, effort and improvement of quality of life.

5 Impact of Internet on Kids, Adolescents and Adults

The Internet is the source of entertainment and learning with fun device for kids to adults. The school-going children and college students use social media for their education, project work and research-related activities. The students of each generation easily get educational application form, online test, online admission, projects through online, online classes, research tools, reading techniques, e-learning etc. educational institutions, colleges and universities spend a large amount of money for digital setup, teaching and reading. Learning of kids, adults, Gen-X, Gen-Y and Boomers all easily access information related to their studies and retrieve and utilize these as and when required. There are specific applications like: Webx, Zoom, Google Meet, etc. which do not need boundaries of institutions. Students can easily learn their lessons related to normal courses, extracurricular activities, technology, science, arts and sports or games, etc. Internet helps in educating pupils according to their needs, helps to enhance their knowledge and develop their skills. The main barrier in the education is lack of information and expenses associated with education. Internet has made the education affordable for students and they can get updated information regarding their studies. They can learn and relearn their lessons through YouTube tutorial videos and other apps. It is also interactive for teachers, students and peers group by using new technique of teaching and learning through animation, colourful slides and images. Similarly, learning through multimedia by different apps like BYJU's (from pre-kindergarten—class 12), ABC mouse.com (age 2–8), Epic (age 2–12) and many others for kids rather than traditional method of classroom notes, pen–paper or slates create more fun and interest among them. For them, reading is just like a play not a task and they compete with their friends to complete it. Parents can also aware of the performance of their children and conscious for their studies. Internet spreads education across the globe for each generation and every community.

6 Internet Service Providers (ISP) in India During This COVID-19 Lockdown

During this pandemic, almost all the ISPs in India are trying their best possible way to provide unlimited Internet service to their customers and companies. It is also challenging for those as there is heavy competition. In India, Internet Service Provider (ISP) provides broadband speed which is growing during this time.

According to Global Digital Report 2019, total number of Internet users are 4.4 billion that means YouTube users account for 45% world's entire online population. According to Alexa, YouTube has the second position in the world, just after the Facebook. Users are spending 13 minutes 18 s in a daily basis on this site as of late 2019 with compare to 8 min 41 s in the 2018. Alexa has also declared that each visitor on an average involves in YouTube 6.5-page views. The following data table shows about the monthly active users in the social platform which was released on January 2019:

Social platforms	Active users (in millions)
Facebook	2271
YouTube	1900
WhatsApp	1500
FB Messenger	1300
Instagram	1000
Tiktok	500
Twitter	326
LinkedIn	303
Skype	300
Snapchat	207

Internet sources

7 Objective and Methodology of Primary Data Collection

We wanted to study, taking a large sample, as part of this research objective, the 'digital consumption pattern and diversity' of Internet users. The specific research objective is to understand 'role of different demographic attributes' with 'Internet use pattern' of different categories of users.

For this, anybody using Internet, by any computing device, including smart phones, formed the target population for this study. We limited the primary data collection to states of Eastern India, particularly Odisha and its neighbouring states. A convenience sampling approach was adopted and a structured questionnaire was canvassed, by both physical and online survey, guided and administered by specially trained investigators, for this research project. We circulated around 1400 printed questionnaires and sent soft copies to over 400 contacts, from different categories of Internet users, from different age groups.

Finally, (a little over) 1030 printed questionnaires and 320 online questionnaires were received, duly filled in. So, for our statistical analysis of primary data, the sample size is 1350. After filtering, editing and data validation, descriptive statistical data analysis was carried out by using SPSS (version 24).

8 Data Analysis

The following tables first give a summary statistics of 'uses-profile' of respondents and their response to relevant key questions, as applicable to the specific objective of this research paper. Table 1 and Fig. 1 give statistics about usual mode of connecting to Internet, highest being 'through personal mobile data', with 96% of sample responses. WiFi (which includes both at own homes and/or office connections) is also used in high frequency (with 42% respondents using it).

Next, we studied the purpose, why and how Internet is used (by our target population, represented by a sample of 1350 respondents). As respondents had the 'multiple-choice' option in questionnaire, we have (as shown in Table 2 and also in Fig. 2) many 'purpose' with very high percentage of use, exceeding 70% of cases. Most common use (or purpose of using Internet) is found to be 'communication' and 'entertainment' purposes (with 95% and 91%, respectively).

Table 1 Connection source/medium: how they connect to Internet

Connection source	f (no.)	%
Personal mobile data	1301	0.96
Broadband connection	260	0.19
Shared WiFi	565	0.42
Public WiFi	247	0.18

Percentages are calculated by diving respective frequencies by 1350, the sample size

Fig. 1 Connection source/medium

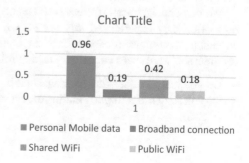

Table 2 Purpose of use of Internet

Purpose of use	f (no.)	%
Entertainment_a	1229	0.91
Communication_b	1289	0.95
Video_c	535	0.4
Education_d	846	0.63
Work and research_e	990	0.73
Personal finance_f	505	0.37
Current events_g	907	0.67
Online booking_h	758	0.56
Online shopping_i	1028	0.76
Video-audio download_j	849	0.63

Percentages are calculated by diving respective frequencies by 1350, the sample size

Fig. 2 Purpose of use of Internet

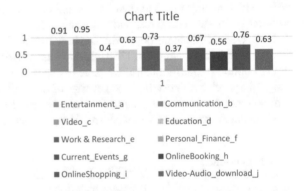

We asked respondents, to know if they 'felt necessity of recharging net-pack' same day, when day-limit got exhausted/exceeded. Majority response to this question, as shown in Fig. 3, was '3-less important', while there were cases/sections of users who felt it (recharging) '1. very important' and '2. important'. What we understand from this observation, whenever there is a 'need', users would not mind 'recharging the net-pack' again same day.

We had checked first, if users (our respondents) exceeded daily data limit, and how often they experienced this. As shown in Fig. 4, many people/respondents exceeded 'sometimes' (coded as 3). Next, to understand, how deeply/intensely users use Internet, we tried to check it by asking following two questions:

- Do you feel like checking messages/mails, even late at night? (Fig. 5)
- Do you sometimes forget 'pending urgent work', while using Internet 'intensely'? (Fig. 6).

Fig. 3 Importance of recharging net-pack

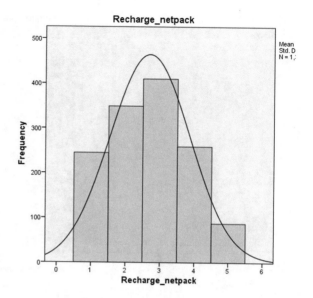

Fig. 4 Exceeding daily data limit

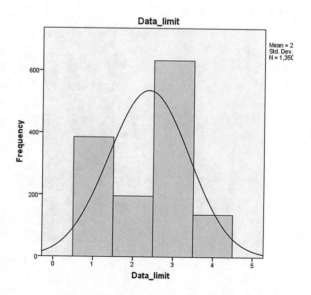

It is interesting to observe (as shown in respective figures) that there are plenty of such examples, where people (represented by the respondents of our sample) 'sometimes' check messages late at night and also have forgotten (sometimes) 'important work'.

As part of our exploratory study, we had asked for opinion on 'availability in plenty' of 'adult contents' in Internet, these days, and if it would badly affect 'social life/relation' of 'individuals'. It was intentionally an indirect question, asking their

Fig. 5 Checked net at late-night

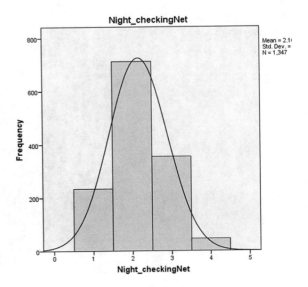

Fig. 6 Forgot work due to Internet

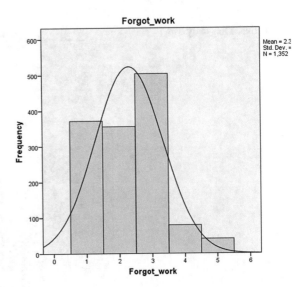

opinion about 'use by other individuals, like members of their family'. Majority respondents confirmed with their response, as '1. Strongly agree', '2. Agree', saying, 'yes, it will disturb social/family relation' (Fig. 7).

Fig. 7 Will adult contents/websites disturb social life?

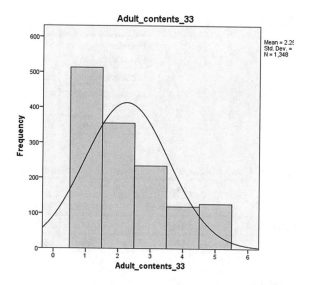

9 Bi-variate Analysis

We have studied 'statistical significance' of possible association/dependence of a few inter-related factors, by using Pearson Chi-square tests.

Hypothesis 1 People can watch (Internet) videos, while doing their (other) routine jobs, irrespective of age groups.

Table 3 gives the bi-variate table, showing 'very high frequency' of such 'multi-tasking', only for 'young and adults', within age group of 18–40 years. (Young) 'Age' has 'significant' relation/association with the other attribute (of watching and working, together), as shown in Table 4.

Hypothesis 2 (H02): 'Gender' has no-role/association with 'Internet-induced forgetfulness'.

Cross-tabulation was done to check (data in Table 5), if there is any 'significant relation' between 'gender' and 'Internet-induced forgetfulness'. People are often found to be highly engrossed in the fascinating world of Internet and some times

Table 3 Cross-tabulation: Age versus 'watch-while-working'

		1. Usually	2. Sometimes	3. Never	Total
Age_2	<18	35	61	17	113
	18–40	407	572	91	1071
	41–60	26	56	31	113
	>60	15	24	6	45
Total		483	713	145	1342

Table 4 Chi-square values

	Value	df	Significance (2-sided)
Pearson Chi-square	44.275	9	0.000
Likelihood ratio	36.750	9	0.000
Linear-by-linear association	4.639	1	0.031
N of valid cases	1342		

Table 5 Cross-tabulation: Gender versus 'forgot my work, while doing Internet'

		Forgot_work_26					Total
		Never	Rarely	Sometimes	Often	Very often	
Gender_3	Male	182	201	258	37	17	695
	Female	189	155	245	42	24	655
Total		371	356	503	79	41	**1350**

Chi-square $= 6.65$; $p = 0.156$; not significant

forget their important work. Does this happen equally with male and female? Has gender anything to do with this 'forgetfulness'? Though we wanted to test it statistically, as per available data, it could not be found/established to be significant, as 'p' value exceeded 0.05.

Hypothesis 3 Gender has a role/association with SM-time spent.

Following bi-variate table (Table 6) gives data on frequency of use of social media (SM) sites/applications by both male and female categories. As the data show, and as the significance (Chi-square) test verifies, 'time spent on social media (SM) sites/applications' is not associated with 'gender', as per sample data.

Hypothesis 4 'Profession' of user has an association/relation with 'watching-while-working' attribute of Internet users.

To test above hypothesis, bi-variate table was generated (as presented in Table 7), linking the said two attributes of Internet users ('profession' and 'watching-while-working'). As shown in next Chi-square table (Table 8), it is found to be statistically significant.

Table 6 Cross-tabulation: Gender versus 'time spent of social media/SNS'

		Time spent social networking sites				Total
		Frequently	Sometimes	Once/day	Rarely	
Gender_3	Male	172	342	130	54	698
	Female	176	295	124	55	650
Total		348	637	254	109	**1348**

Pearson Chi-square $= 1.958$; $p = 0.581$; not significant

Table 7 Cross-tabulation: Profession versus 'watch-while-working'

		Watch_while_work_16			
		1. Usually	2. sometimes	3. Never	Total
Profession	1. Student	364	521	82	968
	2. Service	57	107	30	194
	3. Home maker	18	17	11	46
	4. Business	28	44	19	91
	5. Others	16	23	3	42
Total		483	712	145	1341

Table 8 Chi-square values (for data as given in Table 7):

	Value	df	Significance (2-sided)
Pearson Chi-square	32.361	12	0.001
Likelihood ratio	29.365	12	0.003
Linear-by-linear association	6.408	1	0.011
No. of valid cases	1341		

Hypothesis 5 Social media helps significantly countering the 'isolation' of senior citizens.

It is now seen in society that retired/senior citizens isolated, even within the family they live with. It is both social and psychological isolation. To cover this important issue, we had two specific questions: 1. If people fee and are aware that senior citizens 'feel isolated', and 2. If Internet-based social media (SM) engagements help senior citizens countering such 'isolation'. Table 9 gives the bi-variate data, while Table 10 gives the result of Pearson Chi-square test. As shown in Table 9, there exists a significant relationship between the two attributes. Internet is a boon for the senior citizens, fighting 'social isolation'.

Table 9 Cross-tabulation: 'Isolated senior citizens' versus 'social media usefulness'

		Social media helps emotional support				Total
		1. No/NA	2. Disagree	3. Agree	4. Strongly agree	
Lonely_ Seniors	1. No/NA	74	20	93	27	214
	2. Disagree	23	46	77	30	176
	3. Agree	93	67	373	70	603
	4. Strongly agree	71	44	168	73	356
Total		261	177	711	200	1349

Table 10 Chi-square values (for data as given in Table 9)

	Value	df	Significance (2-sided)
Pearson Chi-square	92.209a	9	0.000
Likelihood ratio	82.827	9	0.000
Linear-by-linear association	15.751	1	0.000
No. of valid cases	1349		

10 Conclusion

During the COVID-19, many people used Internet for a much longer duration, for the purpose of entertainment, shopping and chatting and communicating with their near and dear ones, along with their routine jobs. This pandemic has forced the people to stay at home in the past few weeks and it is very difficult particularly for the kids, school and college going young generations. 'Digital consumption', particularly during this COVID-19 lockdown period, has been seen as a 'catalyst' for digital world. Digital and social media (DSM) were viewed differently and used differently, often converting it as an 'enabler' and 'medium' of official communications and e-learning requirements of concerned users.

The education of kids and young adults is done now by using different DSM platform and the 'e-learning' is now considered as the 'new normal'. Users are also engaged in online playing, shopping, banking, health purpose and also for self-learning. As we find from our survey research, many (Internet) users are finding Internet-based social media as very part of their daily life. They go to bed and again wake-up with Internet, they often sleep with (Internet) connected devices. It is now 'most common' to find 'multi-window', 'multi-tasking' new-generation netizens, who are fortunately 'concerned' about senior citizen's 'isolation' and also take a 'mature view' of adult contents available in Internet.

Long live Internet!

Acknowledgements This paper is part of the ICSSR–IMPRESS funded research project of author. The author acknowledges the support and contribution of Prof. SC Das, Ms Sonalimayee Sahu and Ms Vandana Pattnaik.

References

1. Ahorsu, D.K., C.Y. Lin, V. Imani, M. Saffari, M.D. Griffiths, and A.H. Pakpour. 2020. The fear of COVID-19 scale: development and initial validation. *International Journal of Mental Health and Addiction*
2. Blasi, M.D., A. Giardina, C. Giordano, G.L. Coco, C. Tosto, and J. Billieux. 2019. Problematic video game use as an emotional coping strategy: Evidence from a sample of MMORPG gamers. *Journal of Behavioural Addictions* 2019 (8): 25–34.

3. Borg, K., and L. Smith, 2018. Digital inclusion and online behaviour: five typologies of Australian internet users. *Behaviour & Information Technology, 37* (4): 367–380

4. Choudhary, R. 2020. COVID-19 pandemic: Impact and strategies for education sector in India. ET Government. April 16, 2020, 09:45 IST.

5. Cooper, H., L. Hedges, and J. Valentine. 2009. Research synthesis as a scientific process (eds.). *Handbook of research synthesis and meta-analysis* (3–16). Russell Sage Foundation

6. Grau, S., and S. Kleeiser, 2018. Exploring social media addiction among student Millennials. *Qualitative Market Research: An International Journal, 22* (2): 200–216 © Emerald Publishing Limited. 1352-2752. https://doi.org/10.1108/QMR-02-2017

7. Gregory, C. 2019. Internet addiction disorder: signs, symptoms, diagnosis and treatments for those who may be on their PC or Smart phone.

8. Griffiths, M. 2010. Professional practice internet abuse an internet addiction in the workplace. *The Journal of Workplace Learning, 22* (7): 463–472. © Emerald Publishing Limited. 1366-5626. https://doi.org/10.1108/13665621011071127

9. Gvili, Y., and S. Levy. 2016. Antecedents of attitudes toward eWOM communication: differences across channels. *Internet Research, 26* (5): 1030–1051. https://doi.org/10.1108/IntR-08-2014-0201

10. Hamari, J, and M. Sjoblom. 2017. What is e-sports & why do people watch it. *Internet Research, 27* (2): 211–232. © Emerald Publishing Limited. 1066-2243. https://doi.org/10.1108/IntR-04-2016-0085

11. Hill, W.W., and Beatty, S.E. 2013. A segmentation of adolescent online users and shoppers. *Journal of Services Marketing, 27/5*: 347–360. q Emerald Group Publishing Limited. ISSN 0887-6045. https://doi.org/10.1108/JSM-10-2011-0157.

11. Hunady, J. 2019. The effect of the Internet on corruption awareness and corruption incidence in the EU. *Information Policy* 24 (2019): 75–89. https://doi.org/10.3233/IP-180086 ((IOS Press)).

13. Illingworth, L., D. Williams, and S. Burnett. 2002. The costs & benefits of the internet as a marketing & communications tool: the attitudes, perceptions and experiences within the non-profit environmental sector in Scotland. *Aslib Proceedings, 54* (5): 280–293. © MCB UP Limited. ISSN 0001-253X. https://doi.org/10.1108/00012530210448235

14. Jiang, Q. 2013. Internet addiction among young people in China: Internet connectedness, online gaming and academic performance decrement. *Internet Research, 24* (1): 2–20. Emerald Group Publishing Limited. 1066-2243. https://doi.org/10.1108/IntR-01-2013-0004.

15. Kulesz, O. 2016. The impact of digital technologies on the diversity of cultural expressions In Spain and Hispanic America. DCE/16/10.IGC/INF.4 Paris, 10 November 2016 Original: Spanish

16. Lafferty, B.A., R.E. Goldsmith, and S.J. Newell. 2002. The dual credibility model: The influence of corporate and endorser credibility on attitudes and purchase intentions. *Journal of Marketing Theory and Practice, 10* (3), 1–11, https://doi.org/10.1080/10696679.2002.11501916

17. Serrano, and Munoz-Soro, J.F. 2018. A multivariate study of internet use and the digital divide. *Social Science Quarterly, 99* (4,). © 2018 by the South-Western Social Science Association. https://doi.org/10.1111/ssqu.12504.

Applicational Statistics in Data Science and Machine Learning

Indrashis Das and Anoushka Mishra

Abstract In the domain of data science and machine learning, statistics plays a huge role. When it comes to gaining insights and building quality features out of the data to train any model, statistical tools and techniques along with the concepts of exploratory data analysis assist in doing the same. A data scientist or data analyst is incomplete without the knowledge of statistics because this is the building block of a machine learning or deep learning model which has learned or needs to learn trends and patterns from the features which were built by analysing the data end-to-end, be it in any tabular form or in picture format or video format. Also, as it covers a lot many concepts under statistics like variables, sampling, correlation, outlier treatment and much more, this chapter solely aims to take the reader to a tour of applicational statistics and how it can be combined with exploratory data analysis to easily work on data science and machine learning. Also, data analysis and machine learning are domains that are experiment heavy and need correct statistical methods for correct inferencing. Hence, for these experiments, the different statistical methods in place are discussed here in detail. There are different languages like Python, MATLAB, R and much more which have libraries for statistical mathematics and make simple API calls to do the required experiments within any dataset.

Keywords Statistics · EDA · Data science · Machine learning · Data analysis · Python

1 Introduction

Statistics and exploratory data analysis are interlinked where both hand-in-hand assist in seeking and deriving insights by doing an end-to-end analysis on top of the data.

I. Das (✉)
HighRadius Technologies, Bhubaneswar, India
e-mail: indrashisdas98@gmail.com

A. Mishra
Dell Technologies, Bhubaneswar, India
e-mail: anoushka1900@gmail.com

© The Author(s), under exclusive license to Springer Nature Singapore Pte Ltd. 2021
S. Rautaray et al. (eds.), *Trends of Data Science and Applications*,
Studies in Computational Intelligence 954,
https://doi.org/10.1007/978-981-33-6815-6_4

But very often, budding data scientists and analysts fall in a dilemma to understand which tool or technique to use. Therefore, all those tools and techniques are discussed here such that it can help terminate all such dilemmas that turn up in the mind. For example, many times just by seeing a variable, a person might not be able to identify the variable type for either being a categorical variable or numerical variable. Hence, to vanish such confusions, several scenarios are explained using examples.

1.1   Statistics and Exploratory Data Analysis

Talking about statistics, it is the methodological science for collecting data, defining data, analysing data and inferring insights from data. Now, when we talk of collecting and defining data, the concepts of population and sample along with the concept of variable types pop up. It is very true that before proceeding with analysing the data, there should always be a sanity check so that the data scientist or analyst gets aware of the discrepancies or anomalies in the data before moving for the analysis phase.

The next phase in the process turns out to be analysing data and inferring insights from it. This particular phase is a handshake between exploratory data analysis and statistical tools and techniques which demands experimenting on the data to understand the hidden trends and patterns. With this phase invoked, concepts like measures of central tendency, outlier treatment, probability distributions and many other such concepts show up. Hence, this is a guide that can help you out with all these concepts under applicational statistics.

1.2   Statistical Tools and Techniques

The concepts are organized in an order which a data scientist or data analyst would follow to do the general exploratory data analysis. Further comes up the concept of variables and types of variables. It is important to analyse the data and define the variables before doing modelling. Once we understand the variables, the phase of hardcore data analysis begins. Hardcore data analysis includes describing the required variables for analysis by its mean, median and mode known as the measures of central tendency. Apart from measures of central tendency, other statistical techniques included are various distributions in statistics, outlier treatment, correlation analysis, variance analysis, chi-square analysis, ANOVA, Z-score, bias and variance, and hypothesis testing. Accordingly, Fig. 1 shows how the concepts of probability and statistics fused together along with exploratory data analysis can help in finding meaningful insights from the data. It is very important to find insights from the data as these act as features for training a machine learning model.

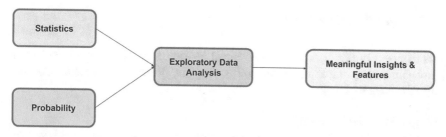

Fig. 1 Use of statistics and probability in data science and machine learning

2 Sampling Techniques

2.1 Population Versus Sample

For dealing with any problem, it is very important to understand the data related to it. All data that satisfies the objective being studied is called a population. Let us say we need to understand the job possibility of students in a particular university. This would not just include students on-campus, but also those involved in foreign exchange programmes and distance education as well. All these students constitute the population of the problem statement. This leads to the accumulation of a large volume of data that subjects our system. To overcome this, a portion of the population referred to as a sample is considered. In the above case, analysing only the on-campus students could give us the required insights about such students only.

Sampling is a statistical technique that focuses on selecting a portion of the population to obtain cost model training. Sampling of a population is not as simple as it seems. While sampling, each section of the population should be considered. Let us consider a COVID-19 dataset representing the probabilities of people from different age groups being affected. Let us say, we train our model with a sample that contains people above 50 years of age. This model may not capture the required pattern since it represents only a specific age group of the dataset. Hence, it is very important to choose a sample that depicts the characteristics of the entire data and satisfies the problem statement. The following are some sampling techniques that help us achieve these objectives.

2.2 Sampling Methods

As depicted in Fig. 2, there are two broad categories of sampling which include probability sampling and non-probability sampling. Below are the techniques used under each one of them.

Simple Random Sampling
It is the technique of randomly selecting a sample from a population where each member has the same possibility of being preferred.

For example, 5 students have been chosen from a class of 100 for a group discussion. In this case, the population is 100 and the sample size is 5. Each student has an equal possibility of being selected. It has been proven that the storage required is only $O(\sqrt{k})$, where k is the sample size as discussed by Meng [1].

Systematic Sampling
This sampling method is applicable when the entire population from which the sample is drawn is available. Under this method, elements of the sample are picked at regular intervals say k, where k can be derived by taking the ratio of the population and the proposed size of the sample. For example, if 50 elements are to be selected from a population of 400 elements, then every 8th (400/50) element is selected, i.e. 8, 16, 24, 32, 40… 384, 392, 400.

Stratified Sampling
This sampling method is applicable when there is a population of varied nature; i.e. subdivisions are possible. The population is divided into groups, and independently from each group, elements are selected to form the sample. For example, if you have 4 groups with 200, 400, 600 and 800 respective sizes and you decide to choose 1/4 as a sampling fraction, then you would have to select 50, 100, 150 and 200 members from the respective stratum.

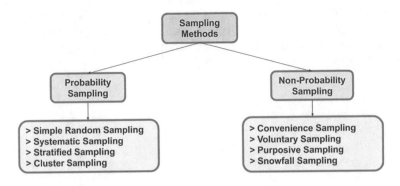

Fig. 2 Types of sampling

Cluster Sampling

This method proposes the division of the population into clusters in such a way that every cluster is homogeneous and has a representation of all the groups in a population. Any one of the clusters is selected through random or systematic sampling. For example, in a survey of students from a city, we first select a sample of schools, then we select a sample of classrooms within the selected schools, and finally, we select a sample of students within the selected classes.

Convenience Sampling

In convenience sampling, selection is made based on convenience and proximity and no emphasis is given on the representation of the entire population. For example, Tata Steel is planning to have some vendors who would provide them raw iron. It selects the top 5 cities based on the accessibility.

Voluntary Sampling

This is an important non-probability sampling. In voluntary sampling people willing choose to be a part of the survey. Generally, such people have a strong opinion about the topic and such samples are always subjected to bias.

Purposive Sampling

Purposive sampling is a sampling technique where the researcher trusts his/her judgement while choosing members of the population to participate in the study. The selection of people is based on a particular profile.

Snowball Sampling

This is a non-probability sampling technique based on chain referral. After selecting the initial subject, these subjects enrol future subjects from their contacts. Snowball sampling is of different types like linear, exponential discriminative, exponential and non-discriminative snowball sampling (Table 1).

3   Types of Variables

There are several types of variables, and it is highly appreciated to define the variables and the types so that it becomes easier to analyse the data. Different variables consist of different values and might be of different data types; i.e. it might be an integer-type or float-type variable or may be a date-type variable as well. Hence, Fig. 3 explains the different types of variables that one can find while analysing the data. The variables which are shown in the diagram above are explained below starting with random variables and then going forward towards categorical, numerical, qualitative and quantitative variables. Each variable is cited with an example so that it gets easier for the reader to understand what the variable is all about and what it tries to explain.

Table 1 Summarizing types of probabilistic and non-probabilistic sampling

Sampling technique	Description	Application
Simple random sampling	Technique of probabilistic sampling where each sample has equal probability of being picked	Technique is mostly used with homogeneous data. Example sample of employee to understand work efficiency
Systematic sampling	Entire population is available to pick samples where samples are picked after a certain interval of k	Example sample of employee to understand work efficiency
Stratified sampling	Equally proportionate samples are picked from each class or stratum	Sample of employee collected from each level of employee to understand their satisfaction with the job
Cluster sampling	Entire population is divided into homogeneous groups or clusters where simple random or systematic sampling is applied	Sampling of people of different strata of life to understand their lifestyle
Convenience sampling	Emphasis on representing the entire population is not taken care of. Rather, based on proximity samples are chosen	Pilots use convenience sampling to understand the weather condition around then to make sure flight is safe or not
Voluntary sampling	People are chosen for surveys. These people have domain knowledge. Such techniques are subjected to bias	Sampling of workers by the union leader to take feedback from workers and understand if the work environment is healthy or not
Purposive sampling	This is based on a researcher's judgements or understanding. The samples are selected based on the judgement of the researcher	A researcher sending his work to other scientist for feedback
Snowball sampling	This is a chain mechanism of selecting samples. Once one sample is picked, the other ones are picked by the previous samples	A survey conduction where a form is circulated and people who have filled further circulate the form to others; in this way data is collected

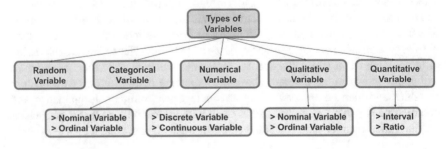

Fig. 3 Types of variables

3.1 Random Variable

A random variable is one that takes values from a whole set of predefined values. Random variables are important to understand probability distribution in statistics. There are 2 types of random variables—discrete and continuous random variables. If a random variable considers integer values in a particular range such as 1, 2, 3, 4..., then it is a discrete random variable else is a continuous random variable. Random variables are different from the normal variables.

For example, $x + 5 = 10$. Here, the value of $x = 5$. This is a normal variable. A random variable on the other hand is defined as $X = \{0, 1, 2, 3\}$ where X can take values 0, 1, 2 or 3 randomly where each one has the same or different probabilities.

3.2 Categorical Data

Categorical data helps to group the information collected based on various classes in the variable. Categorical variables are of two types:

Nominal Variable
A nominal variable is one which does not follow an order while naming a variable. These are sometimes known as labelled or name data. For example, gender has two categories (male and female), season having four categories (summer, winter, monsoon and spring).

Ordinal Variable
An ordinal variable has a clear ordering among its values, for example, the size of T-shirts or shoes.

3.3 Numerical Data

Numerical data is represented in numbers such as temperature, weight, height and length. Numerical data can be further divided into two types:

Discrete Variable
A discrete variable represents countable items. The values can be grouped into lists that can be finite or infinite. The number of buckets of water present in a tank is finite countable, and the number of buckets of water present in an ocean is infinite countable.

Continuous Variable
In continuous variable, the values can take any number from the number line, instead of counting numbers. For example, the heights of people in a class can take values from 0 to 10 where the value can be in decimals as well.

3.4 Qualitative Data

Qualitative data is descriptive in nature. They are also known as categorical data. Types of such data include:

Nominal Variable
It is derived from the Latin word "Nomen" (meaning name), name's variables. Nominal data does not have any numeric characteristics. Such data cannot be manipulated using available mathematical operations.

Ordinal Variable
This is a category of qualitative data where variables are represented as ordered categories and there is some relation among those categories. For example, a product having 4 categories is excellent, good, bad and worst.

3.5 Quantitative Data

This data is numerical, and they can be counted, measured and manipulated.

Interval
This is a continuous data and is not ordered on a scale (e.g. ratings of 1–10). Each value has equal spacing from the value after and before it. For example, the separation between 19 and 20 km is equivalent to that between 6 and 7 km.

Ratio
Such data has an absolute zero value, and the data is continuous. Ratio data and interval data are similar, for example, temperature, which can take to zero degrees (Table 2).

4 Visualizing Data

This section explains the visualizations available for viewing categorical and numerical data. There are separate visualizations for each of these broad categories. The different visualizations are shown below.

Table 2 Summarizing types of variables

Type of variable	Description	Application
Random variable	Takes values from a whole set of defined values. Can be further divided into discrete and continuous variables	Rating of a product in a review analysis dataset
Categorical variable	Refers to the stringed categorical variables where data can be grouped based on classes. Can be further divided as nominal and ordinal variables	Nominal: a column in dataset containing the gender of the people Ordinal: a column having T-shirt size
Numerical variable	Refers to the variables with data in number format. This can also be divided as discrete and continuous variables	Count of boxes in a dataset comes under discrete data A dataset containing temperature of different places as a column is a continuous column
Qualitative variable	These variables are descriptive in nature. Can be divided as nominal and ordinal variables	Label encoding of weather will give nominal qualitative variable If satisfaction of a product (good, average, bad) is label encoded, it comes under ordinal label encoding
Quantitative variable	Numerical data that can be counted, measured and manipulated. Can be further divided as interval and ratio	Cost of a place height of a person comes under quantitative variables

4.1 Categorical Data

Pie chart

Pie charts are helpful when we represent qualitative data, where an attribute or a feature is not numerical. Each part of pie illustrates a different category, and each slice represents a particular trait. The below example in Fig. 4 shows T-shirt size.

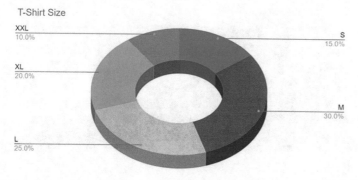

Fig. 4 Pie chart

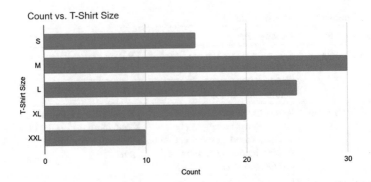

Fig. 5 Bar chart

Bar Plot

This kind of plot helps to visualize qualitative data. Each bar represents the frequency of the individual category, and it helps to compare the frequency count of two categories. As shown in the below diagram, i.e. Fig. 5, this is a typical bar chart that is used for exploratory data analysis on categorical variables.

4.2 Numerical Data

Histograms

These are used to represent quantitative data. The X-axis represents the range of values, i.e. classes, and the classes with maximum frequency have taller bars. Histograms are different from bar plots in the level of measurement of data. Figure 6 justifies the above statement.

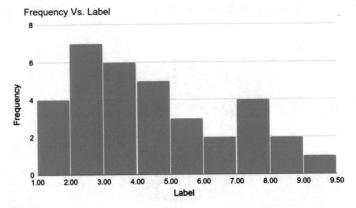

Fig. 6 Histogram

Scatter Plot

As shown in Fig. 7, a scatter plot helps to show trends or correlation among variables. It also helps us to determine the shape and strength of the correlation. It also helps us to determine outliers.

Box Plot

It helps us to understand the distribution of the data as shown in Fig. 8. It helps us to determine the maximum, minimum and interquartile range of the values. It also helps us to find outliers in our data.

P-P Plots

Probability–probability plot is a distribution plot that determines how comparable the two data are. It helps to compare a sample distribution with a theoretical distribution. Figure 9 justifies the above statement (Table 3).

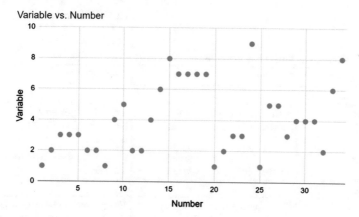

Fig. 7 Scatter plot

Fig. 8 Box plot

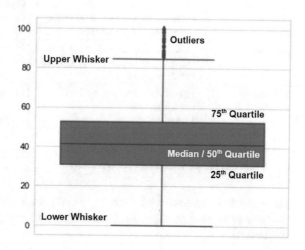

Fig. 9 P-P plot

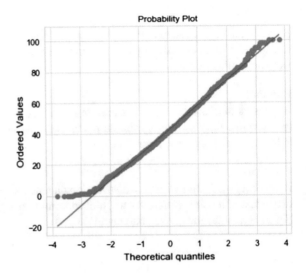

Table 3 Summarizing types of visualizations

Visualization name	Description	Application
Categorical variable		
Pie chart	A circular chart with sections as percentage division of classes	Analysis of percentage of population from different states in a conference
Bar chart	A chart that shows frequency of different classes in a variable	Count analysis of favourite music in a movie
Numerical variable		
Histograms	Show frequency of continuous variables by binning data on the X-axis	Histograms can be used to visualize data of students' performance by providing different buckets
Scatter plot	Shows data points as simple dots or scatters. Mostly used to see data concentration and outliers	Plotting distance travelled with petrol consumption under their relation
Box plot	Used to view outliers in the data. Can also be used to view IQR and other quartiles	Price hike of a place can be determined using a box plot
P-P plot	Shows how comparable two variables are	Comparison of output generated by model with the already existing output column

5 Measures of Central Tendency

There are three measures of central tendency which basically include the mean, median and mode. Apart from this, standard deviation and variance are also included in this section.

5.1 Mean

Mean is the average of all the values in a set. There are different ways of calculating mean, and this includes:

Arithmetic Mean
Arithmetic mean is the most used method. If $X_1, X_2, X_3, ..., X_n$ are the values, then mean of these values is denoted by μ

$$\mu = (X_1 + X_2 + X_3 ... X_n)/n \tag{1}$$

Geometric Mean
Geometric mean is the nth root of the n positive values $X_1, X_2, X_3, X_4...X_n$.

$$G = \sqrt[n]{(X_1 X_2 X_3 ... X_n)} \tag{2}$$

Harmonic Mean
For a set of n values $X_1, X_2, X_3, X_4 ... X_n$, harmonic mean is the reciprocal of the arithmetic mean of the reciprocals of the values

$$H = 1/\left(\sum (1/X_i)\right) \tag{3}$$

5.2 Median

The value in the exact centre, when the data is arranged in ascending order, is known as the median. In case when the number of observations is odd, median is the $(n + 1)/2$th observation and when there are even numbers of observations median is the mean of $n/2$th and $((n/2) + 1)$th observation. For example, consider this set {1, 13, 2, 34, 11, 57, 27, 47, 9} and the arranged set is {1, 2, 9, 11, 13, 27, 34, 47, 57}. The median of this set is 13.

5.3 Mode

As mentioned by Manikandan [2], mode is the value with the maximum frequency. For example, there is a set {1, 3, 3, 3, 6, 7, 7, 7, 7, 9, 10}. In this set since 7 has the maximum frequency as it appears 4 times in the set, the mode is 7. However, there are cases when there is no mode such a situation arises when all the values in the set have a single occurrence.

5.4 Variance

It is the expected value of the squared variation of a variable from its mean value. It estimates how far a set of numbers are from their mean value. Let us consider a sample of n numbers, i.e. $x_1, x_2, x_3, x_4 \ldots x_n$. Hence, the variance would be

For a population,

$$S^2 = \sum (X_i - X_{mean})^2 / n \tag{4}$$

For a sample,

$$\sigma^2 = \sum (X_i - X_{mean})^2 / (n - 1) \tag{5}$$

5.5 Standard Deviation

The square root of variance is called standard deviation. The symbols S and σ represent standard deviation for a sample and population.

For a population,

$$S = \sqrt{\left(\sum (X_i - X_{mean})^2 / n \right)} \tag{6}$$

For a sample,

$$\sigma = \sqrt{\left(\sum (X_i - X_{mean})^2 / (n - 1) \right)} \tag{7}$$

To understand variance and standard deviation in more detail, you can refer to Ahn et al. [3]

6 Distributions in Statistics

A very eminent and influential part of statistics includes distributions which is very intuitive in understanding various variables and fields while building features in the process of machine learning. Basically, distributions speak about how values are distributed for a particular variable, be it more towards larger values, or smaller values, or maybe towards a higher probability for moderate values in case if the distribution is Gaussian or normal.

6.1 Probability Distributions

As discussed by Hernandez [4] in his paper, probability distributions generally depend on two factors of a random variable which includes probability distribution of each random element in the random variable along with particular selection probability of each element. Therefore, probability distributions are functions that basically depict the probability or chance of occurrence of each element in a random variable.

For an example, if we find a probability distribution of scores for a set of students in any examination or test, then the probability distribution formed will always be a Gaussian or normal curve which is the perfect example of randomness, because in a class, there can be higher number of average students scoring average marks because of which probability for average scores would always be high and less number of below average and above average students scoring either too less or too high in any examination leading to lesser probability for such scores.

Further, there are some common probability distributions which are discussed below along with some terminologies associated or linked with the concept of probability distributions.

6.2 PMF Versus PDF

Probability Mass Function

As discussed by Hossain et al. [5] in the paper, to visualize probability distribution for discrete random variables as shown in Fig. 10, probability mass functions are used. Probability mass function helps to define discrete probability distributions. Bernoulli distribution, binomial distribution, Poisson distribution and geometric distribution fall under probability mass function. Also, in case of multivariate probability mass function, a joint probability distribution is used generally.

Fig. 10 Probability mass function

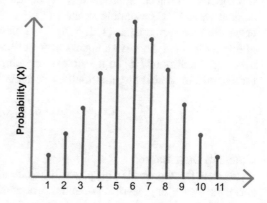

Fig. 11 Probability density function

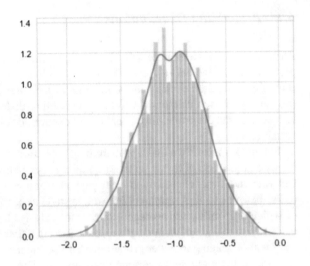

Probability Density Function

In the case of probability density function as shown in Fig. 11, the distribution is used to visualize probability distribution for continuous random variables. Normal or Gaussian distribution falls under probability density function. In the probability density function, for any sample in the sample space, the function returns a chance of occurrence or probability of occurrence for that particular sample.

6.3 Common Probability Distributions

Uniform Distribution

Uniform probability distributions also known as rectangular probability distribution have a constant density over the interval, and hence for all elements in a random variable, there is an equal probability of occurrence for each random element in the random variable. For example as shown in Fig. 12, while we roll a die, there are 6 values that can turn up, i.e. {1, 2, 3, 4, 5, 6}. Now, if we do not add any bias while rolling a die, then there is an equal chance of any value turning up; i.e. all values have a probability of 1/6, be it 1 or 2 or any number from the set. Hence, this will form a uniform probability distribution. Formula for uniform distribution is

$$P(x) = 1/(b - a), \quad a \le x \le b \tag{8}$$

Gaussian Distribution

Dytso et al. [6] described in his paper that Gaussian distribution has been very much accepted by the engineering community because of its flexible parametric form. Gaussian distribution, also known as normal distribution, is the best example of

Fig. 12 Uniform
distribution

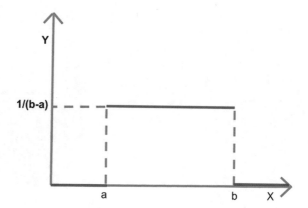

randomness. Gaussian distributions are symmetric about the mean having a bell-shaped curve, with mean, median and mode on the same point in the X-axis. For example, one can see in Fig. 13 that as discussed in the above sections, scores in a test for a class of 80 students are Gaussian in nature. Formula for Gaussian distribution is

$$P(x) = \left(e^{\left(-(x-\mu)^2/2\sigma^2\right)}\right)/\sqrt{(2\pi\sigma^2)}, \quad -\infty \leq x \leq \infty \tag{9}$$

where μ is the mean, σ is the standard deviation and π is the universal constant pi.

Bernoulli Distribution

One can see in Fig. 14 that the Bernoulli distribution is a discrete distribution, which has binary outcomes that include 0 and 1. Here, 1 means success that occurs with probability p, while 0 means failure that occurs with probability $(1 - p)$. For example, the distribution of heads and tails in tossing a coin is a Bernoulli distribution. Formula for Bernoulli distribution is

Fig. 13 Gaussian
distribution

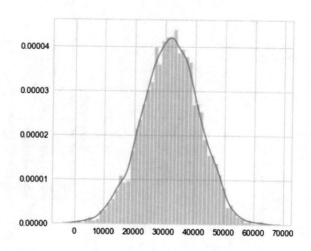

Fig. 14 Bernoulli
distribution

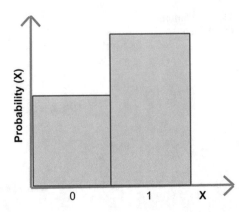

$$P(x) = p^x p^{(1-x)}, \quad x = 0 \quad \text{or} \quad x = 1 \tag{10}$$

where x is the outcome which is either 0 or 1.

Binomial Distribution

As discussed in his paper by Palmisano [7], binomial distribution is the method of modelling likelihood of only two certain outcomes, i.e. by either a success or failure. The conditions for a binomial distribution are that the number of trials is fixed, there is either a success or failure to the trials, there is an equal probability to success in all the trials and the trials are independent as shown in Fig. 15. For example, binomial distribution can be applied to flipping a coin n-number of times. The trials are fixed, i.e. either a head or tail may come, there are two cases possible for getting a head, either it is a success or failure, also, the trials are independent and the probability of success is constant, i.e. probability of getting a head is 0.5. Formula for binomial distribution is

$$P(x) = \left(n! p^x (1 - p)^{n-x}\right) / (x!(n - x)!), \quad 0 \le x \le \infty \tag{11}$$

Fig. 15 Binomial
distribution

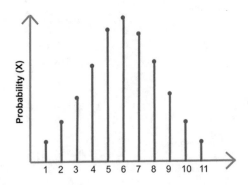

Table 4 Summarizing the formulas for different distributions

Distribution type	Distribution formula
Uniform	$P(x) = 1/(b - a), a \leq x \leq b$
Gaussian/normal	$(x) = (e^{(-(x-\mu)^2/2\sigma^2)})/\sqrt{(2\pi\sigma^2)}, -\infty \leq x \leq \infty$
Bernoulli	$P(x) = p^x p^{(1-x)}, x = 0$ or $x = 1$
Binomial	$P(x) = (n!p^x(1-p)^{n-x})/(x!(n-x)!), 0 \leq x \leq \infty$
Poisson	$P(x) = (\lambda^x e^{-\lambda})/x!, 0 \leq x \leq \infty$

where n is number of trials, x is the specified number of successes and p is the probability of success.

Poisson Distribution

Unlike binomial distribution, in Poisson distribution, it does not model frequencies of success and failures. Rather, in a fixed unit of time and space, it provides an expected number of events that can occur. Poisson distribution helps to formulate if any event was generated in randomness. The random variable is a positive integer that exists from 0 to infinity due to no cut-off. Also, Poisson distribution approximates binomial distribution when n is large and p is small. Formula for Poisson distribution is

$$P(x) = (\lambda^x e^{-\lambda})/x!, \quad 0 \leq x \leq \infty \tag{12}$$

where λ is the number of events occurring per time per space and x is the general number of events (Table 4; Fig. 16).

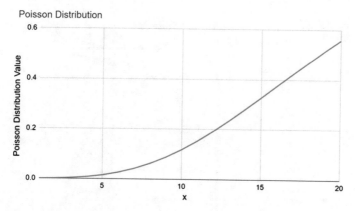

Fig. 16 Poisson distribution

6.4   Kurtosis

Introduction to Kurtosis and Interquartile Range
Kurtosis can be defined as the measure that defines how well are the tails of a distribution spread with respect to the normal distribution. It defines if tails of a distribution contain extreme values or not.

Usually, kurtosis is a quality statistical measure for various domains. For example, a higher kurtosis refers to financial risk as it signifies lesser returns; hence, financers look forward to a higher kurtosis for moderate-level risk.

Usually, kurtosis is one of the finest methods or measures to detect outliers in any variable. Many times, extremely smaller or larger values might signify some meaningful or logical information in the dataset for which such values should not be removed as it might show some important pattern. This is the reason why outlier treatment should be done carefully. In the below sections, these concepts will be discussed in detail. Formula for kurtosis is

$$\text{Kurtosis} = \left(\sum_{i=1}^{n} (x_i - \mu)^4 / n \right) / \left(\sum_{i=1}^{n} (x_i - \mu)^2 / n \right)^2 \tag{13}$$

where x_i is an element in the random variable, μ is the mean and n is the number of samples in the random variable.

As it can be seen from Fig. 17, there are certain terminologies which need to be explained and hence are explained in detail below:

25th Quartile—This can be defined as the lower quartile starting which 50% of the values from the random variable lie in the box plot.

Fig. 17 Box plot and normal distribution

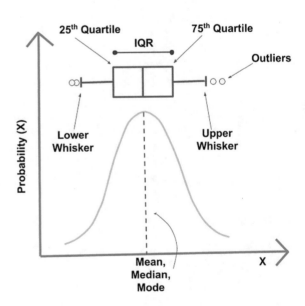

50th Quartile—This can be defined as the middle quartile which is midway to 50% of the values from the random variable in the box plot. This value also signifies the median of the variable.

75th Quartile—This can be defined as the upper quartile till which 50% of the values in the random variable exist.

Interquartile Range—The range of scores from lower quartile to the upper quartile of the box in box plot is known as the interquartile range.

Formula to calculate interquartile range is

$$IQR = 75th\,Quartile - 25th\,Quartile \qquad (14)$$

Lower Whisker—This is the lower boundary beyond which any value is considered as an outlier.

Formula to calculate lower whisker is

$$Lower\,Whisker = 25th\,Quartile - 1.5 * IQR \qquad (15)$$

Upper Whisker—This is the upper boundary beyond which any value is considered as an outlier.

Formula to calculate upper whisker is

$$Upper\,Whisker = 75th\,Quartile + 1.5 * IQR \qquad (16)$$

Types of Kurtosis

Kurtosis is mainly divided into three types as shown in Fig. 18.

Fig. 18 Types of kurtosis

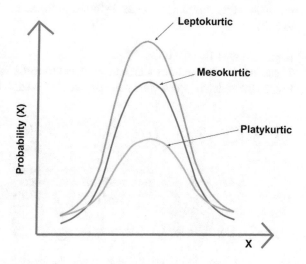

Mesokurtic—Such distributions are in sync with the normal distribution; i.e. they have a moderate height, and the tails are spread lesser. They have a kurtosis of zero or near to zero.

Leptokurtic—These distributions have a higher peak with larger number of outliers because of which the tails are spread wide. Also, these distributions have a positive excess kurtosis.

Platykurtic—These distributions have a flatter peak with lesser number of outliers because of which the tails are also flat. Also, these distributions have a negative excess kurtosis.

6.5 Skewness in Distributions

If one of the tails of a distribution is stretched towards a particular direction be it towards left or towards right, then the distribution is known as a skewed distribution. The prime reason for skewed distributions is due to higher probability of very higher values and lower probability of extreme lower values or higher probability of lower values and lower probability of extremely higher values. The different types of skewed distributions are discussed below.

Left Skewed Distribution
These are the distributions which have its tail towards the negative side as shown in Fig. 19 for which it is also known as negative skewed distribution. Also, higher ranged values being more in the random variable lead to higher probability for such values while lower ranged values having lesser probability due to lesser occurrence in the random variable. Value on the X-axis for which the probability is highest shows the mode of the distribution which is followed by the median and the mean from left to right.

Right Skewed Distribution
These are the distributions which have its tail towards the positive side as shown in Fig. 20 for which it is also known as positive skewed distribution. Also, lower ranged

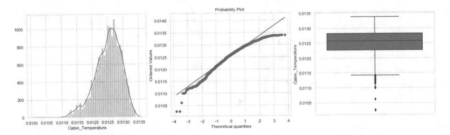

Fig. 19 Probability distribution, P-P plot and box plot for a left skewed distribution

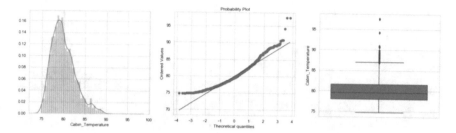

Fig. 20 Probability distribution, P-P plot and box plot for a right skewed distribution

values being more in the random variable lead to higher probability for such values while higher ranged values having lesser probability due to lesser occurrence in the random variable. Value on the x-axis for which the probability is highest shows the mode of the distribution which is followed by the median and the mean from right to left.

Non-skewed Distribution
This is a perfect normal distribution or bell-shaped curve that has the highest peak at the centre of the range of values on the X-axis. The mean, median and mode lie on the same point in the X-axis.

6.6 Scaling and Transformations

Generally, while working on datasets it happens that for some continuous random variable, the distribution is skewed, i.e. either it is left skewed or right skewed. Hence in order to make the random variable Gaussian, transformations are required as models understand Gaussianity well. Also, many times, values in random variables are so large that they need to be scaled for which scaling techniques are applied. The requirement for scaling comes as larger values take time to be learned by the model. The purpose of scaling values is to reduce time complexity while training models (Fig. 21).

Standard Scaling
This is a technique that is used to scale extremely larger values to smaller ranges by keeping the distribution or Gaussianity intact. The transformed values fall in a positive and negative range, i.e. to ± 1 or ± 2 range. Values lesser than the mean fall in the negative range, while values higher than the mean fall in the positive range. Formula for standard scaling is

$$x' = (x_i - \mu)/\sigma \tag{17}$$

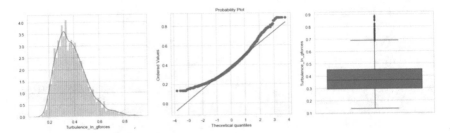

Fig. 21 Probability distribution, P-P plot and box plot for a continuous variable

where x_i is an element in the random variable, μ is the mean, σ is the standard deviation and x' is the transformed value.

This is also known as the z-score method which is discussed in detail in the upcoming sections (Fig. 22).

Min–Max Scaling

In this method, the minimum and maximum values in a random variable are used to scale values from large ranges to a lower range keeping the distribution or Gaussianity same in the random variable. Also, in this method of scaling, the transformed values fall in a range between 0 and 1. Formula for min–max scaling is

$$x' = (x_i - x_{min})/(x_{max} - x_{min}) \tag{18}$$

where x_i is an element in the random variable, x_{min} is the minimum value in the random variable, x_{max} is the maximum value in the random variable and x' is the transformed value (Fig. 23).

Logarithmic Transformation

According to simple mathematics, logarithm is a power to which if a base is raised, then we can get the exact number for which we are trying to find a log for. Hence, as the above statement is intuitive enough, logarithmic transformations are used to make left skewed distributions more Gaussian as the larger values are decreased

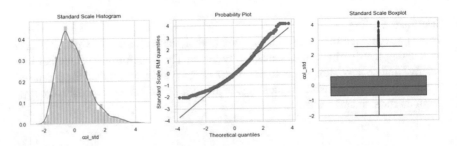

Fig. 22 Probability distribution, P-P plot and box plot for standard scaled data

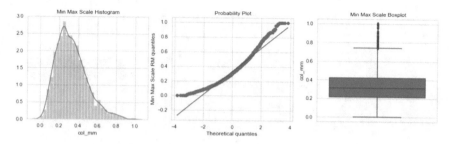

Fig. 23 Probability distribution, P-P plot and box plot for min–max scaled data

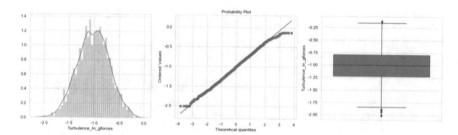

Fig. 24 Probability distribution, P-P plot and box plot for logarithmic transformation

to lower scale by taking a log. For example, log of 1000 with base 10 is 3 which reduces the corresponding value from 1000 to 3. But logarithmic transformations have a disadvantage which is that it does not work for random variables containing negative values and the number zero as well (Fig. 24).

Exponential Transformation
In comparison with logarithmic transformation, exponential transformation can not only be used to Gaussianize left skewed distributions, but be used to Gaussianize right skewed distributions as well. Hence, it is a dual-purpose transformation. Also, another advantage is that it can work for negative values as well. For example, in case of right skewed data, higher degree power can be used to transform values to a larger scale while in case of left skewed data, lower degree power can be used to transform values to lower scale to make the distribution Gaussian (Fig. 25).

Reciprocal Transformation
This is another process of transforming left skewed distributions to a normal distribution. Similar to logarithmic transformations, it can be used to scale a left skewed random variable. But in comparison with logarithmic transformation, this can be used to downscale large negative numbers as well. For example, if a random element is valued 1000 in a random variable, then its reciprocal would be 0.001 which is a higher downscaling of value to increase the Gaussianity of the random variable (Fig. 26).

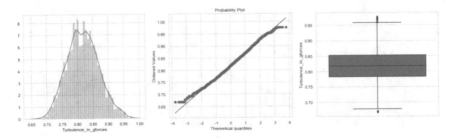

Fig. 25 Probability distribution, P-P plot and box plot for exponential transformation

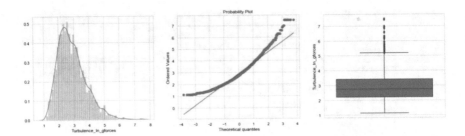

Fig. 26 Probability distribution, P-P plot and box plot for reciprocal transformation

Root-Based Transformation

Similarly, this is also a mechanism to downscale larger values for left skewed random variables. This is exactly similar to logarithmic transformation as in this case as well negative values cannot be downscaled due to generation of complex numbers. Hence, random variables with only positive values can be downscaled. For example, if a random variable has a random element to be valued as 1,048,576, then, on square rooting this number, the transformed value would be 20 which is a very small value in comparison with 1,048,576. Also similarly, if we square root $-1,048,576$, then it will return $20\sqrt{i}$ which cannot be understood by a machine learning model (Fig. 27).

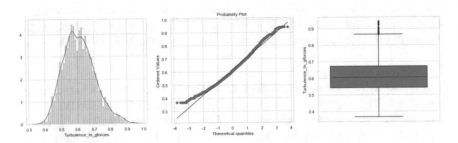

Fig. 27 Probability distribution, P-P plot and box plot for root-based transformation

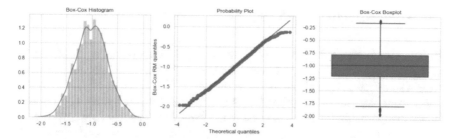

Fig. 28 Probability distribution, P-P plot and box plot for Box–Cox transformation

Box–Cox Transformation

Another effective transformation technique is the Box–Cox transformation. As discussed in his paper by Amir et al. [8], Box–Cox transformation is a transformation that is used to make data work in accordance with assumptions of linear regression and ANOVA. Box–Cox transformation is usually applied to non-normal-dependent random variables. Box–Cox transformation comes from a family of power transformations which is applied to skewed data to convert it to a normal distribution. Also, it cannot be applied to negative random elements in random variables. However, there is another transformation named Yeo–Johnson transformation that can be applied to zero and negative values to random elements in a random variable as well. Formula for Box–Cox transformation is (Fig. 28)

$$y(\lambda) = (y^{\lambda} - 1)/\lambda, \quad \text{if } \lambda \neq 0 \quad \text{or} \quad y(\lambda) = \log(y), \quad \text{if } \lambda = 0 \tag{19}$$

7 Outlier Treatment

Outliers play a significant role in screwing up the data used for modelling. Such entities should be removed from the dataset. But outlier treatment should be done extremely carefully because many times, it might show some important patterns or intuitive information, which when excluded or removed might be lossful for the model in the process of training. For example, a very high value in any continuous column might indicate null values, which when removed from the dataset might be a loss in capturing pattern. Hence, this section specifically talks about outliers which can be either disastrous for machine learning models or a blessing in disguise for the same.

7.1 Understanding Outliers

To define outliers, as described by Yang et al. [9] in his paper, it is nothing but deviations in typical data. So a value that cannot exist typically in a variable, but is present somehow is considered as an outlier.

For example, usually test score data of a class full of students cannot be negative, but suddenly you find a negative value in the variable. Now, there are certain cases possible for the same. Firstly, it can be an error from the teacher's end who entered the data in the portal. In such a case, the data scientist who wants to predict future score range for a student will have to clarify the same from the teacher and take action on it. Now, if the value is not supposed to be negative, then either it can be imputed by some other value, or the record might be dropped or ignored according to the requirement. Secondly, if there was some negative scoring on wrong answers, then in such a case the records should exist without any imputation in the same. Hence, as you saw in the above example, this is how the decision of the data scientist would change depending on the case.

7.2 Detecting Outliers

Below are some techniques for detecting outliers:

Standard Deviation
As explained by Yang et al. [9], earlier the method to detect outliers included mean $\pm 3\sigma$, i.e. three times the standard deviation following the standards of normal distribution. But there the scientists decided to keep an unknown control parameter which any user could define according to its requirements. The formula for calculating outlier score using standard deviation is as follows

$$\text{Threshold}_{\text{Min}} = \mu - a * \sigma \quad \text{Threshold}_{\text{Max}} = \mu + a * \sigma \tag{20}$$

where μ is the mean, σ is the standard deviation and a is the control parameter that can be defined by the user.

Now, smaller the control parameter, more would be the outliers in the data, while larger the control parameter, lesser would be the number of outliers.

Median and Median Absolute Deviation Method
Similar to the above process as described by Yang et al. [9], in this method, mean is replaced by median and standard deviation is replaced by median absolute deviation. Median is considered in comparison with mean because median is robust to outliers and it does not get affected. The formula for calculating outlier score using median and median absolute deviation is as follows

$$\text{Threshold}_{\text{Min}} = \text{median}(x) - a * \text{MAD} \quad \text{Threshold}_{\text{Max}}$$

$$= \text{median}(x) + a * \text{MAD} \tag{21}$$

$$\text{MAD} = b * \text{median}(|x - \text{median}(x)|) \tag{22}$$

where median(x) is the median of the random variable, MAD is the median absolute deviation, a is the control parameter and b is a constant value 1.4826.

MAD or median absolute deviation contains a term median $(|x - \text{median}(x)|)$ which means that for every variable in a random variable, first subtract the median from each value and then find the absolute value of each median deviation. Further, from the set of median deviations, find the median value.

Box Plots and Interquartile Range
Box plots are a significant help to visualize and detect outliers in the dataset. It is considered that any value below the lower whisker and any value above the upper whisker is an outlier. This helps understand the range in which the maximum data exists. As discussed in the above sections, IQR is the range of values in the box plot, i.e. the difference between 75th quartile and 25th quartile.

Scatter Plots
Scatter plots are also beneficial in detecting a lesser number of outliers from the data. When it can be observed that there are lesser outliers in a particular dataset, then scatter plots can be used to detect and make a call for action on the same. Extreme values can be detected via this method, but close outliers would be hard to detect using the same. This method can just be an indication to outliers if any exist in the dataset.

Probability Distribution Technique
Similarly, probability distribution technique is only intuitive enough to know if outliers exist in a variable or not. If there exist any outliers in the dataset, then the distribution is supposed to be skewed, either left or right in accordance with lower value outliers or higher value outliers. But, this method cannot provide exact values to the outliers, which is a demerit in this case (Table 5).

8 Correlation Analysis

Two or more variables are correlated means that if one variable is altered, the other would possibly get altered. Correlation is a statistical technique used for analysing the behaviour of two or more variables. Correlation provides us the measure of the direction and degree of sympathetic movement in two or more variables (Table 6).

The coefficient of correlation which ranges between -1 and $+1$ gives us a measure of the degree of correlation. Also, the below diagram, i.e. Fig. 29, shows correlation matrix using visualization tools like heatmaps.

Table 5 Summarizing outlier detection techniques

Outlier detection technique	Description	Application
Standard deviation	Detect outliers by defining a range based on the standard deviation, mean and control parameter	Outliers in height for a particulate set students
Median and median absolute deviation	Detect outliers by defining a range based on the median absolute deviation, median and control parameter	Outliers in price of a place can be determined using this technique and removed so that prediction of prices of the houses
Box plots and interquartile range	Find outliers by visualizing values beyond the lower and upper whisker in box plots	Price hike of a place can be determined using a box plot and outliers can be determined
Scatter plots	Find extreme outliers by visualizing using scatter plots. Close outliers cannot be detected using this method	Scatter plot can be used to determine outliers in areas, for example ground living area versus sales, and remove them
Probability distribution	Find extreme outliers by visualizing if the distribution of a variable is skewed or not. Here also, it is hard to find the exact values of the outliers, but a range can be defined such that any value outside it would be an outlier	Probability distribution can also be used to determine outliers in case of income versus people of a particular place are outliers can be removed for better results

Table 6 Bucketing correlation values

Degree	Direction	
	Positive	Negative
Perfect	+1	−1
Significant (very high)	+0.75 to +1	−0.75 to −1
High	+0.5 to +0.75	−0.5 to −0.75
Low	+0.25 to +0.5	−0.25 to −0.5
Insignificant (very low)	0 to +0.25	0 to −0.25
Absent	0	0

8.1 Steps for Correlation Analysis

Let us consider two separate data series to determine their correlation.
  The various steps involved are as follows

1. Calculate mean of two series. X, Y

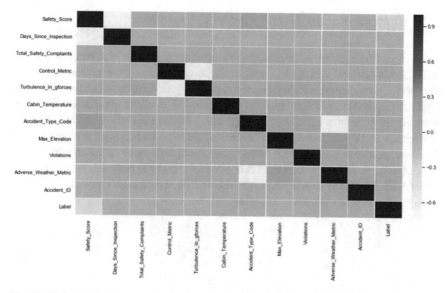

Fig. 29 Example of a heatmap showing correlation between various variables

$$\mu_x = \left(\sum_{i=1}^{n} X_i\right)/n, \quad \mu_y = \left(\sum_{i=1}^{n} Y_i\right)/n \tag{23}$$

2. Calculate variable and standard deviation for each variable
   Sample variance,

$$\sigma^2 = (X_i - \mu_x)^2/(n-1)\sigma^2 = (Y_i - \mu_y)^2/(n-1) \tag{24}$$

Standard deviation,

$$\sigma_{\text{index 1}} = \sqrt{((X_i - \mu_x)^2/(n-1))} \quad \sigma_{\text{index 2}} = \sqrt{((Y_i - \mu_y)^2/(n-1))} \tag{25}$$

3. Covariance between two series is determined.

$$\text{COV}(X, Y) = \sum_{i=1}^{n} (X_i - \mu_x)(Y_i - \mu_y)/(n-1) \tag{26}$$

4. Calculate the correlation coefficient.

$$\text{Correlation} = \text{COV}(X, Y)/(\sigma_{\text{index 1}}\sigma_{\text{index 2}}) \tag{27}$$

8.2 *Autocorrelation Versus Partial Correlation*

Autocorrelation is a representation of the amount of similarity between a given time series and a delayed version of itself. For example let's say we have a time series data of daily temperature for a week we start off by shifting the time series one day now we find the correlation between the shifted time series and the original one. Now we shift the time series by two days and find the correlation between original time series and shifted time series. Similarly, we keep increasing the days and find the correlation. We observe that there is not much difference in the temperature in the consecutive days. The correlation between the original time series and the time series shifted by a day is still very high. As we shift the time series by more number of days, the correlation starts decreasing.

The mean is

$$\mu_y = (1/n) \sum_{i=1}^{n} Y_i \tag{28}$$

The autocovariance function at lag k for k > 0 is

$$\sigma_k = (1/n) \sum_{i=1}^{n} (Y_i - \mu_y)(Y_{i-k} - \mu_y) \tag{29}$$

Autocorrelation function at lag k for $k > 0$ is

$$r_k = \sigma_k/\sigma_0 \tag{30}$$

When there are more than one independent variable controlling the dependent variable in correlation analysis, the correlation between the independent variable and dependent variable is called partial correlation or net correlation, the influence of other correlations is excluded in a partial correlation analysis.

For example, weight loss (dependent variable) is dependent on both dieting (independent variable) and exercise (independent variable). If we consider the effect of dieting on weight loss without considering exercise or vice versa, this measure is known as partial correlation.

r_{xy} is the correlation between x and y.
$r_{xy.z}$ is the partial correlation coefficient for x and y controlling z.

$$r_{xy.z} = \left(r_{yx} - \left(r_{yz}\right)\left(r_{xz}\right)\right)/\left(\sqrt{(1 - r_{yz}^2)}\sqrt{(1 - r_{Xz}^2)}\right) \tag{31}$$

9 Variance and Covariance Analysis

While working with various datasets and doing multivariate analysis on the same, people often get confused between ANOVA, ANCOVA, MANOVA and MANCOVA. But it is necessary to understand how different the above terms are. This section completely focuses to explain and differentiate between these terminologies such that the purpose of each analysis is understood well.

9.1 Analysis of Variance (ANOVA)

Analysis of variance or ANOVA is a method of analysing continuous dependent variables for mean differences under three or more groups. As explained by Akbay et al. [10], there are two types of ANOVA, one-way ANOVA and two-way ANOVA. In case of one-way ANOVA, groups of an independent variable are compared with respect to the mean difference obtained for each group from the continuous dependent variable, while in the case of two-way ANOVA, groups of two independent variables are compared with respect to mean difference obtained for each group from the continuous dependent variable. For example, by level of education if we analyse the mean of test score, then it is a one-way ANOVA while if we introduce country of education to differentiate out level of education and then analyse mean of test scores, it is called a two-way ANOVA.

9.2 Analysis of Covariance (ANCOVA)

As discussed by Rasch et al. [11] in the chapter, ANCOVA combines ANOVA with components of regression analysis. Now, ANCOVA can be understood as multiple linear regression analysis such that there is one categorical independent variable and one continuous independent variable. In this, a dependent variable is compared using both, i.e. the categorical and continuous independent variables. For example, if test score is a continuous dependent variable while level of education is categorical independent variable and hours spent in studying is continuous independent variable, then an analysis of mean test score is considered as an experiment of ANCOVA. The formula for covariance is as follows

$$\text{Cov}(x, y) = \sum_{i=1}^{n} (x_i - \mu_x)(y_i - \mu_y)/n \tag{32}$$

where x and y are two independent variables, μ_x is the mean of independent variable x and μ_y is the mean of independent variable y.

9.3   Multiple Analysis of Variance (MANOVA)

The difference between ANOVA and MANOVA stands that in MANOVA there are multiple dependent variables to analyse while in ANOVA there is only one dependent variable to analyse. Similarly, this also has one-way MANOVA and two-way MANOVA. From the example in ANOVA, if we analyse test scores along with annual income and introduce the same to both one-way and two-way ANOVA, then it becomes a MANOVA experiment.

9.4   Multiple Analysis of Covariance (MANCOVA)

Similarly, MANCOVA is nearly similar to ANCOVA. While it is just that for MANCOVA, there are two independent variables to be analysed. With respect to the example in ANCOVA, if we introduce annual income as a dependent continuous variable to be analysed with test scores, then it can be considered as a MANCOVA experiment.

10   Chi-Square Analysis

As per MacFarland et al. [12], the above analysis method helps to tell how well two columns are related or what is the difference in proportions of categories in two or more variables. The formula used is stated below.

$$\chi^2 = \sum_{i=1}^{n} \left(x_i^o - x_i^e \right)^2 / x_i^e \tag{33}$$

where x_i^o is the observed value and x_i^e is the expected value.

Let us assume that we have education level and annual income slab as two categorical columns (Table 7).

Table 7 Sample dataset for chi-square analysis with observed values

Education level	$10,000–$30,000	$300,001–$100,000	>$100,000	Total
High school	30	15	5	50
B.Tech.	20	40	10	70
M.Tech.	10	40	30	80
Ph.D.	5	45	50	100
Total	65	140	95	300

After seeing this table, we can calculate the table with expected values as shown below using the formula (Table 8)

$$\text{Expected Value} = (\text{Column Total} * \text{Row Total})/\text{Total \# of Observations} \quad (34)$$

Now, calculate the chi-square for each observed and expected value (Table 9). With this, it can be seen that the degree of freedom = (# of column − 1) * (# of rows − 1) = 6 in this case.

For the degree of freedom to be 6 and significance level of 0.05, the critical value is 12.59, and as the total chi-square value is 81.8 which is very higher than the critical value, there exists a high relation between education level and annual income slab which is true in real life as well.

Table 8 Preprocessed dataset for chi-square analysis with expected values from observed values

Education level	$10,000–$30,000	$300,001–$100,000	>$100,000	Total
High school	10.83	23.33	15.83	50
B.Tech.	15.17	32.67	22.17	70
M.Tech.	17.33	37.33	25.33	80
Ph.D.	21.67	46.67	31.67	100
Total	65	140	95	300

Table 9 Intermediate calculations to perform chi-square analysis

Observed value	Expected value	$(O - E)^2$	$(O - E)^2/E$
30	10.83	367.49	33.93
20	15.17	23.33	1.53
10	17.33	53.72	3.1
5	21.67	277.88	12.82
15	23.33	69.38	2.97
40	32.67	53.72	1.64
40	37.33	7.12	0.19
45	46.67	2.79	0.06
5	15.83	117.29	7.41
10	22.17	148.11	6.68
30	25.33	21.80	0.86
50	31.67	335.98	10.61
Total			81.8

Table 10 Example showing Z-score analysis

Score (x_i)	Mean	Score mean	Z-score ($x_i - \mu$)/σ
85	80	+5	+0.54
80	80	0	0
68	80	−12	−1.30
75	80	−5	−0.54
92	80	+12	+1.30

Table 11 Example showing Z-score visualization

52.4	61.6	70.8	80	89.2	98.4	107.6
$\mu - 3\sigma$	$\mu - 2\sigma$	$\mu - 1\sigma$	μ	$\mu + 1\sigma$	$\mu + 2\sigma$	$\mu + 3\sigma$
−3	−2	−1	$Z = 0$	+1	+2	+3

11 Z-Score

Z-score is used to calculate how far off a point is from the mean in terms of standard deviation. The formula used is stated below

$$Z = (x_i - \mu)/\sigma \tag{35}$$

where σ is the standard deviation, μ is the mean and x_i is the value from a field.

Let us assume that we have the score of students where σ is 9.2 here (Tables 10 and 11).

When we add 1 standard deviation to our mean, the Z-score is +1 similarly. And, when we add 2 standard deviations to mean it becomes +2 and when we subtract 1 standard deviation it becomes −1 and so on.

This is a real-world application of Z-score explained by Bandyopadhyay [13].

12 Bias Versus Variance

Bias is termed as the error that occurs on approximating a problem which is extensively complicated, by a simpler model. If we consider a linear regression model, it is assumed that there is a linear relation between the input and the output variables which is unlikely to occur in a real scenario.

For example, consider height versus weight plot where weight is the output column. If we try to fit a linear model, it will undoubtedly result in some error in the estimate since a short-heighted obese person would weigh more. The true function is nonlinear so irrespective of the size of the training data, our linear model will not predict the output accurately. This sum of the vertical distances between the true value and the linear function line where the output column is on the y-axis is called bias. A high bias will cause the model to miss a dominant pattern of the variable.

When the bias is too high, it is assumed that the model does not favour the complexity of data.

When a machine learning model is asked to predict unseen data, and if the predicted values differ from the true values largely, then the model is said to have variance. In a model with high variance, the predicted values are too far from the true values. On the other hand in a model with low variance, the predicted values are away close to the true values which directly affects performance of the model. A medium variance model is accepted which shows a generic medium fit.

12.1 Bias–Variance Trade-Off

Bias–variance trade-off is the fundamental topic for understanding the model's performance. It is not always the case that if a model performs well on the training set then it would perform well on the unseen/testing data as well. If this happens that means the model has a low bias (the model has fitted very well to the training data or it is overfitted) and high variance, the predicted values are far off from the true values while when we test on the training data, it gives a quality performance, due to either output leakage or overfitting.

Low Bias and Low Variance
It means that the error with the training data and the error with test data both are less. This is an ideal case, and we would want your final model to be as close to this as possible.

Low Bias and High Variance
It means although the model has less error with the training error while it had a greater error with the test data, i.e. the predictions were deviated with respect to the true values.

High Bias and Low Variance
It means that the error with the training data is very high and that with the test data is low. In this case even if the output error is low, it takes a toll on the reliability of the model since it has high training error which means the model has not captured the true pattern of the data which makes it unreliable for future predictions.

High Bias and High Variance
It means that there is high error with the training data and high error with the testing data.

When we try to decrease the bias by adding complexity to the function, it so happens that the model starts overfitting the training data and which increases the variance on the testing data.

On the other hand in order to decrease the variance on the testing data, if we try to decrease the complexity of the training data, i.e. if we increase the bias, this may cause the model to underfit. We need to find an optimum ratio between the bias and the variance, in order to obtain a model having optimum bias and variance. This trade-off between the bias and variance in order to obtain the optimum value is called bias–variance trade-off.

One can also refer to Briscoe et al. [14] to get a better understanding of bias–variance trade-off and its complexity.

12.2 Overfitting and Underfitting

Overfitting
The bias is very less in the model, while variance is very high. It has captured the unnecessary patterns in the training data. Equation: $\theta_0 + \theta_1 x + \theta_2 x^2 + \theta_3 x^3 + \theta_4 x^4$.

Underfitting
The bias is high and fails to capture any pattern in the data having a low variance. Equation: $\theta_0 + \theta_1 x$.

Medium Fitting
This is the perfect fit which is needed when neither bias nor variance is very high. Equation: $\theta_0 + \theta_1 x + \theta_2 x^2$ (Fig. 30).

Fitting	Description	Application
Underfitting	When the model fails to capture necessary data	When less number of features are provided to train the model example if we do not provide price per-square fit in housing cost prediction model

(continued)

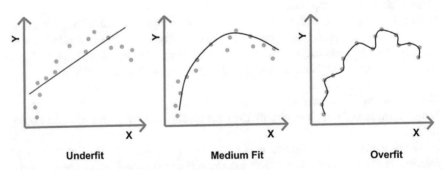

Underfit   Medium Fit   Overfit

Fig. 30 Example of underfitting, medium fitting and overfitting

(continued)

Fitting	Description	Application
Medium fitting	When the model is able to capture all necessary pattern in the training data	When exact features are provided to train example if we provide price per-square, number of bedroom and other important features
Overfitting	When the model capture unnecessary patterns of a training data	When more features are provided to train then what is necessary

13 Hypothesis Testing

An assumption about a feature of the population is called hypothesis, and the test means validating something. Hence, hypothesis testing is the process of validating an assumption about a population feature using statistics.

The steps involved in hypothesis testing are referred below:

Step-1: Specify the null hypothesis. Null hypothesis ($H0$) is a statement of no relationship, effect or the difference between two or more groups or factors.

Example: $\mu \geq 0.15$

Step-2: An alternate hypothesis is specified. Alternate hypothesis is the effect or the difference. The hypothesis that is to be proven is the alternative hypothesis.

Example: $\mu < 0.15$

Step-3: The significance level (α) is set. The value of the significance level is generally 0.05; this means that there is a 0.05 chance that even when the null hypothesis is true, the alternative hypothesis will be selected. A smaller value of α makes it difficult to prove the null hypothesis.

Step-4: The corresponding P-value is calculated from the test statistics. So now we consider a sample data assuming the mean of the sample to be x where σ is the size of the standard deviation of the sample, μ is the hypothesis mean and n is the sample size. Now we calculate the Z statistics. This Z is a representative of what the sample data is indicating,

$$Z = (x - \mu)/(\sigma/\sqrt{n}) \tag{36}$$

One thing that we need to keep in mind is that if the sample size is less than 30 then the Z statistics is called T statistics.

P-value also called probability value is the probability of determining a sample "more extreme" than the one observed in your data, assuming that the null hypothesis is true. The value of P can be calculated from the normal distribution table using the Z statistics value.

Step-5: Construct the acceptance and rejection region. From Fig. 31, the Z value is evident. The area left and right of Z statistics is the probability value. While considering the T statistics, only one side is considered because in smaller samples the distribution cannot be normal as shown in Fig. 32.

Step-6: Drawing a conclusion. P-value less than or equal to significance level (α) is used to reject the null hypothesis; i.e. the alternative hypothesis is preferred. If P-value is greater than significance level (α), we fail to reject the null hypothesis. This outcome is statistically significant.

Fig. 31 Graph showing P-value for Z statistics

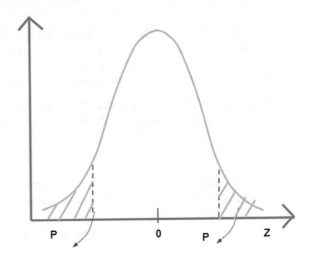

Fig. 32 Graph showing P-value for T statistics

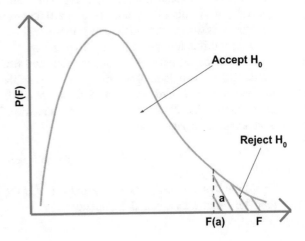

Table 12 Confusion matrix for hypothesis testing

Decision	In reality	
	H_0 is true	H_0 is false
Accept H_0	Ok	Type II error β = Probability of type II error
Reject H_0	Type I error α = Probability of type I error	Ok

13.1 Errors in Hypothesis Testing

Given below is a contingency table (Table 12).

Type I Error
When the P-value is less than the significance level, the hypothesis is rejected. However, there is a possibility of type I error. There is usually no warning when it occurs and the sample error could have overestimated the chance. Even though we do not know the significance of this error, we know the rate of occurrence of this error, i.e. α. If we reduce α, the probability of false positive reduces.

Type II Error
When we perform hypothesis test and the P-value is greater than the significance level, it is not statistically significant. However, there are chances that the effects are present in the population. Hence, it might be a type II error and β is called the probability of type II error. This type II error could have occurred because of small effect size, small sample size or high data variability. $1 - \beta$ is called statistical power analysis, which is calculated, and then β is derived from it. Low variability and large effect size reduce the type II error which increases the statistical power.

Figure 33 shows the graphical representation of type I and type II errors.

14 Conclusion

This chapter is focused for researchers seeking for all concepts related to statistics in an integrated form. As discussed in the above work, one can find concepts beginning with different types of sampling techniques, concepts of different variables and how to visualize them using standard visualization techniques. Further, one can get to know about the measures of central tendency, different distributions, scalings, transformations, outlier detection techniques and much more. Also, the above work discusses different statistical tests like correlation analysis, covariance analysis, Z-score testing, chi-square testing, bias–variance concepts and hypothesis testing.

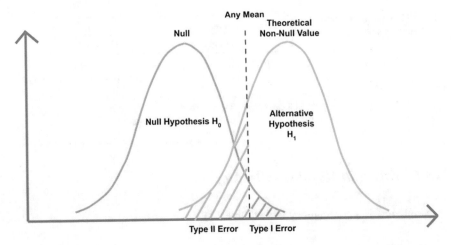

Fig. 33 Graph showing type I and type II errors for hypothesis testing

All the concepts and statistical tests are discussed in detail. They follow the standard approaches with available figures and comparative tables for easy understanding and conceptualization.

References

1. Meng, X. 2013. Scalable simple random sampling and stratified sampling. In *30th international conference on machine learning, ICML 2013*.
2. Manikandan, S. 2011. Measures of central tendency median and mode. *Journal of Pharmacology and Pharmacotherapeutics* 2 (3): 214.
3. Ahn, S., and J. Fessler. 2003. *Standard errors of mean, variance, and standard deviation estimators*.
4. Hernandez, H. 2020. *Probability distributions in groups of random elements*.
5. Hossain, M.F., and A. Joarder. 2007. *Identities based on probability mass functions*.
6. Dytso, A., R. Bustin, H.V. Poor, and S. Shamai. 2018. Analytical properties of generalized Gaussian distributions. *Journal of Statistical Distributions and Applications* 5 (1): 1–40.
7. Palmisano, A. 2018. *Poisson and binomial distribution*.
8. Amir, M., W.M.A. Ahmad, N.N. Naing, and N. Halim. 2008. *An application of box-cox transformation to biostatistics experiment data*.
9. Yang, J., and S. Rahardja. 2019. *Outlier detection: How to threshold outlier scores?*.
10. Akbay, L., Akbay, T., Erol, O., and M. Kilinç. 2019. İnadvertent use of ANOVA in educational research: ANOVA is not a surrogate for MANOVA. *Eğitimde ve Psikolojide Ölçme ve Değerlendirme Dergisi*.
11. Rasch, D., Verdooren, R., and J. Pilz. 2019. *Analysis of covariance (ANCOVA)*.
12. MacFarland, T., and J.M. Yates. 2016. *Chi-square*.
13. Bandyopadhyay, A. 2006. Predicting probability of default of Indian corporate bonds: logistic and Z-score model approaches. *Journal of Risk Finance* 7 (3): 255–272.
14. Briscoe, E., and F. Jacob. 2011. Conceptual complexity and the bias/variance tradeoff. *Cognition* 118 (1): 2–16. ISSN 0010-0277.

Evolutionary Algorithms-Based Machine Learning Models

Junali Jasmine Jena, Manjusha Pandey, Siddharth Swarup Rautaray, and Sushovan Jena

Abstract Machine learning models have found immense applications in various sectors such as energy, stock market, demand–supply chain, logistic management, health and many more, but their efficiency and accuracy depend on a very important factor, i.e., data. Data plays a major role in affecting the performance of these models. Mostly dataset with several features is fed as input to the machine learning models. A dataset containing crisp and relevant features can highly improve the accuracy whereas a dataset containing redundant and irrelevant features can detoriate the same. While performing data collection, basically one tries to collect as much information as it can about a specific domain, but when we need to draw any viable inference from that data, definitely, the inference is drawn, basing on a particular perspective and we may not need all the features of the collected data for this purpose. So, choosing some appropriate features of the dataset, which need to be fed to the models, has always been a crucial task. It is an obvious fact that complexity of feature selection increases with the increase in number of features and choosing some features out of human intuition is, of course, not appreciable. Evolutionary algorithms have been proved highly beneficial for solving such issues, for their stochastic nature. This chapter discusses some recent application of evolutionary algorithms such as genetic algorithm, particle swarm optimization, artificial bee colony, etc. to optimize the parameters of machine learning algorithms, e.g., support vector regression, artificial neural networks, random forest, etc. and their uses in various sectors like engineering, applied sciences, disaster management, finance and economy and health sector.

J. J. Jena (✉) · M. Pandey · S. S. Rautaray
School of Computer Engineering, Kalinga Institute of Industrial Technology (Deemed to be University), Bhubaneswar, Odisha, India
e-mail: junali.jenafcs@kiit.ac.in

M. Pandey
e-mail: manjushafcs@kiit.ac.in

S. S. Rautaray
e-mail: siddharthfcs@kiit.ac.in

S. Jena
Wipro Limited, WIPRO BHDC, Bhubaneswar, Odisha, India
e-mail: sushovanjena@gmail.com

© The Author(s), under exclusive license to Springer Nature Singapore Pte Ltd. 2021
S. Rautaray et al. (eds.), *Trends of Data Science and Applications*,
Studies in Computational Intelligence 954,
https://doi.org/10.1007/978-981-33-6815-6_5

Keywords Machine learning · Evolutionary algorithms · Soft computing ·
Applications

1 Introduction

Computer science and data could not exist without each other. If there is no data, then
there is nothing upon which computation could be done. If there is no computation,
then data could not be exploited and represented in a useful manner. So this field
of computer science and engineering revolves around the relationship of data and
computation. That is why data has been the focus of research interest from the very
beginning. Always there has been some breakthrough technologies coming up for
data handling, storing, utilizing, etc., and thus, data science field has gained popularity
and has got wide range of applications.

One of such application of data science could be found in the area of machine
learning. Learning always happens by extracting and improving knowledge from the
available data. Without data, learning has got no existence. Classification and prediction
are the output that we get by application of learning algorithms. Machine learning
models have got tremendous applications in various fields where classification and
prediction are needed and are major tools of artificial intelligence.

Evolutionary algorithms (EAs) can be treated as intelligent algorithms which
follow randomized or stochastic approach to find solutions to those problems which
cannot be solved by general computational strategies. As the field of data science and
machine learning involves many complex situations where finding a good solution
with better accuracy is very much required, EAs help in enhancing the performance
of the model to a greater extent. EAs can be used to determine optimal feature set
selection which is to be given as input to the model or it can be used to optimize the
hyper-parameters of a model. Figure 1 explains the whole process diagrammatically.

This chapter discusses various field of application of data science and machine
learning where EAs have been successfully used for enhancing the accuracy of the
models. Classification and prediction are the end result of every machine learning
model. So, we have discussed several EAs-based machine learning models used
for different applications. State-of-the-art techniques proposed after 2010 have been
considered, so that readers could get a glimpse of the trend of the current decade.

2 Application Domains

In this work, papers based on application of ML and EAs to five application domains,
such as engineering applications, applied sciences, disaster management, finance
and economy and health, have been discussed. Description about frequently used
abbreviations in this paper has been provided in Table 1.

Feature Selection

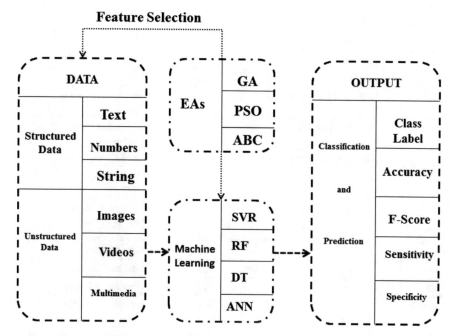

Fig. 1 Hybridization of EAs and ML models

2.1 Engineering Applications

EAs and ML models have found wide range of applications in the field of engineering. Some of such works have been discussed in this section. Table 2 provides the classification of the discussed works as per their field of use.

Behandish and Wu [1] used artificial neural network (ANN) for pump scheduling of complex water distribution system. They used genetic algorithm (GA) for optimization of pump operation and tank levels at the cycle end. Zhao et al. [5] used ANN and artificial bee colony (ABC) to predict the rate of penetration in drilling process. Salcedo-Sanz et al. [9] used coral reef optimization (CRO) with harmony search (HS) operators and extreme learning neural network model (ELM) for short-term wind speed prediction. Mashhadban et al. [14] used ANN and particle swarm optimization (PSO) to study the effects of fiber in self-compacting concrete. Mia et al. [19] found support vector regression (SVR) and GA provided better results in predicting the optimal control parameters for high pressure coolant (HPC)-based machining.

Chen et al. [10] used ensemble long short-term memory, SVR and extremal optimization(EO) algorithm for wind speed prediction. Muralitharan et al. [24] found that GA-based ANN provided better result in short-term load forecasting and PSO-based ANN provided better result in long-term load forecasting for energy consumption

Table 1 Details about frequently used abbreviations in the paper

Terms used	Description	Terms used	Description
ML	Machine Learning	SVR	Support Vector Regression
EAs	Evolutionary Algorithms	EO	Extremal Optimization
ANN	Artificial Neural Network	SFLA	Shuffled Frog Leap Algorithm
GA	Genetic Algorithm	ICA	Imperialist Competitive Algorithm
ABC	Artificial Bee Colony	FFA	Fire Fly Algorithm
CRO	Coral Reef Optimization	BPNN	Back Propagation Neural Network
HS	Harmony Search	SMO	Sequential Minimal Optimization
ELM	Extreme Learning Neural Network	RBF	Radial Basis Feed Forward Neural Network
PSO	Particle Swarm Optimization	BS	Backtracking Search
RF	Random Forest	HE	Hierarchial Entropy
ANFIS	Adaptive Neuro Fuzzy Inference System	kNN	k-Nearest Neighbor
DE	Differential Evolution	TLBO	Teaching–Learning-Based Optimization
MLP	Multi-Layer Perceptron Neural Network	SCA	Sine–Cosine Algorithm
BGO	Bio Geography-Based Optimization	ACO	Ant Colony Optimization
GS	Grid Search	FFO	Fruit Fly Optimization
GVS	Gravitational Search	GRNN	General Regression Neural Network
MDA	Multi-variate Discriminant Analysis	LSTM	Long Short-Term Memory

prediction in smart grid. Chatterjee et al. [15] used PSO-based ANN for faulty structure prediction in multi-storied RC buildings. Cong et al. [28] used SVR and FFA for prediction of traffic flow on the road. Kant and Sangwan [20] used ANN and GA to minimize surface roughness in machining process. Asrari et al. [25] used ANN and shuffled frog leap algorithm (SFLA) for short-term solar power prediction. Ahmadi et al. [6] used fuzzy logic, ANN and imperialist competitive algorithm (ICA) for oil rate prediction of wells. Wan et al. [11] used ELM and PSO for optimal interval prediction of wind power generation. Olatomiwa et al. [12] used SVM and firefly algorithm (FFA) for global solar radiation prediction. Zhou et al. [42] used GA and PSO-based SVM to predict the possibility of underground rock burst. Yu et al. [45] used particle classification optimization-based BPNN for prediction of telecommunication customer churn. Armaghani et al. [43] used PSO-based ANN for prediction

Table 2 Classification of the application of EAs and ML models in the field of engineering

Fields of engineering	Purpose	Methodology used
Water resource management	Pump scheduling [1]	ANN + GA
	Water supply demand prediction	ANN + BS
	Ground water mapping [2]	RF + GA
	Ground water fluctuation [3]	SVM + FFA
	Surface and ground water for irrigation [4]	SVM + GA
Petroleum	ROP in drilling [5]	ANN + ABC
	Oil rate prediction [6]	Fuzzy ANN + ICA
	Permeability prediction of hydrocarbon reservoir [7]	Hybrid PSO + SVR
	Minimizing miscibility pressure in CO_2 crude oil	ANFIS + DE
	Prediction of saturates in oil [8]	ANN + GA
Meteorology	Wind speed prediction [9]	ELM + CRO-HS
	Wind speed prediction [10]	EnsemLSTM + SVR + EO
	Wind interval prediction [11]	ELM + PSO
	Global solar radiation [12]	SVM + FFA
	Rainfall prediction [13]	RBF + Hybrid PSO
Civil	Properties of self-compacting concrete [14]	ANN + PSO
	Structure fault prediction [15]	ANN + PSO
	Prediction of soil bearing capacity [16]	ANN + PSO
	Structural health monitoring [17]	SVM + GA
	Soil coefficient of consolidation [18]	MLP + BGO
Mechanical	HPC-based machining [19]	SVR + GA
	Minimizing surface roughness in machining [20]	ANN + GA
	Roller bearing fault diagnosis [21]	SVM + PSO + HE
	Fault analysis of rotary kiln [22]	SVM + Binary SGO

(continued)

Table 2 (continued)

Fields of engineering	Purpose	Methodology used
	Fault diagnosis of rotating machine [23]	LS-SVM + improved PSO
Electrical and Electronics	Load forecasting [24]	ANN + GA, PSO
	Solar power prediction [25]	ANN + SFLA
	VLSI [26]	SVM + SMO
	E-nose [27]	BPNN + chaosPSO
Transportation	Traffic flow prediction [28]	SVR + FFO
	Bus arrival time prediction [29]	SVM + GA
	Short-term traffic flow prediction [30]	Multi-kernel SVM + Adaptive PSO
Automobile	Automotive fault diagnosis [31]	SVM + PSO
	Car line fault diagnosis [32]	Wavelet SVR + PSO
	Automotive bearing fault diagnosis [33]	kNN + binary DE
Aerospace	Friction coefficient prediction [34]	SVM + hybrid PSO
	Fault diagnosis of gas path in engine [35]	TW-SVM + Hybrid PSO
	Aircraft AC generator [36]	RBF + improved adaptive PSO
Software	Software defect classification [37]	ANN + PSO
	Fault prediction in object-oriented system [38]	ANN + PSO
	Software effort estimation [39]	ANN + PSO
	Quality prediction [40]	ANN + TLBO
	Software reliability prediction [41]	SVM + PSO
Mining	Rockburst possibility prediction [42]	SVM + GA,PSO
	Prediction during blast [43]	ANN + PSO
	Compressive strength of Soft rocks [44]	ANN + PSO

of fly rock and vibration produced from blasting. Moayedi et al. [16] found PSO-based ANN performing better in predicting bearing capacity of shallow footing on two-layered soil. Yang et al. [29] used GA-based SVM for bus arrival time prediction. Akande et al. [7] used a hybrid PSO-based SVR for permeability prediction of hydro carbon reservoir. Kuan et al. [26] used SVM and sequential minimal optimization

(SMO) algorithm for VLSI design. Ling et al. [30] predicted short-term traffic flow using multi-kernel SVM and adaptive PSO. Wu et al. [29] used RBF and hybrid PSO for rainfall prediction. Zubaidi et al. [46] used backtracking search (BS)-based ANN for urban water supply demand prediction.

Naghibi et al. [2] found that GA-based random forest (RF) performed better in ground water mapping. Kisi et al. [3] used SVM and FFA for surveying the ground water fluctuation in Urmia lake. Safavi and Esmikhani [4] used GA-based SVM model for conjunctive use of surface water and ground water of Zayandehrood river basin of Iran for irrigation purpose. Karkevandi-Talkhooncheh et al. [47] used adaptive neuro fuzzy inference system (ANFIS) along with few evolutionary algorithms for minimizing miscibility pressure of CO_2-crude oil and found that DE-based ANFIS performed better. Dong et al. [8] used GA-based ANN to predict saturates of vaccum gas oil. Gui et al. [17] found GA-based data driven SVM performed better in structural health monitoring and damage detection. Pham et al. [18] used multi-layer perceptron (MLP)-NN and biogeography-based optimization (BGO) technique for prediction of coefficient of consolidation of soil.

Zhu et al. [21] used hierarchial entropy (HE), SVM and PSO for roller bearing fault diagnosis. Kadri et al. [22] used SVM and binary ACO for fault diagnosis of rotary kiln. Deng et al. [23] used least squares SVM and improved PSO for fault diagnosis in rotating machinery. Zhang et al. [27] used BPNN and chaos-based PSO for air contamitants estimation using E-nose. Yu [31] used SVM and PSO for fault diagnosis in automobiles. Wu used wavelet SVC and PSO for fault diagnosis in car line. Baraldi et al. [33] used k-nearest neighbor (kNN) and binary DE for fault diagnosis of automotive bearing. Zhan and Li [34] used hybrid PSO and SVM for prediction of friction coefficient between aircraft tire and coating. Du et al. [35] used hybrid PSO and TW-SVM for fault diagnosis of aircraft engine gas path. Cheng et al. [36] used radial basis feed forward (RBF) NN and improved adaptive PSO for modeling of AC generator in aircraft. Dhanalaxmi et al. [37] used ANN and adaptive PSO for software defect classification.

Kayarvizi et al. [38] used ANN and PSO for fault prediction in object-oriented systems. Dan [39] used ANN and PSO for improving the accuracy in software effort estimation. Tomar et al. [40] used ANN and teaching learning-based optimization (TLBO) for quality prediction in component based software systems. Qin [41] used PSO-based SVM for software reliability prediction. Mohamad et al. [44] used PSO-based ANN to predict the unconfined compressive strength of soft rocks.

2.2 Applied Sciences

EAs and machine learning models have got tremendous use in the field of applied sciences. Several such works have been discussed in this section. Table 3 provides the classification of these works according to their fields of application. Liu et al. [48] used real-valued GA (RGA) and SVR to predict water quality in crab aquaculture. Shahlol et al. [49] used ANN and sine cosine algorithms (SCA) for prediction of liver

Table 3 Application of EAs and ML models to the field of applied sciences

Field of science	Purpose	Methodology
Biology	Water quality prediction in crab aquaculture [48]	SVR + RGA
	Liver enzymes prediction in Fish [49]	ANN + SCA
	Prediction of dissolved oxygen [54]	SVM + ACO
	QSAR modeling of peptides [55]	SVM + GA, PSO
	Cell group recognition [56]	SVM + adaptive mutated PSO
	Growth cycle prediction of Spirulina plantesis [57]	SVM + hybrid PSO
	Prediction of O-glycosylation sites [58]	RF + GA tuned PSO
	Human RNA precursor prediction [59]	SVM + GA
Chemistry	Chromatin state detection [50]	SVM + GA
	P2Y12 anatagonist prediction [51]	SVM + GA
	Melt index prediction [53]	RVM + modified PSO
	Glass transition temperature prediction [60]	SVR + hybrid PSO
	Copper removal prediction [61]	ANN + GA,PSO
	Synthetic coolants in spectrophotometry [62]	ANN + PSO,GA
	Modeling TE process [63]	ANN + PSO,GA
	pKa prediction [64]	ANN + improved PSO
	Solubility prediction of carbon dioxide [65]	RBFNN + adaptive chaotic PSO
	Flash point temperature prediction [66]	ANN + PSO
Physics	Crystal lattice parameters prediction [67]	SVM + PSO, ELM
	Lattice constant prediction [68]	ANN, SVR
	Forecasting beta systematic risk [69]	LS-SVM + GA
	Band gap characterization [70]	SVR + PSO
	Auto-ignition temperature prediction [71]	ANN + PSO
	Magnetocaloric effect modeling [72]	SVR + hybrid GA
	Maximum magnetic entropy change modeling [73]	ELM + GS, SVR + GVS

enzymes of fish when fed with nano-selenite. Fernandez and Miranda-Saavedra [50] used GA-based SVM for chromatin state detection. Hao et al. [51] used GA-based SVM for prediction of P2Y12 antagonist. Liu et al. [52] used ant colony optimization (ACO)-based least square SVM for prediction of dissolved oxygen. Jiang et al. [53] predicted melt index in propylene polymerization industry using relevant vector machine(RVM) and modified PSO.

Zhou et al. [55] used SVM with GA and PSO for QSAR modeling of peptides. Wang et al. [56] used adaptive mutation PSO-based SVM for cell group recognition. Nieto et al. [57] used hybrid PSO-based SVM for predicting the growth cycle of Spirulina plantesis. Hassan et al. [58] used GA tuned PSO and random forest for prediction of O-glycosylation sites. Wang et al. [59] used GA-SVM for optimized feature set selection for human RNA precursors prediction. Pei et al. [60] used hybrid PSO-based SVR for glass transition temperature prediction of polymetacrylates based on quantum chemical descriptors. Fan et al. [61] used GA and PSO-based ANN for prediction of copper removal from aqueous solutions by magnetic nanocomposites. Benvidi et al. [62] used PSO and GA based ANN for determination of synthetic coolants in spectrophotometry. Braik et al. [63] used GA and PSO-based ANN for modeling of TE process chemical reactor.

Li et al. [64] used improved PSO-based ANN for pKa prediction values for neutral and basic drugs. Li et al. [65] used chaotic self-adaptive PSO-based RBFNN for super critical solubility prediction of carbon dioxide in polymers. Lazzus [66] used PSO-based ANN and group contribution technology for flash point temperature prediction of organic compounds. Owolabbi [67] used ELM and PSO-based SVM for prediction of crystal lattice parameters. Majjid et al. [68] used NN and SVR for lattice constant prediction. Yuan and Lee [69] used LS-SVM and GA to forecast beta systematic risk. Owolabi [70] used PSO-based SVR prediction of band gap characterization for titanium oxide. Lazzus [71] used PSO-based ANN for auto-ignition temperature prediction. Owolabi et al. [72] hybrid GA-based SVR for modeling the magnetocaloric effect of manganite. Shamshah and Owolabi [73] compared grid search (GS)-based ELM and hybrid gravitational search (GVS)-based SVR for modeling the maximum magnetic entropy change of doped manganite.

2.3 Disaster Management

Disaster management is highly essential to prevent the loss of lives as well as economy losses. EAs and ML models can be efficiently used to handle these problems. Some of such works have been discussed in this section. Table 4 provides classification of discussed works according to their fields of application.

Bui et al. [74] used least squares SVM and ABC for prediction of landslides in Vietnam. Chen et al. [75] found PSO-based ANFIS performed better in predicting landslide susceptibility. Moayedi et al. [76] used optimized PSO-based ANN for mapping of landslide susceptibility. Wang et al. [77] used DES-PSO-based ELM for prediction of intervals in landslide displacement. Li and Kong [78] used GA for

Table 4 Application of EAs and ML models in the field of disaster management

Disaster	Purpose	Methodology
Landslide	Landslide prediction [74]	LS-SVM + ABC
	Landslide susceptibility prediction [75]	ANFIS + PSO
	Landslide susceptibility mapping [76]	ANN + PSO
	Landslide displacement interval prediction [77]	ELM + DES-PSO
	Landslide development prediction [78]	SVM + GA
Earthquake	Earthquake prediction [79]	ANN + PSO
	Seismic response prediction [80]	ANN + PSO
	Earthquake prediction [81]	BPNN + PSO
	Seismic slope stability prediction [82]	ANN + PSO
	Parameter identification of seismic isolator [83]	DE + PSO
Flood	Prediction of flash floods [84]	ANN + PSO
	Susceptibility mapping of flash floods [85]	ELM + PSO
	Flood forecasting [86]	ELM + PSO
	Flood susceptibility prediction [87]	MDA + SVM
Cyclone and Typhoon	Cyclone prediction [88]	GA + SVM
	Typhoon intensity prediction [89]	ANN + PSO
	Typhoon rainstorm prediction [90]	ANN + PSO

parameter optimization of SVM for prediction of landslide development. Su et al. [79] used ANN and improved PSO for earthquake prediction. Nguyen et al. [80] used ANN-based PSO for seismic response prediction build in short response. Abraham et al. [81] used PSO-based BPNN for earthquake prediction in Japan. Gordan et al. [82] used PSO-based ANN for seismic slope stability prediction. Quaranta et al. [83] used DE and PSO for parametric identification of seismic isolators.

Khan et al. [84] used ANN-based PSO for prediction of flash floods with minimization of false alarm. Bui et al. [85] used PSO-based optimized ELM for mapping the susceptibility of flash floods. Anupam and Pani [86] used PSO-based ELM for flood forecasting. Choubin et al. [87] multi-variate discriminant analysis (MDA), regression trees and SVM for ensemble prediction of flood susceptibility. Gu et al. [88] used GA-based SVM for cyclone prediction in tropical regions. Zhao et al. [89] used PSO-based ANN for typhoon intensity prediction. Zhao et al. [90] used PSO optimized neural network for typhoon rainstorm prediction.

2.4 Finance and Economy

EAs and ML models can be efficiently used for decision making in the fields of finance and economy also. Some of such works have been discussed in this section. Table 5

Table 5 Application of EAs and ML models in the field of finance and economy

Sector	Purpose	Methodology
Share market	Crypto-currency forecasting [94]	SVM + PSO
	Stock price prediction [92]	ANN + PSO, DE
	Stock market indices prediction [93]	ANN + PSO, Type-2 fuzzy
	Financial time series forecasting [95]	LPP, SVM + PSO
	Inflation forecasting [96]	SVM + PSO
Banking	Financial distress model [91]	GRNN + FFO
	Bank cash flow prediction [97]	ANFIS + adaptive PSO
	Bankruptcy prediction [98]	SVM + hybrid switching PSO
	Loan risk evaluation [99]	LS-SVM + PSO
	Credit card fraud detection [100]	ANN + PSO

provides classification of the discussed works according to their fields of application. Pan [91] proposed the fruit fly optimization (FFO) and used it along with general regression neural network (GRNN) for predicting optimal parameters of financial distress model. Abdual-Salam et al. [92] used differential evolution (DE) and PSO along with ANN for stock price prediction and found DE performed better. Fang et al. [54] used GA optimized wavelet NN for stock market prediction. Chakravarty et al. [93] used PSO-based ANN and type-2 fuzzy systems for stock market indices prediction. Hitam et al. [94] used PSO-based optimized SVM for crypto-currency forecasting.

Zhiqiang et al. [95] used LPP and SVM optimized by PSO for forecasting of financial time series. Tang and Zhou [96] used PSO-based SVM for forecasting of inflation. Wangand Ning [97] used adaptive PSO-based ANFIS for bank cash flow prediction. Lu et al. [98] used hybrid switching PSO-based SVM for bankruptcy prediction. Cao et al. [99] used PSO-based LS-SVM for modeling of loan risk evaluation. Kamaruddin and Ravi [100] used PSO-based ANN for credit card fraud detection.

2.5   Health

Health sector has wide range of applications of these EAs and ML models. Some of such works have been discussed in this section. As, there are various types of diseases covered under this section, providing classification would be very hectic and huge. So, classification of this section has not been done. Chang et al. [101] used GA-based ANN for prediction of hip bone fracture. Suresh et al. [102] used BPNN and PSO for predicting the length of stay of a patient in hospital. Vieira et al. [103] used SVM and modified binary PSO for mortality prediction of septic patients. Khordifi and Bahaj [104] used ANN with PSO-ACO for efficient prediction of heart disease in patients. Kaya used [105] used hybrid CS-PSO and SVM for diagnosis of Parkinson disease. Subasi [106] used PSO-based SVM for classification of EMG signals for neuro-muscular disorders. Babaoglu et al. [107] used binary PSO and GA-based SVM for diagnosis of coronary artery disease. Kar et al. [108] used adaptive kNN and PSO for cancer subgroup classification from gene expression. Zeng et al. [109] proposed a new switching delay-based PSO and SVM technique for diagnosis of Alzheimer's disease.

3   Analysis and Discussion

In this section, the gaps and issues identified from the reviewed papers have been discussed.

3.1   Issues

Every research article reviewed in this paper uses a hybridization of machine learning algorithms and evolutionary algorithms. In every work, authors claim and prove their proposed methodology to be working better than some of the other methodologies. It is not that the algorithms are poor, but they may not be suitable or made adaptable to a particular problem. So, rather than discussing the issues related to every hybridization algorithm, here we have focused on throwing some lights upon the generalized merits and demerits of some of the mostly used machine learning algorithms, which is provided in Table 6.

One of the common problems faced by the machine learning algorithms is that they have lots of hyper-parameters, which needs to be optimized properly for better adaptation to the problem and enhancing its performance and here the role of evolutionary algorithms come into the play. Every evolutionary algorithm more or less does the same work, i.e., optimizing the hyper-parameters, but again the performance is very much problem dependent, so empirical approach should be adopted to select a

Table 6 Merits and demerits of commonly used machine learning algorithms

ML technique	Key Feature	Merits	Demerits
SVM	Kernel function and hyperplane	• L2 regularization prevent from overfitting • Nonlinear data handled using kernel function • Handles high-dimensional data • Stable	• Choosing appropriate kernel function is challenging • Poor performance with noise • Not preferable for large dataset
ANN	Neurons, weights and activation functions	• Fault tolerant and handles missing data • Suitable for real-time problems • Multi-tasking system • Handles massive data	• Non-reasonable solutions • Too many hyper-parameters • Requires efficient hardware • Not good for sequential data
RF	Bagging technique and ensemble learning	• Feature scaling not required • Used for dimensionality reduction • Handles noisy data	• Computationally expensive • Needs proper hyper-parameter optimization
LSTM	Recurrent Neural network (has a feedback unit which acts as memory)	• Handles sequential data • Handles noisy data • Handles massive data	• Too many hyper-parameters • Computationally expensive
RBF	Nonlinear transformation for better classification	• Increases dimensionality • Faster training • Optimized hyper-parameters	• Slow classification • Computationally expensive

suitable algorithm. The major limitation of evolutionary algorithms is that they are computationally expensive and hence increases the overall time complexity.

3.2 Gap Analysis

A very prominent gap that is found is from the review of the application of evolutionary algorithm-based machine learning approaches is the insufficient applications of variety of evolutionary algorithms. The prominent evolutionary algorithms that are being mostly used are GA and PSO. Very few works have been done using other EAs. According to no free lunch theorem, no particular methodology could be used to solve all type of problems. So, this paves a way for wide area of application and experimentation. Several new algorithms have been proposed which have been proven to

Fig. 2 Popularity of ML models in the discussed application domains during the time period of 2010–2020 inferred from the reviewed works

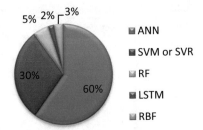

give promising results in various fields. Some of the recently proposed algorithms are gray wolf optimization [110], whale optimization [111], vortex search [112], social group optimization [113] and many more. These may be used along with machine learning models for solving real-time problems.

Figure 2 shows popularity of the ML models in the discussed application domains during the time period of 2010–2020, which has been inferred from the papers reviewed in this work. It was found that artificial neural network was the most used ML model followed by SVM or SVR, RF, RBFNN and LSTM. Models like ANFIS and ELM have also been used but the use is negligible as compared to the discussed models. So it paves a path for the researchers, to do more experimentation with these models.

Figure 3 shows the usage of different EAs in the discussed application domains along with the ML models. PSO was the most used EA for hybridization followed by GA. All other EAs such as ABC, GWO, etc. have been used lesser. So, here also there is scope of research where these algorithms can be explored more. Always, new EAs are evolving with time, so experimentation with some new EAs can also be performed.

Fig. 3 Uses of various EAs in the discussed application domains along with the ML models during the time period of 2010–2020 inferred from the reviewed works

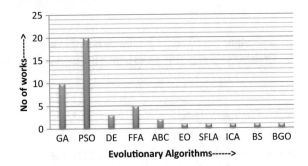

4 Conclusion

Using EAs for improvement of the machine learning models has been found to increase the efficiency many fold. It solves the problem of optimizing the hyper-parameters used in the machine learning models and thus prevents the use of hit and trial methods or random selection. It also helps in preventing the solution to get trapped into local optimum and helps in reaching the global optimum. One drawback associated with the EAs is, they are computationally expensive which increase the time of execution as well as the computational cost, but looking at the improvisation they provide to the models, this may be ignored. So, hybridizing the evolutionary models along with new optimization algorithms may open some new perspectives and paradigms in problem solving. This work will help the researchers in the field of ML and EAs by providing them the information about various application domains as well as works done in the recent decade. This paper gives a clear idea about the models and algorithms that have been mostly used; thus, more experimentation can be done upon the less used or unexplored techniques and modifications can be done to make them more efficient than the existing ones.

References

1. Behandish, M., and Z.Y. Wu. 2014. Concurrent pump scheduling and storage level optimization using meta-models and evolutionary algorithms. *Procedia Engineering* 1 (70): 103–112.
2. Naghibi, S.A., K. Ahmadi, and A. Daneshi. 2017. Application of support vector machine, random forest, and genetic algorithm optimized random forest models in groundwater potential mapping. *Water Resources Management* 31 (9): 2761–2775.
3. Kisi, O., J. Shiri, S. Karimi, S. Shamshirband, S. Motamedi, D. Petković, and R. Hashim. 2015. A survey of water level fluctuation predicting in Urmia Lake using support vector machine with firefly algorithm. *Applied Mathematics and Computation* 1 (270): 731–743.
4. Safavi, H.R., and M. Esmikhani. 2013. Conjunctive use of surface water and groundwater: Application of support vector machines (SVMs) and genetic algorithms. *Water Resources Management* 27 (7): 2623–2644.
5. Zhao, Y., A. Noorbakhsh, M. Koopialipoor, A. Azizi, and M.M. Tahir. 2020. A new methodology for optimization and prediction of rate of penetration during drilling operations. *Engineering with Computers* 36 (2): 587–595.
6. Ahmadi, M.A., M. Ebadi, A. Shokrollahi, and S.M. Majidi. 2013. Evolving artificial neural network and imperialist competitive algorithm for prediction oil flow rate of the reservoir. *Applied Soft Computing* 13 (2): 1085–1098.
7. Akande, K.O., T.O. Owolabi, S.O. Olatunji, and A. AbdulRaheem. 2017. A hybrid particle swarm optimization and support vector regression model for modelling permeability prediction of hydrocarbon reservoir. *Journal of Petroleum Science and Engineering* 1 (150): 43–53.
8. Dong, X., S. Wang, R. Sun, and S. Zhao. 2010. Design of artificial neural networks using a genetic algorithm to predict saturates of vacuum gas oil. *Petroleum Science* 7 (1): 118–122.
9. Salcedo-Sanz, S., A. Pastor-Sánchez, J. Del Ser, L. Prieto, and Z.W. Geem. 2015. A coral reefs optimization algorithm with harmony search operators for accurate wind speed prediction. *Renewable Energy* 1 (75): 93–101.

10. Chen, J., G.Q. Zeng, W. Zhou, W. Du, and K.D. Lu. 2018. Wind speed forecasting using nonlinear-learning ensemble of deep learning time series prediction and extremal optimization. *Energy Conversion and Management* 1 (165): 681–695.

11. Wan, C., Z. Xu, P. Pinson, Z.Y. Dong, and K.P. Wong. 2013. Optimal prediction intervals of wind power generation. *IEEE Transactions on Power Systems.* 29 (3): 1166–1174.

12. Olatomiwa, L., S. Mekhilef, S. Shamshirband, K. Mohammadi, D. Petković, and C. Sudheer. 2015. A support vector machine—firefly algorithm-based model for global solar radiation prediction. *Solar Energy* 1 (115): 632–644.

13. Wu, J., J. Long, and M. Liu. 2015. Evolving RBF neural networks for rainfall prediction using hybrid particle swarm optimization and genetic algorithm. *Neurocomputing* 19 (148): 136–142.

14. Mashhadban, H., S.S. Kutanaei, and M.A. Sayarinejad. 2016. Prediction and modeling of mechanical properties in fiber reinforced self-compacting concrete using particle swarm optimization algorithm and artificial neural network. *Construction and Building Materials* 30 (119): 277–287.

15. Chatterjee, S., S. Sarkar, S. Hore, N. Dey, A.S. Ashour, and V.E. Balas. 2017. Particle swarm optimization trained neural network for structural failure prediction of multistoried RC buildings. *Neural Computing and Applications* 28 (8): 2005–2016.

16. Moayedi, H., A. Moatamediyan, H. Nguyen, X.N. Bui, D.T. Bui, and A.S. Rashid. 2020. Prediction of ultimate bearing capacity through various novel evolutionary and neural network models. *Engineering with Computers* 36 (2): 671–687.

17. Gui, G., H. Pan, Z. Lin, Y. Li, and Z. Yuan. 2017. Data-driven support vector machine with optimization techniques for structural health monitoring and damage detection. *KSCE Journal of Civil Engineering.* 21 (2): 523–534.

18. Pham, B.T., M.D. Nguyen, K.T. Bui, I. Prakash, K. Chapi, and D.T. Bui. 2019. A novel artificial intelligence approach based on Multi-layer Perceptron Neural Network and Biogeography-based Optimization for predicting coefficient of consolidation of soil. *CATENA* 1 (173): 302–311.

19. Mia, M., and N.R. Dhar. 2019. Prediction and optimization by using SVR, RSM and GA in hard turning of tempered AISI 1060 steel under effective cooling condition. *Neural Computing and Applications* 31 (7): 2349–2370.

20. Kant, G., and K.S. Sangwan. 2015. Predictive modelling and optimization of machining parameters to minimize surface roughness using artificial neural network coupled with genetic algorithm. *Procedia CIRP* 31: 453–458.

21. Zhu, K., X. Song, and D. Xue. 2014. A roller bearing fault diagnosis method based on hierarchical entropy and support vector machine with particle swarm optimization algorithm. *Measurement* 1 (47): 669–675.

22. Kadri, O., L.H. Mouss, and M.D. Mouss. 2012. Fault diagnosis of rotary kiln using SVM and binary ACO. *Journal of Mechanical Science and Technology.* 26 (2): 601–608.

23. Deng, W., R. Yao, H. Zhao, X. Yang, and G. Li. 2019. A novel intelligent diagnosis method using optimal LS-SVM with improved PSO algorithm. *Soft Computing* 23 (7): 2445–2462.

24. Muralitharan, K., R. Sakthivel, and R. Vishnuvarthan. 2018. Neural network based optimization approach for energy demand prediction in smart grid. *Neurocomputing* 17 (273): 199–208.

25. Asrari, A., T.X. Wu, and B. Ramos. 2016. A hybrid algorithm for short-term solar power prediction—Sunshine state case study. *IEEE Transactions on Sustainable Energy.* 8 (2): 582–591.

26. Kuan, T.W., J.F. Wang, J.C. Wang, P.C. Lin, and G.H. Gu. 2011. VLSI design of an SVM learning core on sequential minimal optimization algorithm. *IEEE Transactions on Very Large Scale Integration (VLSI) Systems* 20 (4): 673–83.

27. Zhang, L., F. Tian, S. Liu, J. Guo, B. Hu, Q. Ye, L. Dang, X. Peng, C. Kadri, and J. Feng. 2013. Chaos based neural network optimization for concentration estimation of indoor air contaminants by an electronic nose. *Sensors and Actuators A: Physical* 15 (189): 161–167.

28. Cong, Y., J. Wang, and X. Li. 2016. Traffic flow forecasting by a least squares support vector machine with a fruit fly optimization algorithm. *Procedia Engineering* 137 (1): 59–68.

29. Yang, M., C. Chen, L. Wang, X. Yan, and L. Zhou. 2016. Bus arrival time prediction using support vector machine with genetic algorithm. *Neural Network World* 26 (3): 205.

30. Ling, X., X. Feng, Z. Chen, Y. Xu, and H. Zheng. 2017. Short-term traffic flow prediction with optimized multi-kernel support vector machine. In *2017 IEEE Congress on Evolutionary Computation (CEC), 5 June 2017*, 294–300. IEEE.

31. Yu, Z. 2012. Automotive fault diagnosis based on SVM and particle swarm algorithm. *Application Research of Computers* 2.

32. Wu, Q. 2010. Car assembly line fault diagnosis based on robust wavelet SVC and PSO. *Expert Systems with Applications* 37 (7): 5423–5429.

33. Baraldi, P., F. Cannarile, F. Di Maio, and E. Zio. 2016. Hierarchical k-nearest neighbours classification and binary differential evolution for fault diagnostics of automotive bearings operating under variable conditions. *Engineering Applications of Artificial Intelligence* 1 (56): 1–3.

34. Zhan, L., and C. Li. 2016. A hybrid PSO-SVM-based method for predicting the friction coefficient between aircraft tire and coating. *Measurement Science and Technology* 28 (2): 025004.

35. Du, Y., L. Xiao, Y. Chen, and R. Ding. 2018. Aircraft Engine Gas Path Fault Diagnosis Based on Hybrid PSO-TWSVM. *Transactions of Nanjing University of Aeronautics and Astronautics* 35 (2): 334–342.

36. Cheng, R., W. Zhao, H. Deng, and X. Jiang. 2015. Modeling and optimization control for aircraft AC generator brushless excitation system based on improved adaptive PSO. *The Open Automation and Control Systems Journal* 7 (1).

37. Dhanalaxmi, B., G.A. Naidu, and K. Anuradha. 2015. Adaptive PSO based association rule mining technique for software defect classification using ANN. *Procedia Computer Science* 46: 432–442.

38. Kayarvizhy, N., S. Kanmani, and R. Uthariaraj. 2013. Improving Fault prediction using ANN-PSO in object oriented systems. *International Journal of Computer Applications* 73 (3): 0975–8887.

39. Dan, Z. 2013. Improving the accuracy in software effort estimation: Using artificial neural network model based on particle swarm optimization. In *Proceedings of 2013 IEEE International Conference on Service Operations and Logistics, and Informatics, 28 July 2013*, pp. 180–185. IEEE.

40. Tomar, P., Mishra, R., and K. Sheoran. 2018. Prediction of quality using ANN based on teaching-learning optimization in component-based software systems. *Software: Practice and Experience* 48 (4): 896–910.

41. Qin, L.N. (2011). Software reliability prediction model based on PSO and SVM. In *2011 International Conference on Consumer Electronics, Communications and Networks (CECNet), 16 Apr 2011*, pp. 5236–5239. IEEE.

42. Zhou, J., X. Li, and X. Shi. 2012. Long-term prediction model of rockburst in underground openings using heuristic algorithms and support vector machines. *Safety Science* 50 (4): 629–644.

43. Armaghani, D.J., M. Hajihassani, E.T. Mohamad, A. Marto, and S.A. Noorani. 2014. Blasting-induced flyrock and ground vibration prediction through an expert artificial neural network based on particle swarm optimization. *Arabian Journal of Geosciences.* 7 (12): 5383–5396.

44. Mohamad, E.T., D.J. Armaghani, E. Momeni, and S.V. Abad. 2015. Prediction of the unconfined compressive strength of soft rocks: A PSO-based ANN approach. *Bulletin of Engineering Geology and the Environment* 74 (3): 745–757.

45. Yu, R., X. An, B. Jin, J. Shi, O.A. Move, and Y. Liu. 2018. Particle classification optimization-based BP network for telecommunication customer churn prediction. *Neural Computing and Applications* 29 (3): 707–720.

46. Zubaidi, S.L., S. Ortega-Martorell, H. Al-Bugharbee, I. Olier, K.S. Hashim, S.K. Gharghan, P. Kot, and R. Al-Khaddar. 2020. Urban water demand prediction for a city that suffers from climate change and population growth: Gauteng province case study. *Water* 12 (7): 1885.

47. Karkevandi-Talkhooncheh, A., S. Hajirezaie, A. Hemmati-Sarapardeh, M.M. Husein, K. Karan, and M. Sharifi. 2017. Application of adaptive neuro fuzzy interface system optimized with evolutionary algorithms for modeling CO2-crude oil minimum miscibility pressure. *Fuel* 1 (205): 34–45.

48. Liu, S., H. Tai, Q. Ding, D. Li, L. Xu, and Y. Wei. 2013. A hybrid approach of support vector regression with genetic algorithm optimization for aquaculture water quality prediction. *Mathematical and Computer Modelling* 58 (3–4): 458–465.

49. Sahlol, A.T., A.A Ewees, A.M. Hemdan, and A.E. Hassanien. 2016. Training feedforward neural networks using Sine-Cosine algorithm to improve the prediction of liver enzymes on fish farmed on nano-selenite. In *2016 12th International Computer Engineering Conference (ICENCO), 28 Dec 2016*, 35–40. IEEE.

50. Fernandez, M., and D. Miranda-Saavedra. 2012. Genome-wide enhancer prediction from epigenetic signatures using genetic algorithm-optimized support vector machines. *Nucleic Acids Research* 40 (10): e77.

51. Hao, M., Y. Li, Y. Wang, and S. Zhang. 2011. Prediction of P2Y12 antagonists using a novel genetic algorithm-support vector machine coupled approach. *Analytica Chimica Acta* 690 (1): 53–63.

52. Liu, S., L. Xu, D. Li, and L. Zeng. 2012. Dissolved oxygen prediction model of eriocheir sinensis culture based on least squares support vector regression optimized by ant colony algorithm. *Transactions of the Chinese Society of Agricultural Engineering* 28 (23): 167–175.

53. Jiang, H., Y. Xiao, J. Li, and X. Liu. 2012. Prediction of the melt index based on the relevance vector machine with modified particle swarm optimization. *Chemical Engineering & Technology* 35 (5): 819–826.

54. Fang, Y., K. Fataliyev, L. Wang, X. Fu, and Y. Wang. 2014. Improving the genetic-algorithm-optimized wavelet neural network for stock market prediction. In *2014 International Joint Conference on Neural Networks (IJCNN), 6 July 2014*, 3038–3042. IEEE.

55. Zhou, X., Z. Li, Z. Dai, and X. Zou. 2010. QSAR modeling of peptide biological activity by coupling support vector machine with particle swarm optimization algorithm and genetic algorithm. *Journal of Molecular Graphics and Modelling* 29 (2): 188–196.

56. Wang, Y., X. Meng, and L. Zhu. 2018. Cell group recognition method based on adaptive mutation PSO-SVM. *Cells* 7 (9): 135.

57. Nieto, P.G., E. García-Gonzalo, J.A. Fernández, and C.D. Muñiz. 2016. A hybrid PSO optimized SVM-based model for predicting a successful growth cycle of the Spirulina platensis from raceway experiments data. *Journal of Computational and Applied Mathematics* 1 (291): 293–303.

58. Hassan, H., A. Badr, M.B. Abdelhalim. 2015. Prediction of O-glycosylation sites using random forest and GA-tuned PSO technique. *Bioinformatics and Biology insights*, 9 Jan 2015: BBI-S26864.

59. Wang, Y., X. Chen, W. Jiang, L. Li, W. Li, L. Yang, M. Liao, B. Lian, Y. Lv, S. Wang, and S. Wang. 2011. Predicting human microRNA precursors based on an optimized feature subset generated by GA–SVM. *Genomics* 98 (2): 73–78.

60. Pei, J.F., C.Z. Cai, Y.M. Zhu, and B. Yan. 2013. Modeling and predicting the glass transition temperature of polymethacrylates based on quantum chemical descriptors by using hybrid PSO-SVR. *Macromolecular Theory and Simulations* 22 (1): 52–60.

61. Fan, M., J. Hu, R. Cao, K. Xiong, and X. Wei. 2017. Modeling and prediction of copper removal from aqueous solutions by nZVI/rGO magnetic nanocomposites using ANN-GA and ANN-PSO. *Scientific Reports* 7 (1): 1–4.

62. Benvidi, A., S. Abbasi, S. Gharaghani, M.D. Tezerjani, and S. Masoum. 2017. Spectrophotometric determination of synthetic colorants using PSO–GA-ANN. *Food Chemistry* 1 (220): 377–384.

63. Braik, M., A. Sheta, and A. Arieqat. 2008. A comparison between GAs and PSO in training ANN to model the TE chemical process reactor. In *Proceedings of the AISB 2008 symposium on swarm intelligence algorithms and applications, 1 Apr 2008*, vol. 11: 24–30.

64. Li, M., H. Zhang, B. Chen, Y. Wu, and L. Guan. 2018. Prediction of pKa values for neutral and basic drugs based on hybrid artificial intelligence methods. *Scientific Reports* 8 (1): 1–3.

65. Li, M., X. Huang, H. Liu, B. Liu, Y. Wu, and L. Wang. 2015. Solubility prediction of supercritical carbon dioxide in 10 polymers using radial basis function artificial neural network based on chaotic self-adaptive particle swarm optimization and K-harmonic means. *RSC Advances* 5 (56): 45520–45527.

66. Lazzús, J.A. 2010. Prediction of flash point temperature of organic compounds using a hybrid method of group contribution+ neural network+ particle swarm optimization. *Chinese Journal of Chemical Engineering* 18 (5): 817–823.

67. Owolabi, T.O. 2020. Extreme learning machine and swarm-based support vector regression methods for predicting crystal lattice parameters of pseudo-cubic/cubic perovskites. *Journal of Applied Physics.* 127 (24): 245107.

68. Majid, A., A. Khan, G. Javed, and A.M. Mirza. 2010. Lattice constant prediction of cubic and monoclinic perovskites using neural networks and support vector regression. *Computational Materials Science* 50 (2): 363–372.

69. Yuan, F.C., and C.H. Lee. 2015. Using least square support vector regression with genetic algorithm to forecast beta systematic risk. *Journal of Computational Science* 1 (11): 26–33.

70. Owolabi, T.O. 2019. Development of a particle swarm optimization based support vector regression model for titanium dioxide band gap characterization. *Journal of Semiconductors* 40 (2): 022803.

71. Lazzús, J.A. 2011. Autoignition temperature prediction using an artificial neural network with particle swarm optimization. *International Journal of Thermophysics* 32 (5): 957.

72. Owolabi, T.O. 2019. Modeling the magnetocaloric effect of manganite using hybrid genetic and support vector regression algorithms. *Physics Letters A* 383 (15): 1782–1790.

73. Shamsah, S.M., and T.O. Owolabi. 2020. Modeling the maximum magnetic entropy change of doped manganite using a grid search-based extreme learning machine and hybrid gravitational search-based support vector regression. *Crystals* 10 (4): 310.

74. Bui, D.T., T.A. Tuan, N.D. Hoang, N.Q. Thanh, D.B. Nguyen, N. Van Liem, and B. Pradhan. 2017. Spatial prediction of rainfall-induced landslides for the Lao Cai area (Vietnam) using a hybrid intelligent approach of least squares support vector machines inference model and artificial bee colony optimization. *Landslides* 14 (2): 447–458.

75. Chen, W., M. Panahi, P. Tsangaratos, H. Shahabi, I. Ilia, S. Panahi, S. Li, A. Jaafari, and B.B. Ahmad. 2019. Applying population-based evolutionary algorithms and a neuro-fuzzy system for modeling landslide susceptibility. *CATENA* 1 (172): 212–231.

76. Moayedi, H., M. Mehrabi, M. Mosallanezhad, A.S. Rashid, and B. Pradhan. 2019. Modification of landslide susceptibility mapping using optimized PSO-ANN technique. *Engineering with Computers* 35 (3): 967–984.

77. Wang, Y., H. Tang, T. Wen, and J. Ma. 2019. A hybrid intelligent approach for constructing landslide displacement prediction intervals. *Applied Soft Computing* 1 (81): 105506.

78. Li, X.Z., and J.M. Kong. 2014. Application of GA-SVM method with parameter optimization for landslide development prediction. *Natural Hazards and Earth System Sciences* 14 (3): 525.

79. Su, Y.X., J. Shen, D.H. Zhang, and X.F. Hu. 2011. Application of neural networks and improved PSO algorithms to earthquake prediction. *Journal of Computer Applications,* 7.

80. Nguyen, H., H. Moayedi, L.K. Foong, H.A. Al Najjar, W.A. Jusoh, A.S. Rashid, and J. Jamali. 2019. Optimizing ANN models with PSO for predicting short building seismic response. *Engineering with Computers* 6: 1–5.

81. Abraham, A., and V. Rohini. 2019. A particle swarm optimization-backpropagation (PSO-BP) model for the prediction of earthquake in Japan. In *Emerging Research in Computing, Information, Communication and Applications,* 435–441. Springer: Singapore.

82. Gordan, B., D.J. Armaghani, M. Hajihassani, and M. Monjezi. 2016. Prediction of seismic slope stability through combination of particle swarm optimization and neural network. *Engineering with Computers* 32 (1): 85–97.

83. Quaranta, G., G.C. Marano, R. Greco, and G. Monti. 2014. Parametric identification of seismic isolators using differential evolution and particle swarm optimization. *Applied Soft Computing* 1 (22): 458–464.

84. Khan, T., M. Alam, F.A. Shaikh, S. Khan, K. Kadir, M.S. Mazliham, Z. Shahid, and M. Yahya. 2019. Flash floods prediction using real time data: An implementation of ANN-PSO with less false alarm. In *2019 IEEE International Instrumentation and Measurement Technology Conference (I2MTC) 20 May 2019*, pp. 1–6. IEEE.

85. Bui, D.T., P.T. Ngo, T.D. Pham, A. Jaafari, N.Q. Minh, P.V. Hoa, and P. Samui. 2019. A novel hybrid approach based on a swarm intelligence optimized extreme learning machine for flash flood susceptibility mapping. *CATENA* 1 (179): 184–196.

86. Anupam, S., and P. Pani. 2020. Flood forecasting using a hybrid extreme learning machine-particle swarm optimization algorithm (ELM-PSO) model. *Modeling Earth Systems and Environment* 6 (1): 341–347.

87. Choubin, B., E. Moradi, M. Golshan, J. Adamowski, F. Sajedi-Hosseini, and A. Mosavi. 2019. An ensemble prediction of flood susceptibility using multivariate discriminant analysis, classification and regression trees, and support vector machines. *Science of the Total Environment* 15 (651): 2087–2096.

88. Gu, J., H. Liu, X. Liu, and Q. Lü. 2011. Application of genetic algorithm-support vector machine model in tropical cyclone intensity forecast. *Marine Forecasts* 28 (3): 8–14.

89. Zhao, H., L. Jin, Y. Huang, and X. Huang. 2012. An application of ensemble prediction for typhoon intensity based on MDS and PSO-ANN. In *2012 Fifth International Joint Conference on Computational Sciences and Optimization, 23 June 2012*, 885–888. IEEE.

90. Zhao, H.S., L. Jin, Y. Huang, and J. Jin. 2014. An objective prediction model for typhoon rainstorm using particle swarm optimization: Neural network ensemble. *Natural Hazards* 73 (2): 427–437.

91. Pan, W.T. 2012. A new fruit fly optimization algorithm: Taking the financial distress model as an example. *Knowledge-Based Systems* 1 (26): 69–74.

92. Abdual-Salam, M.E., H.M. Abdul-Kader, and W.F. Abdel-Wahed. 2010. Comparative study between differential evolution and particle swarm optimization algorithms in training of feed-forward neural network for stock price prediction. In *2010 The 7th International Conference on Informatics and Systems (INFOS), 28 March 2010*, 1–8. IEEE.

93. Chakravarty, S., and P.K. Dash. 2012. A PSO based integrated functional link net and interval type-2 fuzzy logic system for predicting stock market indices. *Applied Soft Computing* 12 (2): 931–941.

94. Hitam, N.A., A.R. Ismail, and F. Saeed. 2019. An optimized support vector machine (SVM) based on particle swarm optimization (PSO) for cryptocurrency forecasting. *Procedia Computer Science* 1 (163): 427–433.

95. Zhiqiang, G., W. Huaiqing, and L. Quan. 2013. Financial time series forecasting using LPP and SVM optimized by PSO. *Soft Computing* 17 (5): 805–818.

96. Tang, Y., Zhou, J. (2015). The performance of PSO-SVM in inflation forecasting. In 2015 12th International Conference on Service Systems and Service Management (ICSSSM), 22 June 2015, pp. 1–4. IEEE.

97. Wang, J.S., and C.X. Ning. 2015. ANFIS Based time series prediction method of bank cash flow optimized by adaptive population activity PSO algorithm. *Information* 6 (3): 300–313.

98. Lu, Y., J. Zhu, N. Zhang, and Q. Shao. 2014. A hybrid switching PSO algorithm and support vector machines for bankruptcy prediction. *In 2014 International Conference on Mechatronics and Control (ICMC), 3 July 2014*, 1329–1333. IEEE.

99. Cao, J., H. Lu, W. Wang, and J. Wang. 2012. A novel five-category loan-risk evaluation model using multiclass LS-SVM by PSO. *International Journal of Information Technology & Decision Making* 11 (04): 857–874.

100. Kamaruddin, S., and V. Ravi. 2016. Credit card fraud detection using big data analytics: Use of PSOAANN based one-class classification. In *Proceedings of the International Conference on Informatics and Analytics 25 Aug 2016*, 1–8.

101. Chang, Y.T., J. Lin, J.S. Shieh, and M.F. Abbod. 2012. Optimization the initial weights of artificial neural networks via genetic algorithm applied to hip bone fracture prediction. *Advances in Fuzzy Systems* 1: 2012.
102. Suresh, A., K.V. Harish, and N. Radhika. 2015. Particle swarm optimization over back propagation neural network for length of stay prediction. *Procedia Computer Science* 1 (46): 268–275.
103. Vieira, S.M., L.F. Mendonça, G.J. Farinha, and J.M. Sousa. 2013. Modified binary PSO for feature selection using SVM applied to mortality prediction of septic patients. *Applied Soft Computing* 13 (8): 3494–3504.
104. Khourdifi, Y., and M. Bahaj. 2019. Heart disease prediction and classification using machine learning algorithms optimized by particle swarm optimization and ant colony optimization. *International Journal of Intelligent Engineering and Systems* 12 (1): 242–252.
105. Kaya, D. 2019. Optimization of SVM parameters with hybrid CS-PSO algorithms for Parkinson's disease in LabVIEW environment. *Parkinson's Disease* 2: 2019.
106. Subasi, A. 2013. Classification of EMG signals using PSO optimized SVM for diagnosis of neuromuscular disorders. *Computers in Biology and Medicine* 43 (5): 576–586.
107. Babaoglu, İ, O. Findik, and E. Ülker. 2010. A comparison of feature selection models utilizing binary particle swarm optimization and genetic algorithm in determining coronary artery disease using support vector machine. *Expert Systems with Applications* 37 (4): 3177–3183.
108. Kar, S., K.D. Sharma, and M. Maitra. 2015. Gene selection from microarray gene expression data for classification of cancer subgroups employing PSO and adaptive K-nearest neighborhood technique. *Expert Systems with Applications* 42 (1): 612–627.
109. Zeng, N., H. Qiu, Z. Wang, W. Liu, H. Zhang, and Y. Li. 2018. A new switching-delayed-PSO-based optimized SVM algorithm for diagnosis of Alzheimer's disease. *Neurocomputing* 3 (320): 195–202.
110. Mirjalili, S., S.M. Mirjalili, and A. Lewis. 2014. Grey wolf optimizer. *Advances in Engineering Software* 69: 46–61.
111. Mirjalili, S., and A. Lewis. 2016. The whale optimization algorithm. *Advances in Engineering Software* 1 (95): 51–67.
112. Doğan, B., and T. Ölmez. 2015. A new metaheuristic for numerical function optimization: Vortex Search algorithm. *Information Sciences* 1 (293): 125–145.
113. Satapathy, S., and A. Naik. 2016. Social group optimization (SGO): A new population evolutionary optimization technique. *Complex & Intelligent Systems* 2 (3): 173–203.

Application to Predict the Impact of COVID-19 in India Using Deep Learning

Kiran S. Raj and Priyanka Kumar

Abstract The COVID-19 pandemic has hit almost all the parts of the world. Originating in Wuhan, China, to spreading all across the world, it is safe to say now that the pandemic has shocked the people. Despite all the advancements in medicine and science, it is quite frankly a realization to the world that a virus can rip apart everyone's lives [1]. The USA, being the worst impacted by the same [2], countries like India are experiencing the growth of the virus rapidly. From social distancing to living with the virus, we humans are finding out different ways to survive this pandemic and return to normalcy. The importance of understanding the situation reels' core to the lives of everyone. In this chapter, we keep update about the change in events, cases reported, people deceased, people recovered, contact for essential service, the current status of the state/country is essential to provide a system that people could depend on during the time of the pandemic. This research works on the core principle of providing real-time information to the people, by supplementing it with the state-wise report, national reports, essential services, and contacts provided by the state. This chapter also explains the power of AI/ML which will predict how the cases would progress in the coming days. The model used details of a convolutional neural network to predict the spread of the pandemic, hence gives an idea of how the pandemic would spread with a close approximation to the real-world data.

Keywords Deep learning · Time series analysis · COVID-19 forecasting · CNN

1 Introduction

The COVID-19 pandemic has wrecked life in dimensions one cannot perceive. From its inception in Wuhan during December 2019, the virus has spread and affected

K. S. Raj · P. Kumar (✉)
Department of Computer Science and Engineering, Amrita School of Engineering, Amrita
Vishwa Vidyapeetham, Coimbatore, India
e-mail: k_priyanka@cb.amrita.edu

K. S. Raj
e-mail: cb.en.u4cse17430@cb.students.amrita.edu

most of the world. The part of the virus that scares most of the world has to deal with the fact of its ability to spread rapidly. The cases have grown ever since and a total of 952,000 deaths have been reported so far all around the world. The most arduous impact on the people has to do with the fact that lockdown had been imposed in most of the countries which eventually lead to unemployment, scarcity of food, a drop in the overall economy, and most importantly, a very sad time in everyone's life.

The most daring fact about humans revolves around the ability to be resilient and strike back even when the odds are against them. From time immemorial, humans have always overcome obstacles, found solutions, and have come out much stronger than ever before. The research work aims to provide awareness and the threat of COVID-19 among everyone. The application focuses on creating a system that provides the details of how the pandemic would spread in high accuracy, thereby making inferences and taking necessary precautions regarding the same. The saying, "Prevention is better than cure" has become an integral part of what we are currently facing and yet to face.

This work hopes to provide users without sufficient information regarding the pandemic, its spread, and also the contact of necessary resources to be taken during this pandemic. With this system intact, we are trying to create a mentality that would augment the wellbeing of the public by keeping in mind the disastrous effects of the pandemic. It has now reached a situation where scientists and other researchers are instilling the idea of living with the virus as the only way to progress forward, due to the time and other constraints that come to being while developing the vaccine. It is only right to stay updated about the pandemic and move forward with one's life. That is what humans have done since history, defy all the odds, and stop at nothing. With the system in check, government officials and other bodies can have a rough understanding of how the pandemic would spread [3] in the coming days and take necessary precautions to control the spread of the virus. The system can be used as a fundamental basis/blueprint for scripting out the necessary actions.

Our goal is to augment the wellbeing of the public, bring back the confidence among them by taking necessary precautions, and hence move on with life.

Roadmap. The chapter is organized as follows. The aim of the research work for the current pandemic is present in Sect. 2. In Sect. 3, we demonstrate our proposed modules. In Sect. 4, we detail the need for deep learning and the architecture used for the model. Section 5 describes the system implementations and Sect. 6 describes the results and analysis. Finally, the conclusion is presented in Sect. 7.

2 Proposed Work

Some recent works similar to the topic are detailed as COVID-19 trackers. There exist many trackers [4] providing the information on the count of the COVID-19 affected patients, recovered cases, and deceased cases. These works revolve around getting the information from the API and displaying it in a format that the user understands

[5]. Some of the work focuses on predicting the spread of the pandemic using the SIR model.

The proposed system is presented to the user in the form of a web application with Flask in the backend. The user is greeted with a home page detailing cards that have snippets of information regarding the daily impact of COVID-19 in India. These snippets include total number of cases on the current day, deaths on the current day, recovered cases on the current day, and active cases reported on the current day. These pieces of information are updated in real time and are up-to-date for the user to view. The cards are followed by graphical representations. There are four graphs as follows:

- Overview of COVID-19 cases in India (contains total cases, active cases, deaths, recovered cases—with date).
- Total cases of COVID-19 in India, which contains the graph of how the total cases in India are emerging, followed by an artificially intelligent prediction of how the cases will follow for the next 30 days.
- Active cases of COVID-19 in India, which contains the graph of how the active cases in India are emerging, followed by an artificially intelligent prediction of how the cases will progress for the next 30 days.
- Daily cases of COVID-19 confirmed in India, which contains the graph of how the daily confirmed cases in India are emerging, followed by an artificially intelligent prediction of how the cases will progress for the next 30 days.

Figure 1 illustrates the home page of the application. The user is presented with an info card detailing about the present number of cases, recovered cases, daily confirmed cases, and the deaths. This is followed by the visualization of the same where predictions are also incorporated within each of the graphs. This visualization gives a rough idea to the user regarding the spread of the pandemic in the upcoming days.

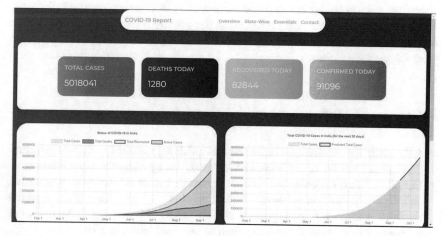

Fig. 1 Home page of the application (till September 15, 2020)

The uniqueness of the graphical representations deals with the ability for users to visualize the impact of COVID-19 for 30 days ahead of time. The prediction of future trends in COVID-19 is modeled with the help of deep learning. The model is trained to predict the future of the impact of COVID-19 in the future with a very close accurate approximation. The predicted values are very close to the real scenario, hence enabling the users to have an idea of how this horrific pandemic would spread, thus taking necessary precautions regarding the same [6]. Predictions are made for the total number of COVID-19 cases in India, active cases in India, daily reported cases in India.

The next tab of the web application holds information regarding the state-wise COVID-19 report. The tab elaborates on the total cases, total active cases, total deaths, total recovered cases, and the last updated time for every state in India. The information present in each cell also contains a daily increase/decrease count in the amount. This section contains the statistical information of the same for any particular state in India.

The next tab of the web application elaborates on the essentials and resources available during this pandemic. It is difficult to comprehend the effects that the pandemic has had in everyone's daily life. Government and health agencies are doing their best to help combat this crisis. Thus, this section of the web application contains details about the contacts, different resources provided by the government, and other agencies in a particular state that would the users in need. With the help of this information, the user is in complete leverage to contact the specific agencies in his/her area when required. Our goal of this chapter is to provide fruitful, genuine and reliable information for the user.

For quick and fast alerts regarding the same, a chatbot built on Telegram equips the user with an efficient way to view the details of COVID-19 in India. The chatbot responds to the queries laid down by the users by fetching relevant pieces of information regarding the query posted.

This system is to be further generalized to any type of pandemic that might hit us in the future. An application to provide the necessary information required to combat the pandemic would be fruitful now and in the coming future. By generalizing the system, it could be a household application that everyone could depend on when a pandemic as such strikes.

3 Proposed Modules

- **Info-card**: The information cards serve as quick snippets of COVID-19 information to the user. These cards contain information regarding the Total Number of COVID-19 Cases, Deaths on a particular day, Recovered Cases on the particular, and the Daily Confirmed Cases. The information cards provide a glimpse of the devastating impacts that the pandemic is having on India. In Fig. 2, the info card is shown details about the cases in each of the specified categories.

Fig. 2 Info-card (till 15th September 2020)

- **Visualizing section**: This section contains a graphical representation of the way COVID-19 has been progressing in India. are namely four graphs on the main page:
  - Real-Time status of COVID-19 in India: This graph contains the history of the entire COVID-19 cases reported in India till date, it includes Total Cases, Total Recovered Cases, Total Deaths, Active Cases.
  - Total COVID-19 Cases in India, along with prediction: This graph contains the total cases of COVID-19 reported in India to date, along with an intelligent prediction of how the cases would increase in the upcoming 30 days in advance.
  - Active COVID-19 Cases in India, along with prediction: This graph contains the active cases of COVID-19 reported in India to date, along with an intelligent prediction of how the cases would increase/decrease in the upcoming 30 days in advance. Active cases are a very crucial factor, emphasis is given on active cases as it is through this can one realize when it would peak and decrease accordingly.

    Active Cases = Total Cases − Total Recovered Cases −− Total Deceased Cases

  - Daily Confirmed COVID-19 Cases in India, along with prediction: This graph contains the daily confirmed cases of COVID-19 in India to date, along with an intelligent prediction of how the cases would increase/decrease in the upcoming 30 days in advance. Daily confirmed cases can be used to conclude how many people are affected depending on the number of people tested.

- **State-wise statistics**: This section details the statistics of COVID-19 impacts in all the states in India. The user is presented with a table which contains the following information:
  - State name.
  - Active Cases in the state.
  - Deaths in the state, due to COVID-19.
  - Recovered Cases in the state.
  - Total Confirmed Cases in the state.
  - Last Updated date and time.

In Fig. 3, illustrates the COVID-19 cases report detailing the case based statistics for each of the states in India. Each of the rows is described with detailed statistics

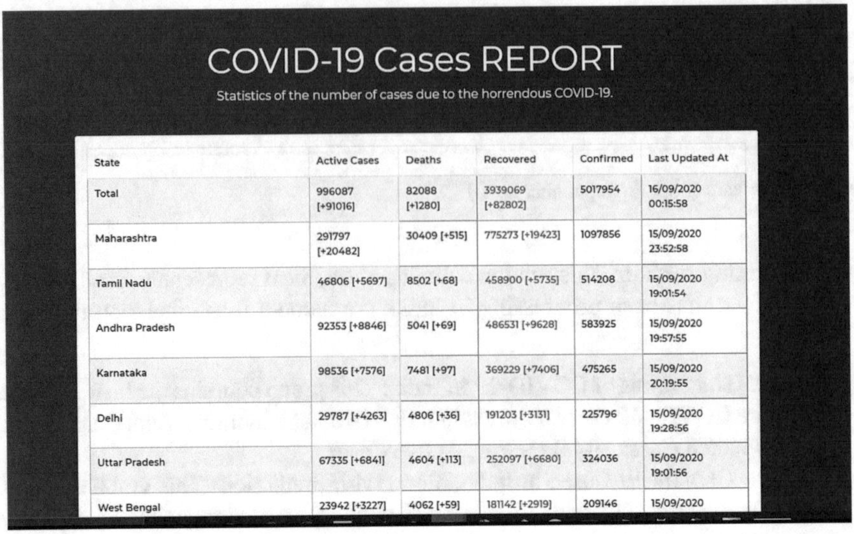

COVID-19 Cases REPORT

Statistics of the number of cases due to the horrendous COVID-19.

State	Active Cases	Deaths	Recovered	Confirmed	Last Updated At
Total	996087 [+91016]	82088 [+1280]	3939069 [+82802]	5017954	16/09/2020 00:15:58
Maharashtra	291797 [+20482]	30409 [+515]	775273 [+19423]	1097856	15/09/2020 23:52:58
Tamil Nadu	46806 [+5697]	8502 [+68]	458900 [+5735]	514208	15/09/2020 19:01:54
Andhra Pradesh	92353 [+8846]	5041 [+69]	486531 [+9628]	583925	15/09/2020 19:57:55
Karnataka	98536 [+7576]	7481 [+97]	369229 [+7406]	475265	15/09/2020 20:19:55
Delhi	29787 [+4263]	4806 [+36]	191203 [+3131]	225796	15/09/2020 19:28:56
Uttar Pradesh	67335 [+6841]	4604 [+113]	252097 [+6680]	324036	15/09/2020 19:01:56
West Bengal	23942 [+3227]	4062 [+59]	181142 [+2919]	209146	15/09/2020

Fig. 3 State-wise statistics COVID-19 in India (the number enclosed between "[ ]" denotes the daily increase/decrease) (till 15th September 2020)

regarding the different categories of cases including Active Cases, Deaths, Recovered Cases, Total Confirmed Cases.

- **Essentials and Resources**: This section details the contact of agencies who would provide a helping hand in cases of COVID-19 related issues and much more in a particular location. The user is presented with a search interface, where the state name is provided as a query and the details of all the government agencies and help groups would be presented. We all owe a big salute to the people who are out there in the most hazardous conditions to help people. Figure 4 describes the Essentials and Resources section, the user is prompted to enter the state name and then to hit 'Click to Select State'. After doing so, the request is sent to the server which fetches the details about the state name and is printed in a tabular format.
- **Telegram Chatbot**: The Telegram chatbot incorporates all the qualities of a reliable, easily accessible source of information. The chatbot will be able to provide real-time information on COVID-19 cases. The user can view national-wide, state-wise information immediately and the chatbot fetches the relevant details such as total cases, active cases, recovered cases, deaths.

Figures 5 and 6 detail the interface of the Telegram chatbot where the user enters the commands such as overview and state-specific details. The details are presented to the user as a text message and the details are presented in the same.

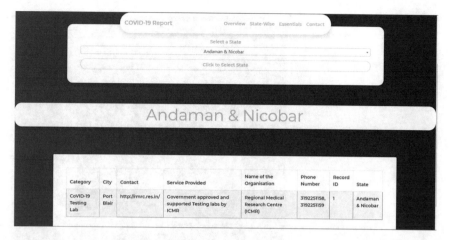

Fig. 4 Essentials and resources

Fig. 5 Telegram chatbot—overview

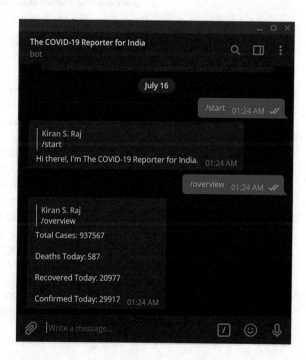

4 Deep Learning

The twenty-first century has been revolving around the developments in technology orchestrating to a much more complicated and sophisticated artificially intelligent system. Most of the innovations that took place in the field had a direct/indirect

Fig. 6 Telegram
chatbot—state-wise selection

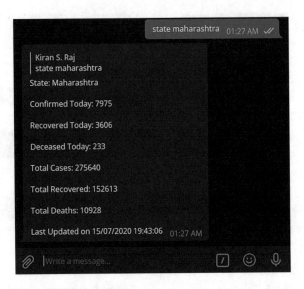

relation to the concept of deep neural networks. The human brain being the source of motivation has helped the computer researchers to craft one of the most innovative architectures based on forward propagation for computing results and back propagation for correcting or improving the predictions, hence the birth of deep learning. The reasons why deep learning has proved to be an important factor when designing the model for the COVID-19 prediction has to revolve around the facts that deep learning models learn complex mathematical function/trends, once stabilized its ability to generalize over most of the input conditions, its ability to tolerate variations from data distributions provided as input to the model, and its ability to extract features from the training set which would then be an important factor to be considered in prediction. Compared to the statistical methods, deep learning architectures outperform them when it comes to its ability to learn all these complicated mathematical functions (multivariate inputs), feature extraction, and the essence of generalizing the predictions over a wide spectrum of inputs [7].

For the current problem statement, convolutional neural networks have been used as the architecture of choice [8]. It can generate feature maps and extract the most important features through convolutional layers, map to the outputs, and generate a prediction that would be a close approximation to the ground truth.

4.1 CNN Model

With the help of the convolutional neural networks on the input data, feature maps are created to extract the components that would play a big role in predicting the progress of COVID-19 with real-life approximations as accurately as possible. With the help of convolutional layers, features are grained out to form feature vectors which are

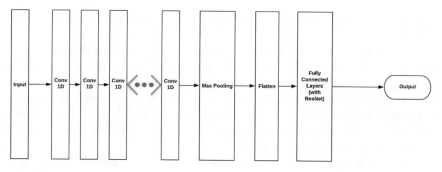

Fig. 7 CNN model used for prediction

then flattened out in a fully connected layered system with a ResNet architecture to provide that dependency on different layers. The results produced simulate the effect of how COVID-19 would progress in the upcoming periods. Figure 7 shows the architecture used in developing the prediction algorithm for COVID-19 cases.

$$x_i^l = \tanh\left(\max_{pooling} \left(\sum_j x_i^{l-1} * w_{ji} \right) + b_i^l \right)$$

where

x_i^l is the map of features obtained after passing through the lth convolutional layer.
w is the weight vector associated with the feature vectors.

This modeling of dealing with the prediction of COVID-19 cases in India with the help of convolutional neural networks helps in predicting values that closely represent the real-world scenario [9]. On further extending the scope to states, the answer to, "How well a particular state is doing?" can be approximated with the help of this approach. On analyzing when the peak would be hit or when the cases have to decrease, one can come to a conclusion regarding the status of COVID-19 in the particular state. This approach can provide insightful information on when people can go out in life returning to normalcy. The results obtained from the approach provide very close approximations to the real-life situation, hence providing a helping tool to have an understanding of how COVID-19 would progress in India.

5 System Implementation

China was the first one to report the case of COVID-19 and as time progressed they were able to flatten the cure. On studying the way COVID-19 progressed in China [10] and analyzing how the trends changed, it is safe to say that the active cases would hit a peak and then drop off. This happens when the active cases are less than the recovered cases, in such a way that the first derivative would be negative and would decrease.

The difficulty in the prediction of such trends deals with the variations and randomness of how the COVID-19 cases would change daily. The factors that would cause such a change include the effects of lockdowns in regions, practices about the prevention of COVID-19 (mask, social distancing), and also the number of tests taken. All these attributes influence the number of cases of COVID-19 reported; hence, it is necessary to understand the underlying trends, variations in the data to predict a close approximation of it in the coming days.

Hence, there requires a system to understand the trend of the temporal quantity to predict in such a way to maintain the correlations between the data and the effective sequence of how the trend follows.

Deep neural networks can estimate nonlinear functions at a very accurate approximation, handle noise present in the data, and good generalization for the time series prediction for a longer period. This is done by taking into account the correlations present within the data and trends, periodical, seasonal nature followed by the COVID-19 data over some time.

Figure 8 displays the process of prediction of the COVID-19 cases in India takes place in the following steps. The first step involves fetching the data details regarding COVID-19 cases from the API. The fetched data is then preprocessed and visualized. The visualization gives a brief idea of how the data points are changing with respect to time. After the preprocessing step and seasonal decomposition, the data is fit to the convolutional neural network model and training. The training would generate the necessary features map and stabilize the weights in the fully connected layers. Then the future trends are then predicted, the convolutional neural networks take into consideration the patterns associated with the changing trends of COVID-19 cases, and also the cases tested. The output from the model is then plotted onto a graph and projected on the web applications for the users to see (Fig. 9).

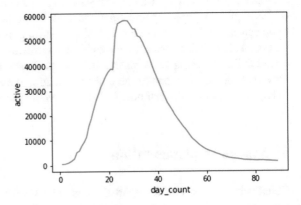

Fig. 8 Active cases in China over the period

COVID-19 CASES ANALYSIS

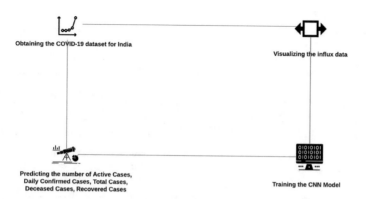

Obtaining the COVID-19 dataset for India

Visualizing the influx data

Predicting the number of Active Cases,
Daily Confirmed Cases, Total Cases,
Deceased Cases, Recovered Cases

Training the CNN Model

Fig. 9 Process of predicting the COVID-19 cases in India

5.1 Decomposition of the COVID-19 Data

COVID-19 Time Series$(t) \rightarrow$ Seasonal$(t) \times$ Trends$(t) \times$ Cyclical$(t) \times$ Irregularity(t)

- **Trends**—represents the persistent increase or decrease of the values along with the overall flow of the data.
- **Seasonal**—occurs when there exists a seasonal in the change of the data over a particular time.
- **Residual**—this denotes the noise/irregularity present in the data that influences the flow over a period of time.

In Fig. 10, the first graph represents the actual COVID-19 progression graph over the days, the second graph represents trend analysis decomposed from the main graph, the third graph represents the seasonal analysis decomposed from the main graph, and the last graph represents residual analysis from the main graph.

6 Results and Analysis

The performance of the trained proposed model is very accurate to real-world data, owing to the architecture underneath and feature map used for training the model. The model also can correct itself from random outliers present along the period; this flexibility enables the model to predict values which would give a rough idea of how the COVID-19 cases would progress in the upcoming days. The following are the error rates providing a rough validation of the accuracy of the model.

From Figs. 11 and 12, the relative absolute error in prediction and the actual cases are represented. From the visualization, we can infer that the error is initially high

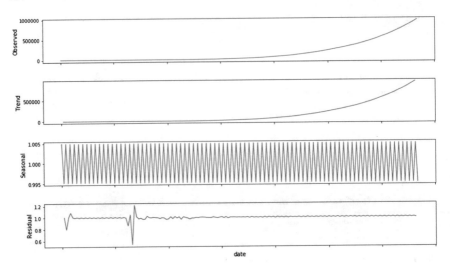

Fig. 10   Decomposition of the COVID-19 data

Fig. 11   Relative absolute
error in the total number of
COVID-19 cases

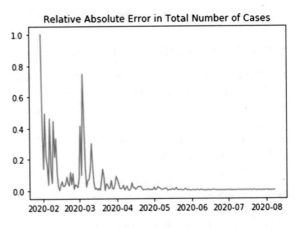

Fig. 12   Relative absolute
error in the total number of
active cases

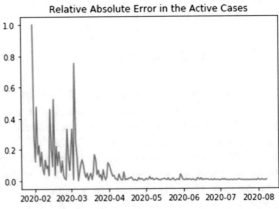

but over time, it drops to the bare minimum, and hence, the accuracy of the model is validated by showing the close approximation of prediction to the real-world data.

The following are the results generated by the model and visualized.

Figure 13 illustrates the current status of COVID-19 in India highlighting the active cases, recovered cases, total cases, and deaths. In Fig. 14, the user can visualize the total cases of COVID-19 in India along with the prediction of the cases for the next 30 days. With the help of Fig. 15, users can visualize the active cases of COVID-19 in India along with the prediction for the next 30 days. Figure 16 illustrated the daily confirmed cases of COVID-19 in India along with the prediction of how it is going to vary in the coming 30 days.

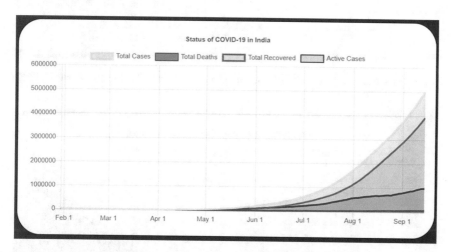

Fig. 13 Status of COVID-19 in India (till September 15, 2020)

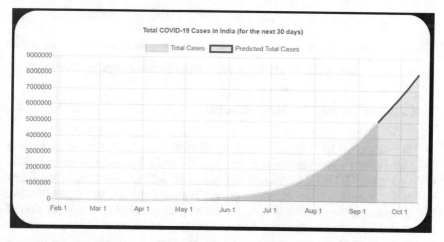

Fig. 14 Total of COVID-19 cases in India along with the prediction (till September 15, 2020)

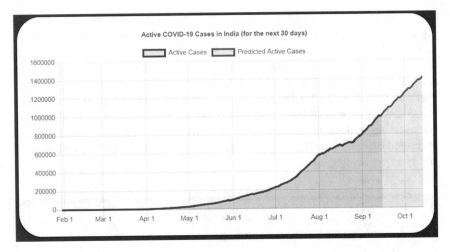

Fig. 15 Active number of COVID-19 cases in India with the prediction (till September 15, 2020)

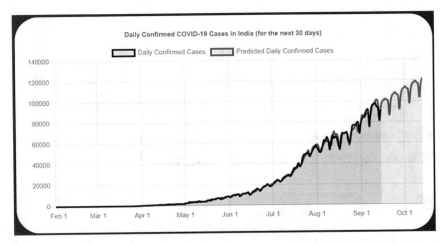

Fig. 16 Daily confirmed COVID-19 cases in India with the prediction (till September 15, 2020

7 Conclusion and Future Direction

In this chapter, we have created a web application that provides a comprehensive and elaborative visualization of the impacts COVID-19 has on India with the help of deep learning. This information is also appended with state-wise statistics, information regarding the essential resources, and Telegram chatbot, which would prove fruitful, enhance the experience from the users' perspective [11]. The prediction of the progress would enable a user to have a rough approximation of how the pandemic would spread across the future days and hence can be prepared in advance for the point of impact. Our goal in this chapter is to provide meaningful information/inference that would help the people understand the extent of severity due to this pandemic and make them aware of how worse it could get with time. In India, where the faces of people are painted with joy and content, we feel that this is our obligation to put forth cutting edge technologies that would help them understand the situation better and bring that smile back onto their faces.

The future scope of this project would be to make the chatbot available in regional languages, generate a score-wise report on how well a particular state is doing to combat COVID-19 in that region and providing a report on the percentage of people affected, detailing out the number of ventilators available in that region and to provide a detailed view on how safe it would be to step out.

For more details, please view the web app (preferably in a laptop):
https://covid-report-india.herokuapp.com/.

References

1. Tang, B., N.L. Bragazzi, Q. Li, S. Tang, Y. Xiao, and J. Wu. 2020. An updated estimation of the risk of transmission of the novel coronavirus (2019-ncov). *Infectious Disease Modelling* 248–255.
2. Marmarelis, V. Predictive modeling of Covid-19 data in the US: Adaptive phase-space approach. *IEEE Open Journal of Engineering in Medicine and Biology.* https://doi.org/10.1109/ojemb.2020.3008313.
3. Berge, T., J.-S. Lubuma, G. Moremedi, N. Morris, and R. Kondera-Shava. 2017. A simple mathematical model for Ebola in Africa. *Journal of Biological Dynamics* 11 (1): 42–74.
4. Binti, Hamzah, Amira Fairoza, Cher Hau, Hafeez Nazri, Dominic Ligot, Guanhua Lee, et al. 2020. CoronaTracker: World-wide COVID-19 outbreak data analysis and prediction. *Bull World Health Organ.*
5. Cleo, Anastassopoulou., Russo Lucia, Tsakris Athanasios, and Siettos Constantinos. 2020. Data-based analysis modelling and forecasting of the COVID-19 outbreak. *Plosone.*
6. Dowd, J.B., L. Andriano, D.M. Brazel, V. Rotondi, P. Block, X. Ding, Y. Liu, and M.C. Mills. 2020. 'Demographic science aids in understanding the spread and fatality rates of COVID-19'. *Proceedings of the National Academy of Sciences of the United States of America* 117 (18): 9696–9698.
7. Ma, J., J. Dushoff, B.M. Bolker, and D.J. Earn. 2014. Estimating initial epidemic growth rates. *Bulletin of Mathematical Biology* 76 (1): 245–260.
8. Yang, C., W. Jiang, and Z. Guo. 2019. Time series data classification based on dual path CNN-RNN cascade network. *IEEE Access* 7: 155304–155312. https://doi.org/10.1109/ACCESS.2019.2949287.
9. Dowell, S.F. 2001. Seasonal variation in host susceptibility and cycles of certain infectious diseases. *Emerging Infectious Diseases* 7 (3): 369–374.
10. Shen, M., Z. Peng, Y. Xiao, and L. Zhang. 2020. Modelling the epidemic trend of the 2019 novel coronavirus outbreak in china. *bioRxiv.*
11. https://www.covid19india.org/.

Role of Data Analytics in Bio Cyber Physical Systems

Utkarsh Singh

Abstract Data science has proved its versatility in all dynamics of the field known to mankind, making decision making faster, and accurate over the past two decades. Coupled with IoT devices and their setups, these have been forerunners in terms of data generation and accurate prognosis. According to advisory firm International Data Corporation (IDC), the number of IoT devices is forecasted to reach 41.6 Billion by 2025, and the data generated from these devices is expected to be 79.4 Zettabytes. One broad sector which has emerged as a gold mine for data generation is the Bio Cyber Physical Systems. Bio Cyber Physical Systems are based on the incorporation of computational elements with biological processes of the human body. The following chapter aims to discuss a new design, implementation of a system based on Bio-CPS, focused primarily on health wearable technologies equipped with state-of-the-art sensors, couple their data with machine learning algorithms to detect real-time health complications primarily in a diabetic person and use of long short-term memory (LSTM) for prediction of such health complications.

Keywords Bio cyber physical systems · Health wearable technologies · LSTM neural network

1 Introduction

The human civilization has been subject to revolutionary changes over the course of its long evolving journey. The pace of changes brought about into the life of people has exponentially increased in the twenty-first century with the advent of technology in cyberworld. People have adapted to these changes in a positive and fast paced response. The most remarkable changes in the twenty-first century is 1. The introduction of Windows XP in the year 2001, this OS single handedly changed the user interface era by bringing in a plethora of changes and making it extremely user friendly for people, 2. In the year 2004, Facebook was founded and brought the

U. Singh (✉)
Bhubaneswar, India
e-mail: singh.utkarsh143@gmail.com

© The Author(s), under exclusive license to Springer Nature Singapore Pte Ltd. 2021
S. Rautaray et al. (eds.), *Trends of Data Science and Applications*,
Studies in Computational Intelligence 954,
https://doi.org/10.1007/978-981-33-6815-6_7

world more closer than ever, coining in the term "social media" and revolutionizing messaging, in 2007, Steve Jobs introduced the world to iPhone, which led to the start of the smartphone era which subsequently has led to changes in almost all aspects of digital technology and later in the year 2009 Fitbit entered the market with an activity tracker. From the year 2010 onwards, most tech giants have invested in improving and coming up with more better specs in that have a big appeal to the customers. One such sector has been the wearable devices sector, with advancement in the field of IoT, semiconductors and the data science playing at the very center of all these revolutionary technologies, it is safe to say that wearable tech is bringing upon a new age of technological revolution. "Bio Cyber Physical System" is the term being used to describe the integration of health wearable tech with the human body and its association with data science. People from all backgrounds have shown interest in these miniature wrist devices which help them keep check on their health without any need for frequent consultations from the doctor. The scope of these wearable devices can easily be expanded to people already suffering from diseases and in turn warn them of severe complications from these diseases or help in mobilizing paramedical forces when the person is in grave danger and the individuals collected data accurately helping in a better diagnosis. In this paper, we propose a system comprising of wearable devices equipped primarily with PPG sensors, EDA sensors, and the data collected from these sensors in detecting diabetic seizures, heart attack, breathlessness and fainting of a person using a multi-channel CNN so as to warn the paramedic services instantly. Later we discuss the possibility of using LSTM model so as to predict future attacks based on the persons activity.

2   Cyber Physical Systems

Cyber physical systems can be defined as an integrated environment of cyber world with the physical components and their processes in various fields such as chemical, mechanical, electrical, and many more. The elements belonging to the cyber division in general interact and coordinate with sensors, which return data from based upon the indications from the physical components. According to the data perceived and real-time decision making by the system, the actuators modify the physical and cyber environment accordingly.

Cyber physical systems are slowly emerging onto the market and are poised to play a vital role over the next decade. The reason for its slow emergence is attributed to the fact that its base concept, and architectural flow is closely related to IoT and industrial IoT. There is a hairline difference between the two terminologies, and leading industry experts have been unable to come up with a distinctive difference between the two.

2.1   CPS and IoT

IoT is defined as the interconnection of various devices equipped with sensors and connected to each other over a network, exchanging data, monitoring the performance of devices and in turn optimizing their processes so as to efficiently run the entire network and generate profits for the users and the industry.

When it comes to find the factor that leads to the difference between IoT and CPS, it can be basically said that CPS concerns itself more with the physical entity and its processes [1].

So, in short, IoT is all about interconnectivity between "Things" over the internet whereas CPS is the amalgamation of physical processes with computer networks.

2.2   Concept Map of Cyber Physical Systems

The above concept map gives a clear idea of the main application sectors of CPS. Each sector of application requires a robust design methodology and security so as to protect the network from foreign attacks. Design methodology revolves around six important factors, specification, modeling, scalability, complexity, validation, and verification [2] (Fig. 1).

For the CPS to be efficient and reliable, there needs to be a feedback system which analyzes and relays back info real time for decisions that are based on the predictions of the model and adapt rapidly to the diverse changes that are occurring in the domain in which it has been implemented. Involving humans and nature into the loop helps the system to accurately identify more setbacks and other thought processes that might not have been detected by the system in the first place. The concept map also

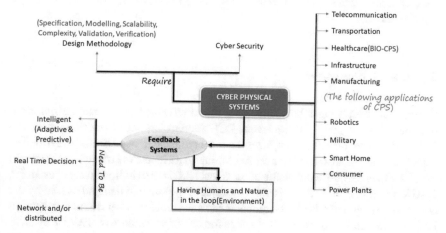

Fig. 1 Concept map of cyber physical systems

points out the main applications of cyber physical systems at the moment. Out of the 10 applications pointed out currently military, healthcare and smart home drive more than 60% of the market. In terms of research military sector leads followed by healthcare division. Healthcare division expands the scope of research with more people coming forward willingly for free medical benefits in return for providing their data to companies which helps in the research and development of such sensors. More people using these devices increases the size of the dataset available for research making the decision making more accurate and increases the chance of a successful diagnosis in the future. The healthcare divisions association with cyber physical systems opens a completely new domain of applications and research, and thus, this domain is termed as 'Bio Cyber Physical Systems' [3].

2.3   Bio Cyber Physical Systems

Biological systems when integrated with CPS lead to the concept of Bio-CPS. At the moment, the most relevant way of integrating CPS to biological processes is available through the healthcare sector. Advancements in the field of surgery, medicine, health monitoring devices have opened a new array of doors for machines to integrate and change the way people used to see the healthcare sector. Many pivotal changes have been brought into robotic surgeries since the first performed in 1985. With a high successful rate and no chances of human error, the medical industry is pushing for such automated technology to perform minimal invasive surgeries. Although there is still a lot to be achieved and understood to make this a reality, the one area where Bio-CPS is taking great leap and stride is health wearables. Health wearables are mostly non-intrusive except a few (pacemakers, etc.).

Robert Bosch in collaboration with IISc Bengaluru has opened up a Bio-CPS research center where they are developing an approach to understand gut biology (cyber-gut) [4].

3   Health Wearables

Health wearables devices have been long used in the medical industry. Monitoring of patients' pulse rate, glucose levels, ECG monitoring have been done since the 1970s–80s, now these devices have been made smaller over time, which makes them portable and also easy to use. According to a survey, the market for wearable devices which monitor vital health aspects reach USD 980 million growing at 21.7% CAGR by the end of 2020. This survey excludes the surge in demand caused by the coronavirus pandemic. In all, the global medical wearable market was at a staggering USD 12.788 billion dollars in 2019 and is forecasted to increase by 19.73% by 2025 standing at a value of USD 37.67 billion. The consumer usage of wearable trackers skyrocketed from 9% in 2014 to 33% in 2018 [5, 6].

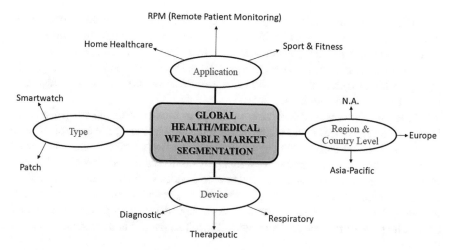

Fig. 2 Segmentation of global health market

The worldwide medical wearable market can be broadly classified into four main segments that are applications, devices, region, and type (Fig. 2).

At the moment, the two primary types of wearable are attributed to Smartwatch/smart band and patch. The Smartwatch/smart band being the most popular of the two as it serves a dual purpose of health monitoring and functionalities of a basic watch. Few more watches also come with inbuilt Wi-Fi, GPS trackers as well as the options to control basic phone activities from the watch. Smart patches are basically designed to deal with details of a person at molecular levels. These patches basically derive data out of the persons sweat. Some patches even have the capability to derive data out of the blood, such as detection of tumor cells.

When it comes to the type of devices, it is basically down to three segments, respiratory, therapeutic, and diagnostic. Smart wearables these days are able to monitor the volume of oxygen in blood and relate it with the heartbeat to give near accurate results for patients suffering from respiratory problems. Diagnostic devices are meant for people suffering from a particular disease and need to maintain record of their activities so as to aid them in their treatment.

The application of wearables can be categorized into RPM, home healthcare, and sports and fitness. Out of these three, the most crucial is remote patient monitoring. This particular monitoring is meant for people suffering from a chronic disease such as kidney failures, heart problems are implanted with a pacemaker or cancer, where the person is undergoing treatment from the confines of his house. RPM allows the devices to send data to the concerned doctor so as to continuously monitor the patient's health. Home healthcare is for people who are in their mid30s-50s and instead of regular consultations and check-ups with a doctor they can keep a tab on their basic health by looking up at their daily activity meter and ECG reports.

North America, Europe, and Asia–Pacific regions are driving the current business on health wearables. North America is leading the demand in consumption of these devices by users followed by Europe and Asia Pacific region. The production of these wearables is at a peak in Asia Pacific region with most tech companies outsourcing.

The increase in the demands of such wearable devices so as to remotely monitor and diagnose include the following reasons: (i) An ageing population; (ii) Improved supply factors; (iii) Better R&D; (iv) Enhanced Functionality, (v) Better integration with IoT system and solutions.

There are various wearable devices out in the medical market and are generally of the following type: (i) Devices which can be integrated into daily life apparels so as to make it the least intrusive or the one which can be worn as a fashion statement as well (this helps in making the patient feel mentally comfortable and also hides the fact that the patient is ill); (ii) Medical implantable devices such as pacemakers, devices to monitor glucose levels (basically needed for patients with extreme health complications and need to be under constant observance); (iii) Sensors placed strategically at specific parts of the body to communicate with an overall body system.

3.1 Fitness Trackers/Smart Watches

As discusses earlier in the paper, the market is dominated by the fitness tracker and watches which offer a plethora of options and functionalities along with the basic features. Smart watches have an edge over fitness trackers because of the functionalities they provide, such as, ECG monitoring, calorie burning according to a specific exercise, manual control of apps linked with the phone, answering calls and reading important texts, and many more [7].

3.2 Types of Sensors

Health wearable devices collect the data through the various sensors and self-input from the users. Wearable devices such as a watch or a wrist band have sensors embedded immaculately across its body. There are various data read by these sensors that ultimately lead to the display of basic health status. Sensors generally found in wearable devices are GPS, accelerometer, gyroscope, pedometer, inertial measurement unit (IMU), piezoresistive or piezoelectric sensor, photoplethysmography (PPG) sensor, and many more. Recently, Samsung's galaxy watch was approved for ECG readings which is recorded by its miniaturized ECG sensor in South Korea.

Each sensor gathers a particular list of feature data that are later on combined together to give readable data to the common user. For example, an ECG sensor gathers data from features such as ECG Peaks, ECG average amplitude, ECG resting, ECG differ mean. Similarly, blood volume pulse feature gathers BVP peaks, average amplitude, and difference mean. The accelerometer in total gathers 15 features, i.e., mean, standard deviation, correlation, kurtosis, and crest factor on each of its 3 axes

(x, y and z). Every sensor gives out near accurate data based on its feature recording and calculations.

The photoplethysmography sensor is perhaps the most vital sensor when it comes to health monitoring. The non-invasive technology is possible through a photodetector and light source. The sensor measures the volumetric variations of blood circulation in the nerves and veins with the help of the photodetector placed at the surface of the skin. The sensor uses a visible infrared light to check the volume of the blood in the nerves, it also is able to determine the oxygen levels in the blood by comparing the color of the blood. Blood color is attributed to hemoglobin which binds with the oxygen, darker blood signifies deoxygenated blood [8]. In total, two information are gathered by the PPG sensor: (1) Heart rate estimation, (2) Pulse oximetry readings. Furthermore, PPGs second derivation wave constitutes vital health information that later on analysis and extensive research may help researchers and physicians in evaluating and further diagnosing cardiovascular diseases such as arterial stiffness and atherosclerosis.

3.3   Activity Log

Data collected by the device is never enough to provide accurate health analysis and decisions, user input is equally important and needs to be near accurate. The two main data that needs to be provided by the user is calorie logging and his/her body composition. Calorie logging is critical for a person suffering from any long-term disease and especially having impact on his/her metabolism. Calorie logging ultimately decides the effect that the person is having, such as weight gain, sleepiness, exhaustion or whether the calorie intake needs to be increased if the stats prove otherwise. Body composition refers to the weight, height, and gender of a person. These three factors help in calculating the body mass index (BMI), which is a primary index of a determining a person's level of fitness [9].

3.4   Advanced Sensors

A company named "Empatica" has manufactured an unobtrusive device which works on the principles of electrical signals. The device is termed as the electrodermal activity sensor and uses skin conductivity to detect a calm or distress condition from the obtained physiological EDA signals. Their research shows that when the results are solely based on the EDA signal processing, the reports come out with a successful accuracy rate of 89% in distinguishing stressful conditions from calm conditions. This capability of EDA sensors can be made to calculate changes in the sympathetic nervous system in the future. The ability to detect a person in stress can be used to predict whether the person is having seizures or strokes if the results are combined with the patients' medical records [10].

Researchers at University of Michigan have developed a wrist wearable proto-type that has the capability to scan for circulating tumor cells (CTCs) in the blood and examining them. This can lead to early detection of tumors and cancer in a person. Researchers have also come up with an inventive way of detecting Alzheimer's in a person by examining their GAIT. Gait is the pattern in which a particular person walks, the pressure that is applied on his foot and his posture while walking. Alzheimer brings a distinctive change in a person's GAIT. Researchers have compared the walking pattern of a person along with the pressure sensors attached in the sole of their shoes to determine whether a person has an onset of Alzheimer or not [11].

3.5 Data Gathering

The wearable devices have a microcontroller placed where the onboard processing of various signals take place and are displayed to the user. The microcontroller gathers input and translates them into decision favorable for the subject wearing it. For a particular individual where constant monitoring and check-ups are necessary, the data can be relayed back to the servers of the hospital where a physician can keep track of his or her records. Due to security concerns, most of the wearable tech companies usually ask for the permission of the individual for storing and analyzing the data for research purposes. Data is generally stored on the individuals' phone which can be received by Bluetooth, Wi-Fi, NFC and is stored in the cloud.

4   Diabetes

Diabetes mellitus is defined as a condition when your blood sugar level is higher than the usual limit. The earliest evidence of diabetes dates back to 1552 B.C. when an Egyptian physician documented symptoms of diabetes. The early prescriptions given to patients in the 1700s and 1800s were to exercise a lot and have to keep fast at regular intervals. There are three main types of diabetes, type 1, type 2, and gestational diabetes. Type 2 diabetes is the most common occurring diabetes with almost 90–95% of the diabetic patients coming under this category. It generally affects people of age 40 and above but recent studies and statistics also show that it is also affecting younger adults and in some cases children as well. Of all the cases in the world type 1 diabetes accounts only for 5% of them and is considered a deadly disease as it reduces the lifespan of an individual by 50% or more. The smaller number of cases in type 1 makes it more difficult for researchers and doctors to find a diagnosis that actually benefits the patient in the long term.

4.1 Complications of Diabetes

Diabetic patients have been steadily on the increase across the world. They are soon going to reach alarming levels that can basically be termed as a health epidemic. Nearly 18 million people lose their life because of cardiovascular diseases and its complications in which diabetes plays a major part of the blame. Around 1.7 billion adults and 115 million children around the world are overweight and 312 million of them are obese. The population with diabetes is around 246 million and is to top 380 million by 2025. 80% of the diabetic people belong to developing countries where health infrastructure is poor, and the treatment provided is not unilateral across the region. India leads the global top ten in terms of the highest number of people with diabetes with a current figure of 40.9 million. There are growing number of cases of type 2 diabetes at a younger age and some even among children who are yet to hit puberty. This scenario is generally common in developing and developed countries. In developing countries, the people diagnosed with diabetes mostly belong to the younger adults, who are in the middle of their productive lives whereas in developed countries this generally affects people post the age of retirement. Diabetes can shorten a person's life by almost 12–14 years and leads to a medical budget two to five times more than that of a regular medical budget. Diabetes is basically attributed to change in a person's lifestyle. Around the clock job, irregular eating habits and other drastic changes are generally the main reasons behind it [12].

Diabetes leads to many complications in the human body and a person not following his diet and prescriptions properly risks a higher chance of death with respect to the other patients. Thus, proper sleep time and daily exercise routine is very important for people suffering from diabetes. High blood sugar can lead to health complications such as heart attacks, heat strokes, seizures, kidney and nerve damage, and many more. Strokes and seizures are the most prevalent among the diabetic patients.

In the coming sections, we will be discussing on the complex problems that the diabetes patient mostly suffers from such as heart attack, seizures, fainting, sun strokes, and the ability of the sensors combined with power of neural networks in-order to detect them in real time and the possibility of predicting such attacks beforehand.

5 Case Studies of Diabetic Complications

Firstly, we will discuss the major life-threatening complications that arise and what data is required from the sensor for the neural network model to recognize the compli- cation accurately. It can be argued that all the sensors with their data can provide far accurate results for a particular complication, but we need to keep in mind not to burden the model with unnecessary data that might delay the response in case of medical emergencies. Of all the data collected by the wearable devices into the system, the most important data is input by the user itself. The calorie logging is extremely important in case of people with diabetes, and food intake regulation is the key to a proper diagnosis.

5.1   Heart-Attack

Diabetes as we know is due to high blood sugar level in the blood. High glucose level in the blood of a diabetic patient eventually degenerates the blood vessel and heart muscles making the particular person more susceptible to heart attack. Failure of a quick response from the medics will ultimately prove the heart attack to be fatal. In this case, notifying health authorities and mobilizing them based on the reading of one particular sensor might turn out to be a false alarm or a malfunction of the device, to increase the certainty of the information being relayed from the watch we take into consideration of the readings from the PPG or ECG sensor, fall detection sensor, accelerometer, and GPS. The GPS sensor will come into effect once it has been established that the person has suffered a heart attack and needs to relay the coordinates to the medical responders. PPG sensor alone is not enough to determine a heart attack, in a circumstance where a person has been frightened all of a sudden, the heart rate tends to shoot up altogether or a scenario might occur where the person has removed the device or the PPG sensor has suffered a malfunction. In this case, we take into consideration the data from the fall sensor. A detection of fall accompanied with irregular or no heartbeat can prove as a strong result to ascertain the heart attack. The above two results can also be accompanied with the accelerometer in case a person on a particular path comes to an abrupt halt. To make the device fool proof, developers can add a prompt on the device which can ask for false alarm from the user. In case the user is not able to verify the false alarm within a particular time interval, the system can be sure that the particular person is having a heart attack (Fig. 3).

5.2   Seizures and Strokes

Both seizures and strokes are correlated in a way to diabetic patients. Recent study has also showed that onset of seizures in late part of life can increasingly lead to a chance of having a stroke. Both stroke and seizures have an effect in the brain and ultimately the nervous system. A stroke occurs when the blood circulation to the brain is stopped suddenly and stroke occurs when there is a surge of electrical activity in the brain.

Diabetic seizures are often caused by extremely low level of blood sugar and can lead to partial or complete paralysis if not taken care of immediately. Now, when a person suffers a seizure, the most notable changes that occur biologically are in the heart rate, respiration rate, and most importantly in the biochemical composition of

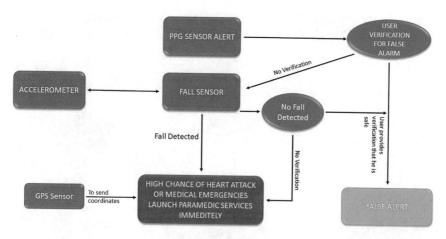

Fig. 3 Flowchart for real-time analysis of a heart attack

a person, i.e., the person starts to sweat profusely. In case of such complications, the data at disposal from the PPG sensor, electrodermal activity sensor, and most importantly the self-input of the persons calorie intake and calories burned by him on that particular day. Most diabetic seizures are attributed to the carelessness of the diabetic patient in following his diet. Electrodermal sensor reading will provide the readings from the sweat of a person. Moreover, electrodermal sensors can also be made to calculate glucose level in blood by analyzing the sweat. In the paper 'Correlation Between Sweat Glucose and Blood Glucose in Subjects with Diabetes' the authors using perfusion method were able to analyze the sweat from the forearms and come to a conclusion from 23 different studies on seven individuals that if sweat is harvested properly by a sensor from a forearm, it can accurately reflect blood glucose levels [13].

In case of a stroke, the chances of a diabetic person of facing this complication increases by 1.5 times, diabetes has the tendency to weaken the biomechanics of a person at all levels. The persons immunity, sustainability, and stamina drop as he/she gets old. The person is more susceptible to stroke who is residing in a hot and humid climate, has a stressful job and most importantly does not follow his diet. Calorie maintenance is of the utmost priority in these scenarios.

Since both the complications have initial common factors and sensors at play, we can categorize the data into one portfolio to detect both the complications. However, the final result will play a major deciding factor in alerting the paramedical forces. By final result, we refer to heartbeat monitoring as in case of a stroke the heartbeat is most likely to falter rapidly and need of professional help and guidance. More importantly, user confirmation itself is going to be vital (Fig. 4).

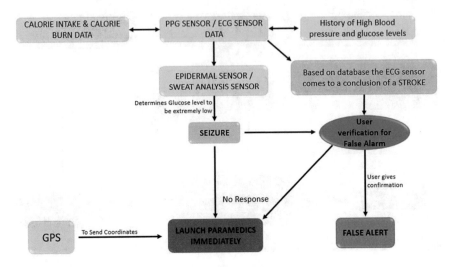

Fig. 4 Flowchart for real-time analysis of seizure or stroke

6 Role of Neural Networks in the Case Scenarios

Machine learning has played a vital role so far in altering the relationship in positive ways between producers and consumers. It has constantly shown strong and progressive potential for the past sixty years. Machine learning, which is a type of artificial intelligence has been in existence since the late 1950s, when Arthur Samuel wrote a progressive learning program in 1959 which made the computer get better at the game of checkers the longer it played. Over the years, machine learning has progressed and brought in numerous changes and new terminologies and concepts across the globe. Deep learning is one of them. Deep learning can be described as a broader part of machine learning dealing with artificial neural networks. Over the past 10 years, deep learning has progressed at a pace unimaginable, marking its dominance in major IT sector firms. Being used in Netflix to suggest you shows on the pattern of your watch to self-driving autonomous cars deep learning has marked its dominance everywhere. The primary difference between machine learning and deep learning is between their data dependencies and subsequent performances. Deep learning cannot perform well on a relatively small amount of data whereas machine learning with their traditional handcrafted algorithms are able to perform satisfactorily in this situation (Fig. 5).

More data dependency of deep learning implies the fact that it needs more high-end hardware for feature learning and training itself as compared to machine learning. It also tries to learn and analyze high level features from a data set and thus takes immense time to train. Neural networks are an integral part of Deep learning. Modeled on the neurological system of the human brain, the neural networks work primarily on the principal of multiple nodes and multi-layer perceptron (MLP). Neural networks are categorized into 3 types: (i) Feed-forward neural network, (ii) Recurrent neural network, (iii) Convolutional neural network.

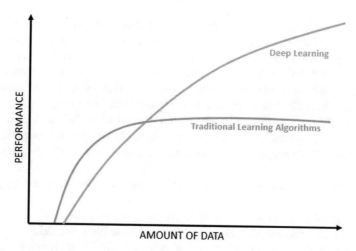

Fig. 5 Data dependency graph

6.1 Convolutional Neural Network

For the above health complications, we will be taking help of convolutional neural network. Convolutional neural networks with their unique convolutional layers provide an edge over recurrent neural networks, feed-forward neural network, and other traditional machine learning algorithms. Figure 6 shows the various layers available in a basic CNN architecture, which according to complexity can be modified and more layers can be subsequently added [14]. When it comes to RNN, each input is evaluated on a single layer and the output is then presented based upon that layer whereas in CNN multiple layers are used to process the data and bring out a more accurate result. Within each convolutional layer the input is transformed before being passed to the next layer.

CNN makes use of filters in between convolutional layers to filter the data, i.e., CNN transforms data with the help of filters. A convolution unit receives its input from

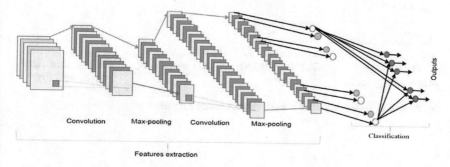

Fig. 6 A basic convolutional neural network architecture

multiple units from the previous layer which together create a proximity. Therefore, the input units (that form a small neighborhood) share their weights. The entire feature extraction process is advantageous for the following reasons: (i) They decrease the number of units in the CNN because of many-to-one mapping architecture which results in lower chances of overfitting the model due to less parameters to learn and in turn making the model less complex than a fully connected network; ii. They consider the context/shared information in the small neighborhoods. This particular feature is valuable in case scenarios where the application of text, speech recognition, data mining, and many more, as related information is generally carried in neighboring units.

CNNs are great at interpreting visual data and data that does not come in a sequence.

In this particular chapter, we will be focusing on multi-channel CNN and the reason for which this model will perfectly work for the given scenarios.

Multi-channel convolutional neural networks basically is a *combination of multiple convolutional networks connected in parallel*, that read data from a source using varying kernel sizes as displayed in Fig. 7 [15].

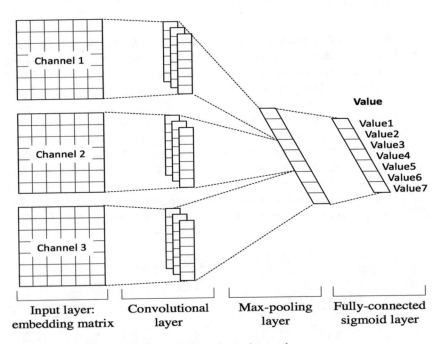

Fig. 7 A basic representation of a multi-channel neural network

7 Multi-channel CNN

As mentioned above, the combination of multiple neural networks in parallel connections will play a very important factor for the case scenarios that we will be dealing with. The detailed flowchart in Sect. 5.1 and 5.2 clearly depicts that at one particular time our proposed architecture would be dealing with a minimum of three or a maximum of 5 different bio-signals at a time. Each bio-signal has a corresponding set of features as well.

The above-mentioned Fig. 8 shows namely a few sensors and various features that are recorded off them. It may also be noted that the number of features may vary upon different sensors, more complex, and advanced the sensor the more features it has to offer. As it is obvious that each and every sensor producing so much data might not be feasible for a regular convolutional neural network to handle and might lead to mixed results not in favor of the physicians and patients alike. Thus, this is the reason for which multi-channel convolutional network is proposed as a model. Separate channels for different bio-signals will help to keep the initial feature learning using CNN discrete from each other thus preventing mixture of information and analysis between individual layers.

Multi-channel CNNs have proved their mettle in different related works as well. In the paper, 'Multi-Channel Convolutional Neural Networks Architecture Feeding for Effective EEG Mental Tasks Classification' the authors aimed to classify mental tasks based on EEG signal processing and its analysis. They put up their multi-channel CNN against various other multi-channel architectures. The result stood in favor of the multi-channel CNN with an astonishing 5% generalization rate [16]. In another research article, 'A Multichannel Convolutional Neural Network Architecture for the Detection of the State of Mind Using Physiological Signals from Wearable Devices' authors used bio-signals from the body to determine the state of mind of a person and ultimately the persons wellbeing. In their research, the model performed effective results based on the classification of the different state of minds for a particular subject. The model achieved an average recall and precision of 97.238% and 97.652%, respectively [17].

It is important to note that despite the fact that the model suiting this situation perfectly, it will still take an enormous amount of data to train upon in-order for it to give out satisfactory results and even more exposure and inputs from various

ECG	ACCELEROMETER	BLOOD VOLUME PULSE	CALORIE COUNT
ECG peak	Mean	BVP Peak	Calorie Intake
Average Amplitude	Standard Deviation	Average Amplitude	Calorie Burnt
ECG resting	Correlation	Mean Difference	Time period between intake
Mean Difference	Kurtosis		
	Crest Factor		

Fig. 8 Features of a few sensors

researchers to make it extremely accurate. Since this is a medical situation in which the model is being applied and lives are at stake, the model would need extensive research, collaboration with plenty patients and third-party members and a dataset with *billions* of data.

8   Complication Prediction Through LSTM

The multi-channel CNN model was used to bring out real-time decision based upon the data collected from the sensors in-order to administer paramedical services on time. It might also be possible to predict such attacks beforehand using a neural network and ultimately be able to save yourselves from the life-threatening complications by getting yourself assessed and taking the proper steps based upon your physician.

LSTM neural networks can be a suitable option for it. *Long short-term memory* neural networks are a special type of recurrent neural network. The problem that used to persist in recurrent neural networks has been easily dealt with in LSTMs, i.e., *long-term dependency problem.* Although in theory, an RNN should not be troubled with long-term dependency, but in practicality it falls short of its expectations. LSTMs were specially engineered to get rid of this problem in recurrent neural networks. The default behavior of a long-short-term memory neural network is to remember information that was analyzed earlier for a longer period of time.

This specialty of LSTM serves as a perfect model for predicting health complications. Basically, this model remembers the analyzed data of the past and also keeps on analyzing the data in real time. Combining both these data, it has a fair enough chance to predict such attacks beforehand. As seen above in Fig. 9 [18], the LSTM repeating module contains 4 interactive layers as compared to the one in the chain RNN repeating module.

In the research paper 'An Effective LSTM Recurrent Network to Detect Arrhythmia on Imbalanced ECG Dataset' the authors use the LSTM network to disentangle timing features in ECG signal to help cardiologists to succinctly diagnose ECG signals. This setup of theirs is very similar to what is proposed in this section. ECG signals need to be remembered by the model for a long duration so as to compare and analyze the older data with the present data and give out satisfactory results. In the above-mentioned paper, the model achieved an accuracy, recall, precision, specificity, and $F1$ score greater than 90%, thus implicating that the LSTM network has immense capability [19].

9   Conclusion

The chapter describes in detail the basic needs for integrating data analytics in Bio-CPS in an effective manner in-order to predict a major complication in a patient and

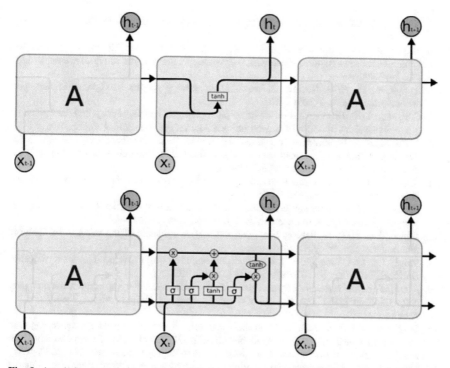

Fig. 9 1st chain structure-repeating module in a standard RNN. 2nd chain structure—repeating module in a LSTM

alert services instantly. The multi-channel convolutional neural network proves to be an extremely efficient model for the situations described above, similar models have brought effective results to other researchers as well. Over the time, with more efficient collection of data and a larger database, the model will cement itself to be near accurate for real-time analysis. With the increase in database, the long-short term memory model will be able to accurately predict complications beforehand.

References

1. [Online]. Available: https://www.researchgate.net/post/What_is_the_difference_between_Int ernet_of_Things_IoT_and_Cyber_Physical_Systems_CPS.
2. CPS concept map [Online]. Available: https://ptolemy.berkeley.edu/projects/cps/#:~:text= Cyber%2DPhysical%20Systems%20(CPS),affect%20computations%20and%20vice%20v ersa.
3. Fass, D. 2016. *Towards a theory for bio-cyber physical systems modelling.*
4. Ananthasuresh, G. 2016. *A BioCPS approach to understand and control gut-biology (CyberGut)*, 28 Nov 2016. [Online]. Available: https://cps.iisc.ac.in/cybergut/.
5. Pvt, B. M. R. C. 2020. Medical Wearable Market Share. *Current Trends and Research Development Report to 2025*, 4 Apr 2020. [Online]. Available: https://www.medgadget.com/

2020/04/medical-wearable-market-share-current-trends-and-research-development-report-to-2025.html.

6. The future of wearable devices in healthcare, 26 Feb 2019. Available: https://www.medica ldirector.com/news/future-of-health/2019/02/new-report-reveals-the-future-of-wearable-dev ices-in-healthcare

7. Aroganam, G. 2019. Review on wearable technology sensors used in consumer sport applications. *Sensors (Basel)* 1–10.

8. Castaneda, D. 2018. A review on wearable photoplethysmography sensors and their potential future applications in health care. *International Journal of Biosensors & Bioelectronics* 1–3.

9. Aivaz, M. 2019. *Importance of tracking calories to maintain a healthy lifestyle*, 31 May 2019. [Online]. Available: https://medium.com/@marudeen_aivaz/importance-of-tracking-calories-to-maintain-a-healthy-lifestyle-1a9ea7379c83.

10. Zangróniz, R. 2017. Electrodermal activity sensor for classification of calm/distress condition. *Sensors (Basel, Switzerland)* 17(10): 1–3.

11. Beauchet, O. 2016. Poor gait performance and prediction of Dementia: Results from a meta-analysis. *Journal of the American Medical Directors Association* 1–3.

12. Tabish, S.A. 2007. Is diabetes becoming the biggest epidemic of the twenty-first century? *International Journal of Health Sciences* 1–3.

13. Moyer, J. 2012. Correlation between sweat glucose and blood glucose in subjects with diabetes. *Diabetes Technology and Therapeutics* 14 (5): 398–402.

14. Alom, M. Z. 2019. *ResearchGate, Mar 2019*. [Online]. Available: https://www.researchgate. net/publication/331540139_A_State-of-the-Art_Survey_on_Deep_Learning_Theory_and_ Architectures.

15. Sharaf, A. 2017. *A multi-channel convolutional neural network for cross-language dialog state tracking, 11 June 2017* [Online]. Available: https://medium.com/@sharaf/a-paper-a-day-19-a-multichannel-convolutional-neural-network-for-cross-language-dialog-state-bb00f5328163.

16. Opałka, S. 2018. Multi-channel convolutional neural networks architecture feeding for effective EEG mental tasks classification. *Sensors (Basel)* 18(10): 3451, 1–2; 20–21.

17. Chakraborty, S. 2019. A multichannel convolutional neural network architecture for the detection of the state of mind using physiological signals from wearable devices. *Journal of Healthcare Engineering* 2019, Human-Centered Systems in Rehabilitation Engineering, 9–16, 18 June 2019.

18. Olah, C. 2015. *Github, 27 Aug 2015*. [Online]. Available: https://colah.github.io/posts/2015-08-Understanding-LSTMs/.

19. Gao, J. 2019. An effective LSTM recurrent network to detect arrhythmia on imbalanced ECG dataset. *Big Data Intelligence in Healthcare Applications Based on Physiological Signals* 2019, 13 October 2019.

Evolution of Sentiment Analysis: Methodologies and Paradigms

Aseer Ahmad Ansari

Abstract With the advent of the digital age, almost everything has come down to better understanding of the data. Natural language processing is equally in pursuit and is rather among the most researched areas of computer science. Post 1980, a major revolution in NLP embarked with the emergence of machine learning algorithms resulting from steady escalation in computational power. Unlike other data, text semantics becomes more complex both because of its contextual nature and daily evolving language usage. While the continuous efforts of improving language representation for logical units interpretation is still prevalent, much to our realization, traditional, and long established recurrent neural networks which were supposed to grasp a bi-directional context of language have been surpassed by attention models in constructing improved embeddings allowing systems to better understand language. Among numerous applications circumventing, understanding sentiment of text has been widespread in fields including but not limited to customer reviews, stock market, elections, healthcare analytics, online, and social media analytics. From binary classification of it to more challenging cases such as negation handling, sarcasm, toxicity, multiple attitudes, or polarity, this research chapter explores the evolution of sentiment analysis in the light of emerging text processing and the transition of text understanding from rule-based to a statistical one with a comparison of benchmark performance from state-of-the-art models over various applications and datasets.

Keywords Natural language understanding · Sentiment analysis · Statistical models · Attention models

1 Introduction

Text data mining, text mining or text analytics belong under the same jargon of computational linguistics. The intent remains the same to derive information of quality from raw text. These raw texts could be all sorts of written resources in

A. A. Ansari (✉)
Deloitte Touche Tohmatsu India LLP, Pune, India
e-mail: a3ahmad.kiit@gmail.com

© The Author(s), under exclusive license to Springer Nature Singapore Pte Ltd. 2021
S. Rautaray et al. (eds.), *Trends of Data Science and Applications*,
Studies in Computational Intelligence 954,
https://doi.org/10.1007/978-981-33-6815-6_8

any human understandable language, though it is necessary for these languages to follow a certain set of rules and these rules are nothing else but grammar. Articles, news, magazines, websites, social media are just a few to count for the sources of this data.

The availability of these resources in present is in excess but in the past when the language remained more voluble and oral or on few occasions on paper, analyzing text was still a distant aim. More profoundly now, but a bit ironic is the fact about the abundance of such resources and computational power though now integrity of data is in question. All in all, the importance of this field of text analytics is evident from its everyday proximity. Even while writing this research chapter, a constant assistance for grammar, usage of words and structures are available through a well designed text analytics engine.

Understanding the paradigm of text analytics is to perform information retrieval and this is accomplished through an understanding of increasing complexity of the science of language in order as follows:

a. Phonetics
b. Morphology
c. Lexical Analysis
d. Syntactic Analysis
e. Semantic Analysis
f. Pragmatics.

All of the above essentially break down the process of natural language processing into smaller and easily accomplished modules. But even after gathering useful information from each of the above, an important step remains which is to make that information interpretable to machines' logical units. Different types of algorithms come into picture at this stage such as regression, tree-based, ensembles, neural networks, or neural turing machines. All these steps combined can serve as the basis for many text mining applications such as entity extraction, text clustering, categorization, summarization, relation modeling among entities, sentiment analysis, and many more.

Sentiment analysis which is at the center of this research can be understood as methods and tools to extract a person's subjective view of a topic, a very simple comprehension of his opinion or attitude which could either be positive or negative. Understanding the sentiment directs us to understand the polarity of it. For most of its occurrence, understanding of sentiment gains mostly from views that are available in a public context or in businesses where customer feedback or opinions is a central element. To our surprise, this practice dates long back when leaders were equally interested in the opinions of their hierarchical underling for maintaining a smooth and transparent functioning of their bureaucracy. Even now this interest in opinion remains an integral part in the functioning of businesses, political agendas. Not so long ago, the case of data scandal for the presidential election which involved Cambridge analytica, pillared entirely on analyzing the data points captured from the views of citizens, essentially text mining.

On this note, it appears logical to point out Prof. Lillian Lee from Cornell University whose work began this great endeavor and is one of the founders of "Sentiment Analysis." Her work on the topic began in the early 2000's with a seminal paper [1] with her colleagues. The problem of document classification by topic was looked for classifying by sentiment. With this, the sentiment analysis began and mostly focused on reviews of products or services which were served on the web.

As for the types of sentiment analysis, there are different methods and ways to identify it differently. First for identifying subjectivity and the second deals with a more comprehensive or nuanced versioning of text called as feature or aspect-based identification. These problems could be high dimensional ones and at the beginning we look at just the positive and negative sentiments but there are datasets out there in ways to think about the problem that would embrace much more of this nuance.

A huge increase in the papers circumventing opinion mining (OM) and sentiment analysis (SA) has been observed since the mid 2000. Besides that fact, another data reveals that this topic has come to be known more profoundly even to the general public. Figure 1 asserts that trend of data from 2004 to 2015 captured from Google Trends and shows the increase in the number of searches for "Sentiment Analysis" and that of the one used generally by the public.

As of the current day, the applications have expanded all the way further to prediction of financial markets or even analyzing potential terrorist cohorts and any such growing organizations activity. Other day to day applications could include social media monitoring, public relations or even serving as an appendage to maintaining an efficient question answering sessions with clients of an enterprise for regular and common enquiries. The essence and the crux of all these applications lies in the fact of how useful they are to institutions and companies in the light that they are able to provide assessment and keen observations to constant questions such as "why isn't a particular service preferred?" or "how can a certain method be more effectively aimed at servicing the target audience?."

Still as the solutions to problems become easily interpretable and accessible, the substratum or the underlying data gains more complexity day by day. And very much like any other application of text mining, this stands as an equally strong deterrent in

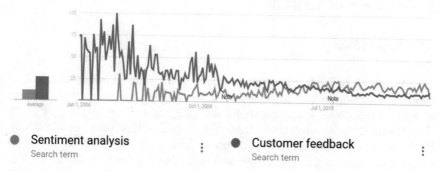

Fig. 1 Google trends data from Jan, 2004 showing search trends of the two strings "Sentiment Analysis" and "Customer Feedback"

efficaciously carrying out sentiment analysis. For one, the complexity of expression and slangs that are used over the web for stating one's point are increasing. Other than that, the semantic meaning of a language alone causes contextual ambiguities and on top of that, people in making opinions, now more than ever use all sorts of available metaphorical devices ranging from satire to rhetoric like sarcasm. This alone makes the comprehension of language by machines difficult. Though in today's date, we have come up with new architectures for ascertaining the meaning and structure of any grammar containing such a complex level of expression but still, this step brings along with it another set of challenges.

The area of sentiment analysis like other applications has become quite vast and hence this research chapter intends to culminate most of the critical methodologies, tools, and perspectives of this topic. Also by means of this, the work intends to help other researchers, groups, and even newcomers to be able to track the evolution as has been brought with advancement in computational ability and varied applications in the field.

The paper makes the following contributions. First, the advent of the topic through most cited, searched, and shared papers in Scopus and Google Scholars is shown. These papers have marked the beginning of "sentiment analysis" and around 2004 for the modern practices in the field. Second, a few applications that came across as unique and useful are covered along with making case for the methodologies adopted through them. Third, a comparison of machine learning to deep learning methods are delineated and how the biologically inspired neural networks overcome and perform better in mapping the contextual nature of language. There are a number of datasets over which the performance of many algorithms have been benchmarked, hence a comparison does assert the quality and performance of those. Fourth, a comprehensive explanation of the attention models that have over performed every other existing algorithm and categorize the state-of-the-art are covered. Furthermore, a deep dive into the limitations of them and the relationship of these models with the traditional ones is reviewed.

2 Foundational Methods

2.1 Supervised

The time ahead of 2000 had adequate data in processible form and so was such available in documents present online. From everyday usage of search engines where results get categorized by topic, part of the attempt by research groups have become to cluster information based on interest and other relatable subject matter automatically. But at the same time, an increase in discussion forums have developed online which include discussion groups and review websites. These reviews involve all sorts of subjective viewpoints from users ranging between strongly positive and strongly negative sentiments.

Keeping the growth of such prospects in consideration, the work of Pang et al. [1] focused at mining the sentiments and thereafter labeling those articles from reviews accordingly. This was aimed at providing a discrete and concise recapitulation to other users and indeed those summaries would add an extra feature to the already present detailed reviews. From hereon, sentiment analysis was looked at from the perspective of other plausible applications such as business intelligence or recommender systems. One such peculiar way to use categorization of sentiment was applied potentially in filtering of messages to be able to either recognize them or discard them entirely. In [1], they applied three different machine learning techniques to classify sentiments. One challenging angle mostly associated with sentiment detection which differentiates it from typical topical classification is the semantic and lexical manner in which it is written. Classification of text on the basis of topics would merely require identifying a few relatable words, but sentiment on the other hand can in a fine way be incorporated even with the use of correspondingly opinionated words. For example, "Yeah Sure! Let's all dance on the street." This example itself without knowledge of its preface is difficult to differentiate for its sentiment that whether it constitutes sarcasm or no sentiment at all. Thus, sentiment-based classification requires interpretation of more than just the topics. Among a few other works by [2–4] for determination of subjective genres like "editorial" differentiates itself from works of [5, 6] which intends to find the subjective language by identifying the features. But in many such works, besides their ability to detect subjectivity or even classify genres at document level, they still are unable to determine the actual subjectivity or opinion the text holds. One attempt in the work of [7, 8] tried to make use of limited knowledge-based methods for classifying the orientation of subjectivity from singular phrases and words, though the idea behind was to make use of pre-selected seed words or semantic heuristics of the grammar. But this idea was based on the assumption that human beings can as easily distinguish between polarities in text constituting sentiments as compared to differentiating between text categories by topics and since a corpus of definite well-defined words exist that mostly express same sentiments, thus comprehending the litany would be as simple as creating such a list of words and depend on them alone. But this hypothesis was tested and proved to be only close to the randomly chosen baseline for majority class. Two graduate students identified both negative and positive set of words in the IMDB movie review dataset to indicate corresponding sentiment. The tabulation when used for a uniformly distributed dataset containing 700 reviews for each polarity found the accuracy at 58% and 64% for the two as is shown in Table 1 [1]. This accuracy reached 69% when combined with frequency matrix and some introspection.

Thus, an examination to comprehend the effectiveness of machine learning algorithms was done assuming the treatment of the problem like that of the topic-based categorization. Maximum entropy classification (ME), Naive Bayes (NB) and support vector machines (SVM) were chosen as three standard algorithms. The performance as shown in Table 2 are threefold cross validated scores. As was evident about Naive Bayes and its assumption of conditional dependency but despite this simplicity which doesn't always hold universally, its performance was surprisingly well with 81.5% accurate when uni-grams were considered along with part-of-speech

Table 1 Words annotated by two humans corresponding to their sentiments and their accuracy

	Proposed word lists	Accuracy (%)
Human 1	Positive: fantastic, brilliant, phenomenal, excellent, dazzling, Negative: awful, suck, terrible, hideous, unwatchable	58
Human 2	Positive: mesmerizing, cool, spectacular, exciting, moving, excellent, awesome, gripping, thrilling, badass Negative: sucks, stupid, slow, boring, bad, cliched	64
Human 3 + stats	Positive: best, great, still, superb, love, beautiful Negative: stupid, waste, worst, bad, boring, ?, !	69

Table 2 Three fold cross validated score for all three algorithms

Feature type	Total features	SVM	NB	ME
(uni)grams + (bi)grams	32,330	**82.7**	80.6	80.8
(uni)grams	16,165	**82.9**	81.0	80.4

tags. SVM eventually stood tall both with and without (bi)grams with accuracy as mentioned in Table 2.

Maximum entropy having outperformed Naive Bayes not always but surely in few cases of standard classification of text on topic has more significance in cases when conditional dependency is not met and thus unlike naive bayes would potentially perform better. Its estimate of $P(c|d)$ for its role of selecting c^* given document (d) is different in its consideration of the exponential form:

$$P_{ME}(c|d) := \frac{1}{Z(d)} \exp\left(\sum_i \lambda_{i,c} F_{i,c}(d, c)\right)$$

Unlike that of Naive Bayes,

$$P(c|d) = \frac{P(c)P(d|c)}{P(d)}$$

where $P(d)$ plays no role in selecting c^*.

Further, the promising large margin classifiers SVM have outperformed the other two and turned out to be largely effective even in the traditional text classification, empirically crediting the possibility to being non-probabilistic and their way of separating the document vectors not just by any hyperplane but by the maximum margin possible.

But these benchmarks when compared on varied applications have shown varied results. But the area of customer reviews has had improvement a step ahead as proposed by Hu et al. [9] wherein features are mined only about what opinions concerns the product or service. Besides the work on summarizing those reviews, the authors kept a narrow window in extracting opinions and their polarity if any, of only such features. They broke down the work in three smaller steps of which this section is concerned only by the first two, i.e., interpreted features upon which customers review and then identifying the opinions and whether they are positive or not. But even here, a constant problem of long reviews was an obstacle. Moreover, many such reviews contained only a few sentences that were concerned with the product or service and this in turn makes it difficult for both the customers to get the complete view and for the manufacturers to keep tabs on those. Also, such length at times becomes a source of biased view. In detail, the steps after product or service feature have been mined, opinions relating to them are identified and consecutively its semantic orientation. At this point, a simple natural language processing method to identify the adjective words gives way for the attitude of the opinion such as a process as simple as bootstrapping using WordNet [10]. Previous suggested methods using algorithms could have worked here, but at this point, the intention is not just to derive the document level opinion but at the sentence level. Major support at this level is provided by part-of-speech tags. Noun and noun phrases in reviews allows to connect the context and hence their presence makes them a crucial part. For instance," The chocolates are quite sweet." The above sentence is easy to interpret with the noun phrase being explicitly mentioned making it easy to understand what the adjective refers to. "Though smooth, it's caffeine content is higher than usual." Here, the *texture* is being referred, but it is not stated directly making it difficult to focus on the feature. Hence, a requirement of *previously referred* is essential which can only be interpreted through parsing the grammar.

The understanding of sentiment even with the help of features still misses a notion. This is the *degree* of the sentiment which varies from mild to strong. In [7], two approaches were evaluated for interpreting word association with different statistical measures. Latent semantic analysis (LSA) and pointwise mutual information (PMI) conducted with a labeled set of ~3600 words containing both negative and positive verbs, adverbs, nouns, and adjectives. In case, a *mild degree* is disregarded the algorithms are able classify the words with 95% accuracy, though with the inclusion of degree the accuracy on the entire test set was 82.8%. A method was introduced to infer the intensity and orientation of the word form automatically through statistical association between opposite polarity words. This is based on the hypothesis that semantic orientation and intensity is related to the orientation of its neighbors. One of the applications through the motivation from [8] was the identification of close antonyms and differentiate between near synonyms. Both of them have strong associations, but it is worthwhile to note that the same synonyms and opposite antonyms might not always share the same magnitude. Table 3 [7] showcases a few examples with ranging semantic orientation.

It is important to accentuate the bold face since they make up for the dominating orientation of the whole text. There are numerous applications suggested in their works such as search engines summary statistics, "flames" filtering newsgroups [11], generating sentiment timelines, appropriating a response in chat bot systems or survey analysis of open ended questions.

Table 3 Examples from the paper with different orientations

	Positive review	Average SO
1.	I love the local branch, however, **communication** may break down if they have to go through head office	0.1414
2.	Bank of America get my business because of it **extensive** branch and ATM network	0.1226
3.	This bank has exceeded my **expectations** for the last ten years	0.1690
	Negative review	Average SO
1.	Do not bank here, their website is even **worse** than their actual locations	−0.0766
2.	Use Bank of America only if you like the feeling of a stranger's **warm**, sweaty hands in your pockets	0.1535
3.	If you want **poor** customer service and to lose money to ridiculous charges, Bank of America is for you	−0.1314

2.2 Unsupervised and Semi-supervised

Let's say if you are considering a vacation at a place in India, our first step would be to go over to a search engine and get ideas. After getting results for the query say "places to visit in India, Bhubaneswar," a search engine such as google would have laid around 5000 results. However, it would be more important to know which of the following results are recommended. A very similar work by Turney in [12] works its way around for unsupervised classification of sentiment orientation. An average semantic orientation is considered from the phrases and based on the average which would either be positive or negative, the review is recommended. By measuring the pairwise similarity in words and phrases, the PMI-IR ascertains the sentiment orientation of the phrase.

An amazing framework and opinion mining lexical resource was presented by Andrea and Fabrizio [13]. The main aim of the work was to target the subdiscipline of computational linguistics, i.e., opinion mining by using a self devised lexical resource which annotates a triplet of scores by making use of an automated classifier to a synset of a WORDNET. These triplet scores belong to positive, negative, or objective. Surprisingly, a set of eight ternary classifiers work associated with equitable accuracy and each score evaluates totally different classification behavior. One of the astounding aspects associated with the concept of these researchers was that they made it freely available.

A major assumption on which this works is based on the presence of multiple and different sentiment oriented properties of the same term in different senses. As expressed in notation, an automated classifier Φ denotes to each *synset s* of WORDNET a triplet of numerical scores $\Phi(s, p)$ **(for p in P = {positive, negative, objective})** which indicates the closeness of the term to the three properties. A synset comprises terms with the same parts-of-speech tag and each is associated with a *sense number* and characterized by a *gloss*. The scores for the term range between 0.0 and 1.0 and the sum of all the terms in the synset is 1.0, indicating that every term corresponds to a non-negative score for the three and hence contains at least a certain degree of all the three properties. Analogous to the intuition presented in the work of Kim and Hovy [14] stated the presence of a certain degree of both positive and negative PN-polarity, SENTIWORDNET for instance to the *sense* "may be computed or estimated" for the adjective [estimable] attribute objective score of 1 and 0 for both positive and negative. But another *sense* "deserving of respect and high regard" would receive a scores 0.75, 0.0, 0.25 for positive, negative, and objective, respectively. One difference wordnet holds compared to other methodologies is the introduction of graded evaluation instead of the binary categorization of sentiment orientation. This is more helpful in sentiment or opinion mining applications. An average length sentence in conversation or reviews contains more than enough nuances and hence binary classifier for instance might label any document containing low SO-polarity as objective and if the document is made of many such, the classifier will miss out on the subtle elements, but on the other hand a lexical resource with graded feature might provide with enough evidence so as to not miss out on those nuances. Thus, SENTIWORDNET works meticulously by making use of eight such ternary classifiers to form that triplet score.

Now, the *classifier committee* of SentiWordNet will contain eight individual synet classifiers and a semi-supervised method is used to decide PN-polarity. A rather radical behavior of the *committee* is observed with different combinations of train set and learner.

1. Smaller number of iterations produces smaller training sets for positive and negative and classifiers of those have high precision but low recall. On increasing the iterations, results in recall increase along with "noise" which leads to decrease in precision.
2. Learners like SVM that make use of prior probabilities are more prone to classifying items into classes that have more positive items. This is probably because of their sensitivity to relative cardinalities in training sets.
3. Though both of the above nuances do not affect accuracy overall, their effect imbalanced when positive, negative are considered individually.

The results of the WordNet are evaluated in a more automatic manner since a full evaluation would require manually labeling all the three properties and since its unavailable, the quality of SentiWordNet is found by using classification for general inquirer [15]. Figure 2 below shows the graphical representation for the three opinion properties.

SentiWordNet 3.0 [16] was presented no long after with improvement in algorithm and addition of a randomly generated hike to even refine the scores. The database of gloss corpus also shifted from Extended WordNet to Princeton WordNet Gloss Corpus. The extra step of random walk along with the semi-supervised learning makes use of manually disambiguated *glosses*. And, these are acquired from Princeton's which were presumably more accurate. But to generate the SentiWordNet 3.0, a prior step of weak-supervision is also appended. As for the random walk, it is an iterative process that views WordNet 3.0 as a graph and then processes the positives, negatives, and objective values which change at almost each iteration.

Fig. 2 Graph representing
opinion orientation for a
synset as adopted by
SENTIWORDNET

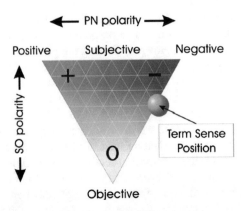

This is iterated till the process actually converges. From the point of view of graph, directed links are assumed to exist in manner from synset s1 to synset s2 given s1 (the definiens) is present in the gloss of s2 (the definiendum). All based on the intuition that if the majority ascertain a definite term as positive, then the possibility of the term being positive is high. In simple explanation, positivity and negativity are seen as "flowing through the graph." Evaluation of SentiWordNet 3.0 is done by synset mapping which makes use of WordNet sense mappings, synset term matching and gloss similarity in order. All in all, the results of SentiWordNet 3.0 are absolutely better and improve the accuracy by 19.48% for positivity and 21.96% for negativity. The ranking as well is improved with the aid of random-walk with 17.11% for positivity and 19.23% for negativity rankings.

As of today, multiple new Wordnet model have been developed in different languages such as Arabi WordNet, BulNet(Bulgarian version), EuroWordNet (numerous european languages), GermanNet, IndoWordNet (18 different languages and their knowledge bases), PolNet (also plWorNet, both are Polish language models by two different universities), OpenDutchWordNet and a few more.

3 Applications

There are other papers that contributed to the foundations of sentiment analysis before the modern sentiment analysis took speed. Some of those works contributed by Maite and Julian in [17], Theresa et al. [18], Cambria et al. in [19], Kushal et al. [20], Mike et al. [21] and a few others made the substratum concrete for applications to stand upon. Besides, one of the then sources from which applications took speed was Twitter.

The kind of content type that was allowed by Twitter was called MicroBlogs and could easily be inferred as the most famous tool for communications. Even other websites provided features for microblogging such as Tumblr and facebook. Content and genre of these microblogs covered almost everything from daily life updates of individuals to opinions and reviews of abstract topics and also those that were prevalent issues. Because of this flexibility, many users transitioned from say, old school blogs to these platforms.

The vastness of the services gave an option to companies to analyze and mine data from the opinions that users held. For instance, any manufacturer could ask its users for preferences or how the services are in effect. Even political parties and other organizations used the platform to inquire about their programs effectiveness or how many people supported it. Though earlier sources for retrieving opinion related corpus was suggested in quite a broad manner by Pang and Lee [1], not many had strongly considered blogs and or even microblogging for that matter.

Alexander and Patrik [22] used these microblogging services as a goldmine for creating a corpus for SA and OM and particularly Twitter for the following:

1. Since users from all different countries who hold different perspectives put their opinions on topics, this makes Twitter a valuable pool of point of views.
2. The size of the corpus on Twitter also is expanding at a daily pace which keeps it an updated source for mining.
3. The users constitute common citizens to influencers, celebrities, political representatives, CEOs of multinational giants and even Presidents, Prime Ministers of Countries. These people use the platform and post their views almost on a daily basis and thus creating Twitter a diverse collection of text.

Text collected using Twitter API formed data in classes of three semantic orientation positive, negative, and neutral. Also, emoticons were decoded for the same purpose. A class of emoticons constituting both sad and happy ones were collected and used. An assumption in the work based on the Twitter ethics and regulations for messaging and posting for 140 character length was that each text would mostly constitute one sentence and the whole of it will be representative of the one semantic orientation.

Based on the analysis of the corpus which was restricted to english language, a trend was mined for both subjective and objective messages. It is also worthwhile to mention that the corpus's frequency distribution followed the Zipf's law. Now, as for the part-of-speech tags, they weren't distributed evenly in objective and subjective sets of texts. This mostly gave way for the following observations:

1. Proper and common nouns such as NNS, NPS, NP were more frequently found in objective texts. This was different from the personal pronouns that mostly constituted the subjective ones.
2. Subjective texts authors refer to themselves mostly in (first person) and the audience in (second person). But objective texts have authors referring themselves in third person.
3. Simple past tense is used in subjective instead of past participle like in objective.
4. As for the adjectives, emotions and opinions in subjective text are expressed through superlative adjectives but more often comparative adjectives state the facts.

Above all, bi-grams receive the best performance mostly because of their balanced ability for coverage in uni-grams and also to capture the sentiment expression present in tri-grams. Though an increase in accuracy was made possible by removing the frequently occurring common n-grams particularly those that did not indicate any subjectivity of any sort and are common. Two strategies were introduced. First way involved calculating the "entropy" of the distribution of n-grams. Uniformity being common in the ones with high entropy were dropped, since their contributions were minimal. Another way made use of "salience," the value of which remains between 0 and 1 and a low value n-grams would be dropped. But even then, the accuracy of the method increased only to a certain point with increasing dataset size.

The performance of automatic classification using Twitter data are limited by the size and quality of data. Experiment suggests no statistical significant difference between different classification models. A multilingual Twitter data analysis for classification suggests that a sufficiently large dataset will eventually reach the inter-annotator agreement, and hence, a periodic monitoring is necessary for self and inter-annotator agreements to improve the data and eventually the model performance. Igor, Miha, and Jasmina [23] quantify the quality of labeled tweets. They focus on estimating the performance after an agreement between human annotators. Their hypothesis is that consensus among annotators to a certain level acts as an upper bound on the performance. So, the experiment involves an estimation of performance in case of manually annotated Twitter posts with models built on agreed annotated training data. The evaluation metric used is that of Krippendorff's Alpha-reliability (Alpha). Two huge corpora containing 1.6 M annotated tweets is the largest available online and also comprises 13 languages. Some of them were: German, English, Albanian, Russian, Polish, Hungarian, and others. Now, it is evident that determination of the sentiment orientation is a challenging task mostly because all individuals hold subjective views. And because of the following reasons, the human annotators also contradict among themselves:

1. Inherent Difficulty of classifying the text.
2. Different domains using similar vocabularies.
3. Topic drift, which is technically brusque shift in topic at discussion.
4. Faulty and imperfect annotaters themselves.

Furthermore, two strategies confirm the confidence interval for the agreements. This involved a certain amount of tweets to be duplicated which was about 15% and to be deliberately annotated again either by the same annotator, i.e., *self-agreement* or by a different one, i.e., *inter-annotator agreement*. And as mentioned earlier, the inter-annotator agreement sets the upper bound of the model performance. Comparative analysis of them and a SVM-based classifier on 13 languages suggest that alpha reaches 0.6 for self-agreement with a few exceptions mostly because of inferior level annotators. The alpha wavers a lot for the inter-annotator agreement and entirely and always lower than that of self. A gap in performance still exists, though alpha increases to 0.516 from 0.422 but still remains below inter-annotation agreement because of less annotations mostly. An improvement in performance with alpha increasing for Russian, Slovak dataset, while for Bulgarian and Sulvanian performance limit is almost reached. In conclusion, the results assert the requirement of multiple or more than one annotator.

Now since the application of sentiment analysis does most good in understanding individuals point of view in reviews, posts, blogs, microblogs, etc. On the web, a derivative of the regime holds an equally important notion and performs several linguistic functions. This common expression frequent among individuals is *Vulgarity*. Mostly, the social media world constituting pragmatic and socio-cultural aspects have been found to use vulgarity in their posts. Though its occurrence is quite common but a standard has been set where a language constituting vulgar words and appearing 0.5% to 0.7% in daily speech during conversation [24, 25] or around 1.15% on Twitter posts [26] can be called vulgarity. Thus, interpreting this form of expression is utterly important in text of any natural language and for many computer scientists, it became imminent to model this nature of language separately to understand its pragmatics and for many psychologists to understand the stand of profanity in a socio-cultural context. Works of Isabel et al. [27] performed a large scale assessment over social media particularly Twitter for an empirical analysis of vulgarity. It is possible to restate a thought without the use of vulgar words, but the fact that they are used indicates a specific intent. An example suggested in the work of Isabel et al., "I am stupid as f**k" will automatically articulate more on the anger or irkness, then simply put "I am stupid" which could convey depending on the context a variety of semantic orientation. Also, the usage of such words is dependent on the demographics and the context of socio-cultural topics and because of this dependency comprehending the pragmatics individually is supposed to improve the performance of sentiment analysis applications. Nine annotators tagged on a five point scale 6.8 K tweets to conduct and assess the experiment and hypothesis. Since demographics itself constitutes education, age, income, gender, political ideology, and religious inclinations all of these also interact with vulgar expression. This has been asserted by other researchers over time. Some grouped its usage into abusive, emphatic, idiomatic, dysphemistic, and cathartic. Others grouped somewhat similarly into emphasis, signaling group identification, aggression, and to express emotion. Thus, analyzing demographics and vulgarity together using statistical measures such as Pearson correlation gender and age were considered as basic traits. The following observation were found:

1. Females make less use of vulgar posts on Twitter as compared to males. Based on the data, vulgar words are composed in about 3.332% tweets of males and 3.060% for females.
2. The biggest affector is Age. Among young people, the usage is much higher compared to older in the posts.
3. Income and education(higher) are inversely correlated with the amount of posts containing vulgar words. Education is relatively more correlated than income.
4. Very similarly faith of the individual is also inversely correlated.
5. Liberal youths and individuals are also strongly correlated and hence are more likely to have vulgar posts.

Table 4 MAE by classes for strategies adopted

Model	Very Negative	Negative	Neutral	Positive	Very positive	Overall
Bi-LSTM	1.084	0.049	1.008	1.978	3.0	0.791
+Masking	1.979	1.0	**0.028**	**1.030**	**2.0**	0.898
+Token insertion	**1.0**	**0.002**	0.996	2.0	3.0	0.791
+Concatenation	**1.0**	**0.002**	0.992	1.993	3.0	**0.759**

Besides the fact that data assessed the above observations, other researchers previously in their works had through surveys, questionnaires observed almost the same relationship. McEnery [28], Wang et al. [29] Jay [30], Sylwester and Purver [31] and a few others reported similar observations in written text, speech, social networks, and surveys. As for the modeling, introducing explicit information about vulgarity into the classification models did show some promising results. A base architecture made up of bi-directional LSTM with Relu activation made use of three strategies namely masking, token insertion, and concatenation to produce the results using mean absolute error as shown in Table 4. With masking the error decreased for Neutral, Positive and Very Positive category but no so much as for Very Negative and Negative by using Token insertion and Concatenation along with vulgar dependent features.

The best (lowest) MAE was found by concatenation which was followed by token insertion. Thus, this experiment showed an increase in sentiment analysis performance when vulgar dependent features are used.

Various other emotions have now become useful for detection such as *joy, sadness, disgust, fear, surprise, anger, neutral*. As we have seen so far, the integrity of datasets and annotaters has proved crucial while also observing that separately auditing emotions such as vulgarity tends to improve the models and classifiers performance for applications. Many researchers have claimed the unavailability of large scale datasets which contain a spectrum of emotions and sentiments labeled. Thus, a work by Soujanya et al. [32] proposed a multi-party emotional conversational database called MELD cotating 13,000 utterances and 1433 dialogues from a popular sitcom "Friends." Besides text, it even contains visual and audio. The primary reason for this dataset is to assert importance for recognizing emotions using contextual and multimodal information. This dataset appears to have evolved from a previous dataset called EmotionLines [33]. A few changes can be observed in Table 5. For instance, annotation as in the 3rd row of the following table is changed to Anger from non-neutral in this database.

Table 5 Different annotations for same utterance as observed in MELD and EmotionLines

Utterance	Speaker	MELD	Emotion lines
I'm so sorry!	S1	Sadness	Sadness
Look!	S1	Surprise	Surprise
This guy fell asleep!	S1	*Anger*	Non-neutral

Besides differences in emotions, additional analysis has been done considering the role of context, inter-speaker influence, emotion shifts, and contextual distance. The applications that can make use of the dataset include emotion classifiers which can eventually be used as emotional receptors. Also by making use of the datasets' multimodality, it can be further used for training such dialogue systems.

One more peculiar application which required heavy research was put forward by Myle et al. [34]. This work pondered back at the online reviews for businesses and manufacturers and how deceptive reviews have become a deterrent in their progress and so in light of the same it became an equally essential notion to detect sentiments that are *deceptive*. Businesses tend to intentionally write sugar coated reviews about themselves in order to sound appealing and this hype they believe encourages and promotes individuals to pursue their services. These kinds of reviews are called *positive deceptive opinion spam*. On the other hand, there are businesses that use slanderous and belittling remarks in reviews for their competitions and in pretty much the same manner these negative reviews are called *negative deceptive opinion spam*. Though a past study by Ott et al. [35] did observe the positive deception but negative deception for berating reviews remained highly unstudied. This work by Myle et al. [34] created and analyzed the very first dataset of negative deceptive reviews for inquiring the relationship between deception and sentiment following the procedure of Ott et al. [35] and created 400 HITs from chicago hotels with 20 reviews from 20 hotels each. The performance of detecting deception is reported by three human judges, a SVM classifier modeled using *n*-grams. Interesting performance is achieved with equitable recall, precision and $F1$-score with 88.4% and 86.0% accuracy for positive and negative, respectively. Furthermore, interactions are found and explained for sentiment and deceptions as following:

1. The positive deceptive reviews are made of less discrete spatial information which substantiates the inexperienced users. The same pattern was mined for negative reviews too.
2. Another observation was the high usage of verbs for negative deceptive reviews that asserts an imaginative style of writing instead of actually experienced.
3. Exaggeration was common in negative reviews comprising terms like terrible and disappointment.

This paper marked the creation of the first gold standard corpora that was freely available of negative deception opinion scam with best performance achieved by using automated classifiers.

4 Comparative Study

Machine learning methodologies have been used long enough by researchers. No doubt it has been the most promising entity in the area of artificial intelligence which is owed to its accurate learning paradigm. ML techniques involve learning from attributes but more importantly the selection of those attributes is in human hands.

In the case of natural language processing, features of text are uni-grams, bi-grams, tri-grams, and other probability-based occurrences. Few more features have been proposed that have been both generic and application centered usages like word count, document length, or counting different parts-of-speech. But specifically for computational linguistics tasks as the discipline gained momentum, a big focus was over rules and lexicons. Obviously, the idea was to make the machine understand the rules of a language very much like a compiler. Hence, lexicon-based or rule-based approach was adopted at a point which made use of dictionary meaning. Though with developing complexity of languages, the method appears amateur but the performance of its derivative techniques have been reported to be valuable. Next step was to utilize learning a representative approach called deep learning (DL) which comprises multilevel representational learning. These involve multiple simple but non-linear modules that transform a raw input to higher representation hence allowing varied complexity to be learned. Soon after the momentum in the natural language processing progressed DL techniques observed high performance throughout all applications in the field. Figure 3 shows a hierarchical difference between machine learning and deep learning.

It is still important to note that irrespective of the methodology adopted whether machine learning, deep learning, lexicon-based, or even hybrid, it is invective to clean the text before. Punctuations, white spaces, non-character symbols, stop words are all that are required to be pre-processed. Reviews on online platforms, posts on Twitter, facebook, tumblr, and other such platforms involve templates, hashtags, links, and other such symbolic text which mostly remains inessential in modeling SA. Figure 4 delineates the taxonomy of deep learning methods for SA.

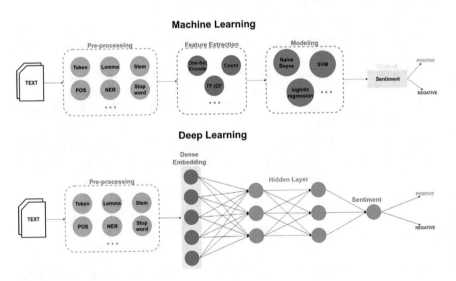

Fig. 3 High level difference between machine learning and deep learning

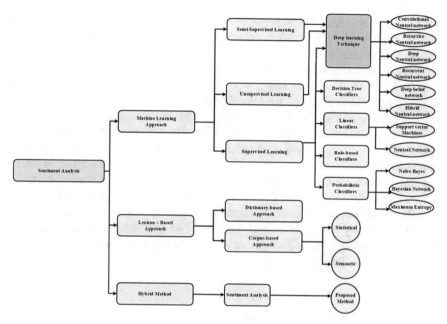

Fig. 4 Taxonomy of deep learning and hybrid methods for sentiment analysis

4.1 Convolutional and Recurrent Neural Network (with LSTMs)

Recurrent neural networks (RNN) have performed absolutely well with low short-term memory (LSTM) [36] cells since their inception in 1997 and many theories and experiments have been performed since then in NLP. This is the prime reason that many researchers and practitioners in the field make use of their cost-effective and easy access in experiments and to production as well. The outstanding feature that separates these networks from the traditional ones is their ability to remember past computed information which they use ahead in time. Though it is quite clear that RNNs with gating mechanisms like LSTMS or gated recurrent units (GRU) [37] are great at modeling units in sequences, convolutional neural networks(CNN) are also widely explored for many applications in the discipline. A general architectural difference between the two is that CNN are hierarchical while RNN are sequential. Since in the field of SA, the orientation is mostly interpreted using key words and phrases, it wouldn't be incorrect to think of a CNN(hierarchical) working out the conjecture for us. And in the similar way, a RNN would be best suited for a sequence modeling task. But the research and literature and natural language processing doesn't quite entirely confirm the assumption. For instance, recurrent neural network's performance on document level sentiment classification has been quite apt, while gated mechanisms with CNN have worked well as compared to LSTM on language modeling tasks. This goes against the assumption and so to say, an accord on architecture selection is not yet achieved for natural language problems. An experiment conducted by Wenpeng et al. [38] involved various tasks among which Sentiment classification was carried out on stanford sentiment treebank (SST) [39] considering CNN, GRU, and LSTM. GRU came out top with an accuracy of 86.32% as compared to 82.38% and 84.51% for CNN and LSTM, respectively. Now as we know so far, CNN would perform better owing to their ability to mine position invariant features, and hence, sentiment classification would have them suited best but currently RNNs outperform.

Based on the quantitative analysis, the performance appears almost inexplicable, but a qualitative analysis suggested a reasoning behind the performance. Error analysis of the two architectures suggested that for the sentence of long length in which sentiment was determined by whole of it instead of by key phrases or words, GRU correctly classified them while CNN did not. For instance, sentences that have positive phrases but are accompanied by negation in the initial or later part of the sentence are missed out by CNN while GRU because of its architecture capture those and hence perform better there. Thus, when it comes to handling, long sequences CNN shouldn't be the first choice.

4.2   Word Embeddings/Representations

The success of deep neural networks for natural language models have been gradual in the community at the beginning. At some point, it became difficult to interpret from the long established statistical inferences of text for many models and architecture to increase performance. The n-grams model, TF-IDF, frequency distributions all computed conditional probabilities of the words. But ahead of 2000, the representation of words took a stronger focus among the community and that escalated the progress in the discipline for deep learning models. The problem was approached to construct a context for words that interpret relation between input and output word vectors and then incline the training of such representations. It was in 2013 that Timos Mikolov and others at google created an embedding for words implemented with two efficient models which is now known as Word2Vec [40, 41]. The two models, SkipGram which was trained to predict the context words given a set of words and a continuous bag of words (CBOW) model which predict the opposite, i.e., the center word when context words were given.

A year later in 2014 Stanford researchers proposed another word embedding called global vectors (GloVe) [42]. A major difference between the two was the consideration of frequent occurrence of context words, weight of which wasn't considered by Word2Vec. GloVe on the other hand articulates this co-occurrence and does not simply disregard them as additional examples.

One other limitation of both the embeddings was overcome in 2016 by Facebook's proposal of FastText. The fact that neither of the embeddings considered *unknown words* was solved by building embedding using characters. Both word2vec and GloVe created the embeddings using words and they formed both the context and the center/focus. The use of characters to create embedding were used on a lower level to create word embeddings and this gave two major advantages: *less training time* and *generalization of unknown words.*

4.3 Deep Belief Networks

A semi-supervised learning method called fuzzy deep belief network (FDBN) was proposed by Shusen et al. [43] for sentiment classification. A previously used DBN was trained by SSL using the training data after which all the class reviews were considered for designing a fuzzy membership function. The abstraction capability of DBN was utilized by FDBN along with fuzzy classification for handling the data. It is the supervised learning stage on the FDBN architecture that increases the classifications performance. Later to take over the upsides of both FDBN and active learning, an active FDBN (AFD) SSL method was suggested. The experiential validation on 5 sentimental classification datasets delineated the effectiveness of AFD and FDBN methods.

4.4 Rule-Based and Other Classifiers

A part of the rule-based classifiers have been mentioned in Sect. 2 of this research chapter. The work on opinion mining and sentiment analysis on web reviews at the inception mostly utilized machine learning algorithms at the center. SentiWordNet's performance is mostly owed to its humongous spectrum for considering all synsets of all wordnets for a numerical score, which considers positive, negative, and objective labels. With the latest version out, it scores words on 7 categories namely strong negative, weak negative, negative, neutral, positive, weak positive, and strong positive. Accuracy of 97.5% was found with fewer error rates on online reviews(political) and books via Kappa measures. Other classifiers such as SVM and K-nearest neighbor used on English testing dataset(ETD) for emotion classification by Vo et al. [44] found a tuned accuracy of 72% and 64%, respectively. This dataset comprised 25,000 documents with 12,500 each positive and negative reviews.

From among the datasets that are widely accepted in the community Table 6 shows the best methodologies adopted by researchers in solving the problems around sentiment analysis. The fact that most of these datasets are easily available and accessible allows all researchers, students, and others across the spectrum to be relieved of any privacy laws.

Noticeably, it is the deep learning methodologies that give the best performance across varied datasets for as many varied topics even when the datasets size varies across all ranges thus offering increasing complexities for problems.

Table 6 Best methodologies adopted for various different problems on sentiment analysis by researchers

Research work	Dataset	Authors	Years	Method(s)
Twitter sentiment analysis	SemEval 2016	Alharbi el al. [56]	2019	CNN
Comparative review of sentiment analysis based on deep learning	SemEval workshop and social network sites	Do et al. [57]	2019	CNN, LSTM, GRU, and hybrid approaches
Aspect-based sentiment analysis	Twitter, SemEval 2014	Yang et al. [58]	2019	Co-attention-LSTM, co-attention-MemNet, co-attention-LSTM + location
Deep learning for financial sentiment analysis	StockTwits	Sohangir et al. [59]	2018	LSTM, doc2vec, and CNN
Joint aspect and polarity classification for aspect-based sentiment analysis	SemEval 2017	Schmitt et al. [60]	2018	CNN, LSTM
A product opportunity mining approach based on topic modeling and sentiment analysis	Twitter, Facebook, Instagram, and Reddit	Jeong et al. [61]	2017	LDA-based topic modeling, sentiment analysis, and opportunity algorithm
Sentiment analysis for recommender system in the cloud	Amazon	Preethi et al. [62]	2017	RNN, Naïve Bayes classifier
A study of sentiment analysis in Thai	Twitter	Vateekul et al. [63]	2016	LSTM, DCNN
Sentiment analysis using AdaBoost combination	Movie reviews and IMDB	Gao et al. [64]	2016	CNN

5 Latest Developments and State-of-the-Art

5.1 Transfer Learning and Language Models

The state-of-the-art performance was achieved using the recurrent neural networks architectures with memory units such as LSTM. Several application areas reached the top of their performance. But the era of transfer learning had not begun then. The performance boost as was brought in computer vision by ImageNet did not actually initiate until 2016 in NLP. Transfer learning began providing architectural advances to practitioners allowing them to make use of foreign knowledge through different tasks to make faster convergence and improve performance even more. In simpler

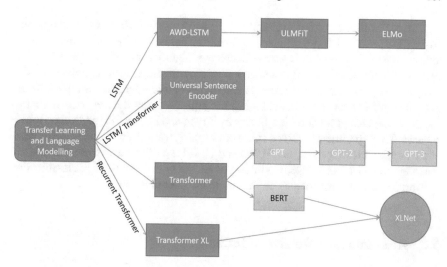

Fig. 5 Different transformers and language models

words, transfer learning allows a neural network architecture to learn and fine-tune itself on a general task and then use those tuned weights for specific tasks with minimal tuning. Primarily, the use of transfer learning is in situations where a dearth of training data is observed. Also, it supposedly is better if the training data for the problem relates to the data used for pre-training. These pre-trained networks use a semi-supervised learning technique and train the architecture to a language model which is then followed by supervised learning with a labeled dataset. Figure 5 shows brief developments in transfer learning and language models.

The *Universal Language Model Fine-Tuning* also called ULMFit formed the foundations of transfer learning. It made use of AWD-LSTM to achieve state-of-the-art performance through it and was on a lower level 3 layers stacked neural network. The three primary steps it utilized were, a pre-training to construct the language model followed by fine tuning the model using a supervised task which used two author proposed methods namely a discriminative tuning and slanted triangular learning rates and finally post tuning the classifier for specific task which involved concatenation of last hidden layers in order to preserve information.

Another in the line of transfer learning foundation were *Embeddings from Language Model* called ELMo which overcame a traditional limitation of word representation to remain the same irrespective of their context. The author of the method made these embeddings a function of the entire sentence which constitute the given word by training a bi-directional LSTM with coupled language model (biLM) on a huge corpus. The calculated representations are more dense because of their dependency on the internal layers. The final representations are calculated on top of two-layer biLM. ELMo has enhanced the performance of traditionally existing models allowing them additional help for processing information with its ability to embed context while generating it.

The performance gain of ULMFit and ELMo were in a study by Enkhbold and Joshua [45] utilized to investigate sentiment analysis. But the experiment was to analyze the performance in Japanese Language. Significant work has been done and proposed for the English language but lack of research in japanese mostly is owed to its sparse nature. With no whitespaces and higher ambiguities, the task of sentiment classification is even more challenging for japanese. The dataset used is that of Japanese Rakuten product review binary by Zhang and LeCun [65] and yahoo movie reviews. The experiment involved entire pre-training from scratch. Final performance and analysis of it suggested that ELMo and ULMFit perform well even with a small language model subset. Also, finetuning of ELMo with a bi-attentive classification network improves performance on the sentiment classification.

5.2 Attention and the Transformer

The need to understand long sentences isn't entirely an essential element in sentiment analysis since the dominant dependencies in most cases happen to be some keywords and phrases. But with longer and bigger documents comes the issue of long term dependencies. The presence of small negative verbs causes the strong positive adjective ahead in a sentence to be interpreted as negative. But even these problems are well handled by recurrent neural networks by utilizing gating mechanisms or memory units called LSTM or GRU, and similarly, the concept of bi-directional seems to have improved the scenario constituting longer dependencies. Still the performance, though better than any other past methods, doesn't appear satisfying.

The problem was addressed at ACL 2014 with the question whether it is possible for one hidden state to store all the global information from dependencies. It was mentioned in the workshop that it isn't possible to capture the entire meaning of a document in a vector. It was then in 2017 by a team of google brain and research individuals along with university of toronto in their work [47] proposed self-attention and the transformers model. The aim of the proposed architecture is not so different from the traditional attention mechanism and while the transformer is based upon the encoder-decoder architecture there are many architectural advantages that lead to better performance and computations.

A fundamental flaw in RNN which was due to its sequential computation ability didn't allow parallelizing the calculations, while also constraining batch size. This made the calculations harder as the length of the sentence grew. The transformer model made improvements in this aspect while also improving the overall performance with the addition of self-attention and positional encodings.

Transformers did overcome the sequential processing but still the sequence of text had to be fed to the architecture through some manner. Thus, positional encodings completed this step by injecting information about absolute and relative position while keeping it distinct for all the tokens. But the chief operator that made the most impact was the mechanism of self-attention or intra-attention. The ultimate aim of this step is to consider the encoding of all other words while encoding a specific word and thus naturally giving it a contextual encoding. This is accomplished by learning three different smaller embeddings called *query, key and value*. The central idea for matrices/ embeddings is to learn how much other words value a specific word, and then finally, calculate the representation based on that valuation. This concept is further bolstered by multi-head attention, the idea of which is to collect different contexts based on different random initialization of query, key, value matrices. The final embedding is calculated from the multiple calculated contexts.

Other than this attention, the architecture demands in its decoder a mechanism to learn and to be able to predict what comes ahead in a sentence. Masked self-attention is the mechanism that helps achieve this notion completely by masking what comes ahead (to predict) in a sentence, logical intuition of which would be to not allow the learning of the next word that is to be predicted.

These concepts combined inside the stacked encoder-decoder along with a few other experimentally proven advantageous step layer normalization has set the precedent for a new era of language understanding and generating. One among the most prevalent transformer-based architecture to become state-of-the-art is the BERT [48].

5.3 Transformers-Based Architectures

It is only appropriate to mention of bi-directional encoded representations from transformers (BERT), the publishing of which created a super stir in the whole community because of its state-of-the-art performance in numerous NLP tasks. The novel technique proposed by the paper's authors of masked language modeling actually allows bi-directional training which lets a deeper context development about the language as compared to unidirectional language models. At its core BERT makes use of the vanilla transformer to learn contextual relation among words and subwords, but since the primary task of BERT is to create a language model, it is composed solely of encoders. While other sequential architectures make use of one direction learning at a time which is either left-to-right or right-to-left, BERT inputs and processes the entire sequence at once. The step to masking language is done prior to feeding where about 15% words are masked, and for computing loss only those many masked words prediction are considered. Because of its extra large size, the pre-training process is quite huge, but since its inception various methods have been introduced to shrink the transformer models while keeping most of the performance. Quantization does the work by reducing the number of bits required for storing the model, although it's hardware dependent it can result in 4 times reduction with faster speed. Another method called pruning, attempts at removing attention heads based on their usage. In a study by Voita et al. [46] suggested about 80% attention heads can be removed. One more promising method called distillation trains a new model to predict the weights of the larger model's layers which results in about 100 times smaller and 15 times faster.

BERT's success even knocked the OpenAI GPT [49] off the GLUE leaderboard. The generative pre-training model (GPT) by OpenAI in 2018 had a downfall primarily due to its traditional usage of language modeling methods like predicting next in contrast to BERT's masked world prediction. But this demerit of GPT actually was converted to its merit when in 2019 OpenAI proposed the new GPT-2 which made use of scaling power by training with 1.5 billion parameters to begin generating coherent text put together beautifully. Yet again in 2020, openAI proposed GPT-3 [50] which was even bigger with 175 billion training parameters allowing the model an amazing feat of capabilities to perform almost any task.

Other language models such XLNet [51], Microsoft Corp's Turing-NLG algorithm [52], Nvidia's Megatron [53], Transformer-XL [54] and few others have drawn the line for amazing performances in the new era of computational linguistics.

5.4 Limits of Transfer Learning

GoogleAI recently in their study explored the limits of transfer learning which they presented in [55] and the different factors that supposedly cause variation. These included self-supervised objective, architecture, dataset, fine-tuning strategy, extra computation, and the attention masking for the foundational architecture. The architecture proposed is scaled up to 11 billion parameters and is perfectly amenable to transfer learning. The model adapts to a unified text-to-text learning framework that allows the model to be used by developers and individuals for different tasks by merely *prefixing* the inputs. Figure 6 shows an example of the frameworks use of prefixes.

The advantage of this unified model is their ability to perform multi task learning, pre-training on the same objective, and then fine-tune to many different tasks. The study involved experimenting over different architecture such as encoder-decoder, encoder stack, decoder stack, and even datasets for which they constructed a new "colossal clean crawled dataset" and others like Wikipedia dataset, Toronto Books corpus.

The performance of T5 has been state-of-the-art on datasets such as GLUE, SQuAD, CNN/Daily Mail. Also its benchmark of 88.9 stands quite close to that of

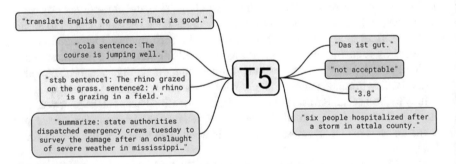

Fig. 6 Framework allowing text-to-text learning by making use of prefixes

human which is 89.8 on SuperGLUE. The agenda behind the model is entirely to show the culmination of the best past ideas proposed into an extremely fine performing model. The model claims state-of-the-art on the twenty established NLP tasks.

6 Conclusions

This research chapter thus has provided in summary the milestone methods and paradigms as the area of sentiment analysis evolved from the use of a traditional, rule or lexicon-based system to models that have on some levels achieved almost human-like language understanding. This chapter aids the practitioners and the newcomers in the field to capture a quick glance at the various applications of sentiment analysis as they and their requirements emerged while discovering the bottlenecks experienced during research and consequently removed. It is also worth mentioning that the credit to the inception of this application field can not be associated to one single author or a team but is rather the amalgamation of many who had keen insight in the ability to utilize the power of natural language processing toward understanding and deciphering the truthfulness of human sentiments and opinions as they are expressed in daily lives in areas such as businesses, markets, and on social media. The newer models though have experimentally proved to be state-of-the-art but their size leading to difficult interpretability would act as an impediment in their utilization in real-world. Also with the newer models coming up an equally strong research is initiated to reduce the size and shape them for specific tasks without losing performance. Models like DistillBERT, RoBERTa, ALBERT are bringing about advantages that appear as a silver lining to their huge architecture in the field of sentiment analysis and opinion mining.

References

1. Pang, Bo, Lillian Lee, and Shivakumar Vaithyanathan. 2002. *Thumbs up? Sentiment classification using machine learning techniques.*
2. Jussi Karlgren, and Douglas Cutting. 1994. *Recognizing text genres with simple metrics using discriminant analysis,* 1071–1075.
3. Brett Kessler, and Geoffrey Nunberg, and Hinrich Schuetze. 1997. *Automatic detection of text genre.* https://doi.org/10.3115/976909.979622.
4. Aidan Finn, Nicholas Kushmerick, and Barry Smyth. 2002. *Genre classification and domain transfer for information filtering.* https://doi.org/10.1007/3-540-45886-7_23.
5. Vasileios Hatzivassiloglou, and Janyce Wiebe. 2003. *Effects of adjective orientation and gradability on sentence subjectivity.* https://doi.org/10.3115/990820.990864.
6. Janyce Wiebe, Theresa Wilson, and Matthew Bell. 2001. Identifying collocations for recognizing opinions. In *Proceedings of ACL/EACL 2001 Workshop on Collocation.*
7. Turney, Peter D.2003. *Measuring praise and criticism: Inference of semantic orientation from association.* Canada: National Research Council.
8. Vasileios Hatzivassiloglou, and Kathleen McKeown. 2002. *Predicting the semantic orientation of adjectives.* https://doi.org/10.3115/979617.979640.

9. Hu, Minqing, and Bing Liu. 2004. *Mining opinion features in customer reviews*. Chicago: Department of Computer Science, University of Illinois.
10. Miller, George A. 1995. *WordNet: A lexical database for english*.
11. Spertus, E. 1997. Smokey: Automatic recognition of hostile messages. *AAAI/IAAI*.
12. Turney, Peter D. 2002. *Thumbs up or thumbs down? Semantic orientation applied to unsupervised classification of reviews*.
13. Esuli, Andrea, et al. 2006. *SENTIWORDNET: A publicly available lexical resource for opinion mining*.
14. Soo-Min Kim, and Eduard Hovy. 2004. *Determining the sentiment of opinions*. https://doi.org/10.3115/1220355.1220555.
15. Stone, Philip, Robert Bales, J. Namenwirth, and Daniel Ogilvie. 2007. The General Inquirer: A computer system for content analysis and retrieval based on the sentence as a unit of information. *Behavioral Science* 7: 484–498. https://doi.org/10.1002/bs.3830070412.
16. Baccianella, Stefano, Andrea Esuli, and Fabrizio Sebastiani. 2010. *SENTIWORDNET 3.0: An enhanced lexical resource for sentiment analysis and opinion mining*.
17. Taboada, Maite, Julian Brooke, Milan Tofiloski, Kimberly Voll et al. 2014. *Supervised sentiment analysis in Twitter using Skipgrams*.
18. Wilson, Heresa, Janyce Wiebe, and Paul Hoffmann. 2005. *Recognizing contextual polarity in phrase-level sentiment analysis*.
19. Cambria, E. 2013. *New avenues in opinion mining and sentiment analysis*.
20. Dave, Kushal. 2003. *Mining the peanut gallery: Opinion extraction and semantic classification of product reviews*. USA: NEC Laboratories.
21. Thelwall, M. 2012. *Sentiment strength detection for the social web*.
22. Pak, Alexander and Patrick Paroubek. 2010. *Twitter as a corpus for sentiment analysis and opinion mining*.
23. Mozetic, Igor, Miha Grcar, and Jasmina Smailovic. 2016. *Multilingual twitter sentiment classification: The role of human annotators*.
24. Timothy Jay. 2009. The utility and ubiquity of taboo words. *Perspectives on Psychological Science* 4. https://doi.org/10.1111/j.1745-6924.2009.01115.x.
25. Mehl, Matthias, Samuel Gosling, and James Pennebaker. 2006. Personality in its natural habitat: Manifestations and implicit folk theories of personality in daily life. *Journal of Personality and Social Psychology* 90: 862–877. https://doi.org/10.1037/0022-3514.90.5.862.
26. Wang, Wenbo, Lu Chen, Krishnaprasad Thirunarayan, and Amit Sheth. 2014. Cursing in English on Twitter. *Proceedings of the ACM Conference on Computer Supported Cooperative Work, CSCW*. https://doi.org/10.1145/2531602.2531734.
27. Cacholaz, Isabel, Eric Holgatey, Daniel Preotiuc-Pietro, and Junyi Jessy Liy. 2018. *Expressively vulgar: The socio-dynamics of vulgarity and its effects on sentiment analysis in social media*.
28. McEnery, Tony. 2005. *Swearing in English: bad language, purity and power from 1586 to the present*, 1–248. https://doi.org/10.4324/9780203501443.
29. Jay, T.B. 1980. Sex roles and dirty word usage: A review of the literature and a reply to Haas. *Psychological Bulletin* 88 (3): 614–621. https://doi.org/10.1037/0033-2909.88.3.614.
30. Jay, T. 1992. *Cursing in America: A psycholinguistic study of dirty language in the courts, in the movies, in the schoolyards and on the streets*. John Benjamins Publishing Company. https://doi.org/10.1075/z.57.
31. Sylwester, Karolina, and Matthew Purver. 2015. Twitter language use reflects psychological differences between democrats and republicans. *PLoS ONE* 10: e0137422. https://doi.org/10.1371/journal.pone.0137422.
32. *MELD: A multimodal multi-party dataset for emotion recognition in conversations*. 2018.
33. Sheng-Yeh Chen, Chao-Chun Hsu, Chuan-Chun Kuo, Huang Ting-Hao, and Lun-Wei Ku. 2018. *EmotionLines: An emotion corpus of multi-party conversations*.
34. Myle, Ott, Claire Cardie, and Jeffrey T. Hancock. 2013. *Negative Deceptive Opinion Spam*.
35. Myle, Ott, Yejin Choi, Claire Cardie, and Jeffrey Hancock. 2011. *Finding deceptive opinion spam by any stretch of the imagination*.
36. Hochreiter, and Schmidhuber. 1997. *Long short-term memory*.

37. Cho, Kyunghyun, Bart van Merrienboer, Caglar Gulcehre, Dzmitry Bahdanau, Fethi Bougares, Holger Schwenk, and Yoshua Bengio. 2014. *Learning phrase representations using RNN encoder-decoder for statistical machine translation.*
38. Yiny, Wenpeng, Katharina Kanny, Mo Yuz, and Hinrich Schutzey. 2017. *Comparative study of CNN and RNN for natural language processing.* Munich, Germany: CIS, LMU; USA: IBM Research.
39. Socher, Richard, A. Perelygin, J.Y. Wu, J. Chuang, C.D. Manning, A.Y. Ng, and C. Potts. 2013. Recursive deep models for semantic compositionality over a sentiment treebank. *EMNLP* 1631: 1631–1642.
40. Mikolov, Tomas et al. 2013. *Efficient estimation of word representations in vector space.*
41. Mikolov, Tomas. 2013. Distributed representations of words and phrases and their compositionality. *Advances in Neural Information Processing Systems.*
42. Pennington, Jeffrey, Richard Socher, and Christopher D. Manning. *GloVe: Global vectors for word representation.*
43. Zhou, Shusen, Qingcai Chen, and Xiaolong Wang. 2014. Fuzzy deep belief networks for semi-supervised sentiment classification. *Neuro-Computing* 131: 312–322.
44. BalaAnand, M., N. Karthikeyan, and S. Karthik. 2018. Designing a frame-work for communal software: Based on the assessment using relational modelling. *International Journal of Parallel Program.*
45. Bataa, Enkhbold, and Joshua Wu. *An investigation of transfer learning-based sentiment analysis in Japanese.*
46. Voita, Elena, David Talbot, Fedor Moiseev, Rico Sennrich, and Ivan Titov. 2019. *Analyzing multi-head self-attention: Specialized heads do the heavy lifting, the rest can be pruned*, 5797–5808. https://doi.org/10.18653/v1/p19-1580.
47. Ashish, Noam, et al. 2017. *Attention is all you need, NIPS.*
48. Devlin, Jacob, Ming-Wei Chang, Kenton Lee, and Kristina Toutanova. 2018. *BERT: Pre-training of deep bidirectional transformers for language understanding.*
49. Jeffrey Alec. 2018. *Language models are unsupervised multi task learners.*
50. Benjamin Tom, et al. 2020. *Language models are few-shot learners.* arXiv:2005.14165 [cs.CL].
51. Zhilin et al. 2019. *XLNet: Generalized autoregressive pre training for language understanding.* arXiv:1906.08237 [cs.CL].
52. Turing-NLG: A 17-billion-parameter language model by Microsoft, Turing-NLG: A 17-billion-parameter language model by Microsoft.
53. Shoeybi et al. 2020. *Megatron-LM: Training multi-billion parameter language models using model parallelism.* arXiv:1909.08053 [cs.CL].
54. Zhilin Zhang, et al. 2019. *Transformer-XL: attentive language models beyond a fixed-length context.* arXiv:1901.02860 [cs.LG].
55. Noam Colin, et al. 2019. *Exploring the limits of transfer learning with a unified text-to-text transformer.* arXiv:1910.10683 [cs.LG].
56. Alharbi, A.S.M., and E. de Doncker. 2019. Twitter sentiment analysis with a deep neural network: An enhanced approach using user behavioral information. *Cognitive Systems Research* 54: 50–61.
57. Do, H.H., P. Prasad, A. Maag, and A.J. Alsadoon. 2019. Deep learning for aspect-based sentiment analysis: A comparative review. *Expert Systems with Applications* 118: 272–299.
58. Yang, C., H. Zhang, B. Jiang, and K.J. Li. 2019. Aspect-based sentiment analysis with alternating coattention networks. *Information Processing and Management* 56: 463–478.
59. Sohangir, S., D. Wang, A. Pomeranets, and T.M. Khoshgoftaar. 2018. Big data: Deep learning for financial sentiment analysis. *Journal of Big Data* 5: 3.
60. Schmitt, M., S. Steinheber, K. Schreiber, B. Roth. 2018. *Joint aspect and polarity classification for aspect-based sentiment analysis with end-to-end neural networks.* arXiv arXiv:1808.09238.
61. Jeong, B., J. Yoon, and J.-M. Lee. 2019. Social media mining for product planning: A product opportunity mining approach based on topic modeling and sentiment analysis. *International Journal of Information and Management* 48: 280–290.

62. Preethi, G., P.V. Krishna, M.S. Obaidat, V. Saritha, S. Yenduri. 2017. Application of deep learning to sentiment analysis for recommender systems on cloud. In *Proceedings of the 2017 International Conference on Computer, Information and Telecommunication Systems (CITS), Dalian, China, 21–23 July 2017*, 93–97.
63. Vateekul, P., and T. Koomsubha. 2016. A study of sentiment analysis using deep learning techniques on Thai Twitter data. In *Proceedings of the 2016 13th International Joint Conference on Computer Science and Software Engineering (JCSSE), Khon Kaen, Thailand*, 1–6.
64. Gao, Y., W. Rong, Y. Shen, and Z. Xiong. 2016. Convolutional neural network based sentiment analysis using Adaboost combination. In *Proceedings of the 2016 International Joint Conference on Neural Networks (IJCNN)*, Vancouver, BC, Canada, 1333–1338.
65. Zhang, Xiang, and Yann LeCun. 2017. *Which encoding is the best for text classification in Chinese, English, Japanese and Korean?*

Healthcare Analytics: An Advent to Mitigate the Risks and Impacts of a Pandemic

Shubham Kumar

Abstract Healthcare analytics is a broad term for a specific facet of analytics which sweeps in a wide sash of the healthcare industry, providing macro- and micro-level insights on patient data, risk scoring for chronic diseases, hospital management, optimizing costs, expediting diagnosis and so on. An efficacious way of implementing healthcare analytics on this sector is to combine predictive analysis, data visualization tools and business suites to get valuable information that can both deliver actionable insights to support decisions, as well as reduce variations to optimize utilization. In order to device a more potent approach to prepare for the next pandemic, the clinical data has to be improved which would provide more granulated information to augment treatment effectiveness and success rate. Confirmatory data analysis combined with predictive modeling can be used to furnish unexpected gaps in terms of getting the required clinical data, which is vital for the success of this approach. The healthcare analytics domain has got the potential to monitor the health of the masses to identify disease trends and can provided enhanced health strategies based on demographics, geography and socio-economics. Hence, healthcare analytics is indeed an advent to counter the imminent threats of outbreaks such as epidemics and pandemics provided it gets the necessary course of action to fulfill its grail.

Keywords Healthcare analytics · Big data · Data analytics · Business intelligence · Seaborn

1 Introduction

1.1 Healthcare Sector

Making itself as the prima facie of the human development the healthcare domain is certainly one of the largest and most prominent industries around the globe. This

S. Kumar (✉)
Deloitte Consulting India Private Limited, Hyderabad, India
e-mail: shubham.sbkr@gmail.com

industry goes in tandem with the survival lifestyle and living quality of the modern humans and hence reflects a direct reverberation of people living in different continents, countries or cities of the world. If one has to define it, healthcare is the diagnosis, cure and prevention of any mental or physical impairments such as an injury, disease or illness. It is performed, grafted and delivered by trained practitioners who could be present in different domains across medicine, psychology, physiotherapy, dentistry, chiropractic, nursing, pharmacy, allied heath and other care providers. This industry is responsible for end-to-end steering of the medical procedures which ensure the correct treatment of patients. This sector has been present in one or the other forms, since ancient times and today, it is the fastest growing sector in the world. The delivery of this sector depends on how the internal segregated group of professionals' function while being aligned together. The healthcare industry also interreacts with different domains to provide augmented and optimized services.

1.2   Analytics Domain

If one has to define analytics, it can be expounded as the application of a progression of phases, which are marked by algorithms or transformations, to generate cognizance from a processed set of datasets. It can be further segregated into in process of fetching, processing, analyzing and interpreting data to gain more knowledge or insight about it. In recent times, especially after the invariable expansion of stored data, analytics has gained a paramount status across multiple domains, sectors or industries. From gaining insights about the nature of crime and criminal psychology to boosting the sales figures of a supermarket, this technology has penetrated across myriad sectors and is considered predominantly utilitarian. It is gaining more prevalence due to the advancement of the quantum computing and storage technology. In the previous generation of information technology, the processing of data was slow and there were limited space to store a set of data. This was a major hurdle in the growth of analytics sector. But, with the augmented ability of machines and systems to store, read and process humongous amount of data in a very less period of time, that hurdle has been crossed. Though there could be a myriad solutions coming out of the analytics domain, it is mostly segregated into three categories which are—descriptive analytics, predictive analytics and prescriptive analytics. At times, the combination of one or more than one of these categories is used to provide an appropriate solution. The first category, descriptive analytics, tries to lay down the already present fact which has been hidden in the vast layers of a dataset. Here, the already captured data is just altered in dimension to get to the internal layer of information which is aggregated as a commodity. The function of predictive analytics starts where the descriptive analysis ended. It uses the results provided by the descriptive analytics as a base to try and predict the future or imminent events. It does not give an exact description of a future event but merely provides an idea of what could be imminent. The final category, prescriptive analytics uses both the descriptive and predictive analytics to suggest the next course of action depending upon the outcomes of a stimulating various potential scenarios.

1.3 Application of Analytics in Healthcare Domain

Analytics has a ubiquitous impact on almost all the industries and healthcare is a sweeping example. To express in a nutshell, analytics has left an ineffaceable imprint on the healthcare industry. There are many live examples which bolster this affirmation, such as the omnipresence of fitness bands and the optimized use of the incoming data and information. This piece of technology has indeed revolutionized healthcare and changed the approach of dealing with diseases, physiological, outbreaks, epidemics and even a pandemic. All sectors across the society are benefitting from this gigantic volume of healthcare data which include the private and the public sector. Such a synergistic domain presents a colossal elbow room for professionals across different sectors to learn this component of advanced analytics, in order to transmogrify business challenges into business successes.

2 Background

Mass health casualty has always been a worrying topic for mankind and there are various accounts of the battle of survival between humans and diseases, which took the form of an epidemic or pandemic. Since Neanderthal's times, humans have fought and survived various ailments of varying potency. Throughout the extensive era of humans, they have been ravaged by the sporadic but ferocious presence of plagues and epidemics which in some scenarios, changed the course of history. The oldest account of a prehistoric pandemic is of circa 3000 B.C when an epidemic completed eradicated a prehistoric village in China. Currently known as "Hamin Mangha," this prehistoric site is among the most primitive and most preserved across the globe [1]. The current records based on a lot of anthropological study stipulate that the epidemic engulfed the population so quickly that there was no time for the proper burial of people who died. Similarly, the plague of Athens is also considered one of the gravest in the history. This epidemic started just after a huge war started between Athens and Sparta which lasted for five long years. According to a lot of studies, this plague is said to engulf as many as a hundred thousand people which was a sizable chunk of population, back in 430 B. C. In the modern era, The Black Death which traveled from Asia to Europe between 1346 and 1353 was among the worst pandemics mankind have face [1]. According to some studies, this even wiped out half of the population across Europe. Some of the other examples for epidemics which succeeded these events were the London Plague, yellow fever, etc. The last pandemic of world which was in presence, volume, effect and destruction comparable to COVID-19, was the Spanish Flu of 1918. It is estimated 500 million people fell victim to this pandemic [1]. The people were from almost all parts of the globe ranging from the South Seas to the North Pole. It is estimated that one-fifth of the population who came in contact of this diseases ultimately died. Because of its potency, some indigenous communities were shoved to the verge of extinction. The unfurling and destructiveness of this pandemic was augmented by the confined and restricted conditions of soldiers during World War I. The poor nutrition which was a byproduct of war also contributed to its lethality. There is an interesting story about the name of this disease. Contrary to popular belief, the outbreak did not start in Spain. During that era, Spain had minimum censorship for press and reporting which resulted into the open publication of the accounts of this outbreak, and consecutive circulation in the Spanish newspapers and magazines. This let to people erroneously believing that it originated from Spain, and hence, the name Spanish Flu became prominent [1].

Talking about the current pandemic, COVID-19 (Coronavirus disease 2019) is a viral infectious disease which allegedly started in December 2019 in the Wuhan district of China. It is caused by SARS-CoV-2 and people suffering from it get flu like symptoms along with breathlessness which is usually an indication of lung infection [2, 3]. In 2020, the World Health Organization (WHO) declared COVID-19 as a pandemic. As of September 2020, over 30 million people had been infected globally with over 1 million deaths. There is no recognized account of the origin of this disease while a lot of studies indicate that it got transferred from bats in Wuhan Market of the Hubei province of China. This disease spreads from close proximity and surface transmission, though there are proven studies which indicate the spread though air transmission [2]. This disease sent the world into a state of panic and hysteria. Mass lockdowns were initiated in different cities across the globe along with the practice of social distancing and quarantine, to curtail the spread of this disease. As of now, the SARSr-CoV virion causing this disease has been observed to be destroyed by the application of a household soap which bursts it protective bubble, provided it is present outside a human body [2, 3].

3   Research on Pandemics and Their Impacts

Once the definition of a pandemic is delineated, then comes the part where its morbidity is accessed. Pandemics are essentially widespread, general cases of global outbreaks of infectious ailment which have multifaceted effects on one or more essential aspect of human survival. They are capable of causing sudden and imminent disruption across innumerable fields such as health, economics, and politics and so on. The impact usually lasts for years and the recovery takes an extra amount of physical and physiological resources. In the modern era, the likelihood of the adverse effects of a pandemic has increased by multiple folds. The prime reason for this change is the augmented and optimized travel options present in the modern era which acts as the peripheral carrier to the disease. Medical science has to change its approach while dealing with a natural hazard of this dimension. Also, specific policies were brought into mitigate the effect of nascent outbreaks which had the potential to become a pandemic. Also, a decent amount of effort was put on to expand the sector which would work toward the sustainable development in creating the preparedness and heath capacity for an imminent outbreak. In the wake of the current century, the world had to deal with an outbreak which had the potency to become a pandemic, namely Severe Acute Respiratory Syndrome (SARS) [4]. In this case, a huge setback was received in the form of delayed reporting which directly contradicts the whole purpose of having a goal of being prepared. The sheer damage caused by this outbreak in a very short span of time forced the World Health Organization to bring major changes to the International Health Regulations (IHR) [4]. Here, the substantial necessity molded innovation and the changes helped in rapid yet outstretched testing of the affected individuals and potential targets. The specific standards delineated for detecting, responding and reporting helped the medical community in being prepared for the imminent danger. The world achieved a short-lived success later in the decade when the effects of 2009 influenza pandemic were successfully mollified with the help of rapid response and planning ahead of time. But, this triumph was short-lived when the next decade exposed the shortcomings of International Health Regulations (IHF). The significant aperture and slits were exposed when the world faced many such outbreaks of similar nature, one example of which can be the 2014 Ebola Epidemic. It was now evidently clear that there were significant challenges for proper screening and detection of the ailment, availability of isolation facility, basic care facility for the affected masses, tracing of potential contacts, coordination, mobilization and so on and so forth [4].

In order to be ready for the next pandemic, the risks, impacts and mitigation have to be delineated. A separate listicle has to be maintained for knowledge and technological gaps which would help the mankind to be prepared and equipped both physically and psychologically to combat an outbreak. Here, analytics plays a crucial role in gathering and asserting the required information on the above three verticals. There are a number of steps which has to be implemented to fully understand the potency of the risks involved in a potential pandemic outbreak. It is a well-known fact that pandemics have occurred often in the history, but in the current epoch, it is more likely to happen because of the surging exposure of viral disease particularly from animals. If one had to segregate the types of risks involved in an outbreak, then it would be broadly classified into two categories—spark and spread. The first category denotes the instantaneous risk at the time when an outbreak comes into existence. Whereas, the second category denotes the risk when an outbreak tries to unfurl and outspread, thus taking a form of an epidemic or in worst case, a pandemic. The population across the globe is not uniform in terms of access to healthcare and exposure to virion. A common myth was exalted till the year 2020 that Central and West Africa are at high instantaneous risk and they do not possess a steady amount of preparedness when compared to the rest of the world. This myth was debunked by the advent of novel Coronavirus also known as COVID-19 when the outbreak started from oriental region of China and then spread across occidental Europe before engulfing the entire globe. In such an unpredictable scenario, medical science would have to take aid of probabilistic modeling and analytical tools to gauge the risk involved and to evaluate and estimate the potential trammel of the pandemic. Exceedance probability curves are one such utilitarian piece mechanism which has the capability to amalgamate the probability of a threat and its corresponding economical loss. Taking an example of one such potential threat, influenza, which has the capability to become a pandemic that can cause 6 million deaths worldwide, it was observed with the help of the EP analysis that in a given year, the probability of the influenza taking the form of a pandemic comes as 1%, which is a potent risk [2, 5, 6].

The healthcare analysis domain can also help in laying down the potential impact which a pandemic might cause in a stipulated period of time. It is observed that an outbreak in the range of a pandemic can cause colossal and outstretched growth in mobility and mortality. It has also been observed that in lower and middle income countries, this kind of an outbreak can have excessive mortality impact. There are a number of verticals and channels through which a pandemic can cause economic turbulence which includes short- and long-term setbacks. The short-term setbacks are usually caused due to the fiscal shock a market receives during the spark period. The long-term setbacks are the result of damage to the perennial economic growth during the spread phase and beyond that. Talking about the physiological impacts, a pandemic may induce behavioral changes which lead to widespread panic and agitation. They are widely interconnected to the long-term setback on the economic market as the fear results to a state of reluctance toward investments and purchase. In certain specific countries, a pandemic can also give rise to political contortions which in turn results to mob violence, mass clashes and a state of constant tension between the citizens and the government. In the view of all these medical, social, economic, physiological and political impacts, healthcare analytics can play a vital role as a study and chieftain combat mechanism by helping with the personalized evidence to enable pro-active decisions.

4 Development of Healthcare Information System and Healthcare Analytics

In order to understand the current implementation of healthcare analytics, it origin and consecutive journey needs to be traced. The example of the evolving healthcare information system has to be taken in order to understand the current state of healthcare informatics. The earliest record of a healthcare information system dates backs to 640 BC where the traces can be found in form of case studies. They portrayed the course of illness though observation of patients symptoms. In order to perform this practice, openness and honesty was required by the patient and observer. It was not a very successful practice and frequent deaths were encountered [7].

In modern times, the roots of information technology in healthcare sector were laid in 1960. For this study of the journey of information technology in healthcare sector, the rise of IT in the healthcare domain of the USA would be observed as model country for the world. The 60 marked the enactment of Medicare and Medicaid into practice.

In July 1965, President Lyndon Johnson signed the landmark bill which is prevalent till today and is referred as Medicare [8]. These two laws gave a strong ground for healthcare information system to emerge. Medicare is a federal program which is intended to provide healthcare coverage to US citizen with greater than sixty-five years of age or having disability regardless of the income. Similarly, Medicaid is state and federal program which provides health coverage for people with extremely low income. Courtesy to these two program, the practice of cost-based reimbursement started, and hence, the US government started keeping a lot of medical and peripheral records. In this decade, significant healthcare expansion also took place. There were a good amount of financial needs and capturing revenues specifically to the healthcare sector. The state of information technology was considerably primitive in this decade. The work was done on mainframe computers. Since mainframes were bulky in size, they were not penetrating through an environment brimming with manual records. Lack of portability and esoteric technical specification thwarted the use of mainframe computers in this decade. Central processing of data was a common practice of this decade which changed in the next ten years. Also, the presence of only a few vendor-based products also contributed in limiting the use of informatics technology services in the field of healthcare. There were presence of administrative and financial systems which mostly started because of the prevalence of Medicare and Medicaid. Talking about the infrastructure, there were presence of large hospitals and medical centers which required presence of information technology solutions. All the solutions which performed centralized data processing on mainframe computers were developed and maintained in-house [8, 9].

The 1970 marked a colossal growth in hospitals and medical centers and there were huge opportunity blooming for this sector. The rising Medicare and Medicaid expenditure exhorted a requirement for a digital solution of higher efficiency [9]. There was a growing need of cost containment in this sector which could only be achieved by replacing mainframes by a less bulky structure and a higher penetration rate among masses. This requirement was fulfilled by the advancement of minicomputers which were like small mainframe computers. Since the portability issue was solved, minicomputer became a lot popular. This decade also marked the dawn of computer network which meant that different computers present in hospitals, independent clinics and the government offices could now be connected and information could be shared instantaneously. This was the decade which also marked the availability of turnkey systems though vendor community. Turnkey were highly personalized computer systems which were customized to function for a specific application. They were mostly industry specific and healthcare industry had a surging demand for such systems. The term turnkey was gleaned from the objective that user can make the system ready by just turning a key. In contained, all the hardware and software required to function for a specific industry or requirement. Because of the presence of such highly personalized devices and solutions, there were an augmented interest in clinical applications. The boom of information system reduced the costs to an extent that it was affordable for medium-sized practitioners and hospitals to use and benefit from these solutions. The shared system which marked its dawn in this decade is prevalent and used till now [8, 9].

The 1980 brought the computers to the vicinity of masses and it had a significant impact on the healthcare industry as well. In this decade, the Medicare program introduced the diagnosis-related group which revolutionized the way information system was perceived in the healthcare industry. The diagnosis-related group was initially implemented across the USA by an organization called as Healthcare Financing Administration (HCFA) [9]. It was aimed at bolstering the cost controlling for inpatient services which was associated with and billed to Medicare. The organization Healthcare Financing Administration (HCFA) started to use DRGs after the development of the preferred plan for provider. This particular instance was started in 1989 and gradually was carried to other HMSA plans. A diagnosis-related group can be defined as a system whose motive is to classify hospital cases into various groups for better processing and storing of information. There were a total of 467 groups when this system began, and usually, the last group denoted cases which cannot be placed into any other groups. The initial motive behind developing this group-based system was to persuade the US congress to allow its use for reimbursement process in Medicare and Medicaid. Before this methodology was adopted, the reimbursement process was "cost-based" and DRGs were developed to make it "group-based" and hence add a certain layer of automation to it. This initiative was very successful and is used till now in determining how much the Medicare program would pay the hospital for carrying out a service on a patient. By dividing the cases into groups, a lot of time and resources were saved since the cares in each groups were clinically similar in nature. Also, same group cases tend to use the same level of hospital resources. Apart from DRGs, once major changed which was a harbinger of the state of information technology in healthcare sector was the emergence of personal computers. The unveiling of PCs brought a sea of change with itself. Networking became more sturdy and structured with the segregation of local area network (LAN) and wide area network (WAN). Later in the decade, the increased use of personal computers led to the decentralization of data processing. The billing system was also introduced which became more common in the next decade. All of the abovementioned development in the healthcare information system led to expansion of clinical information systems in hospitals [8, 9].

The 1990 was a turning point in terms of information technology across the globe and it had multifaceted effects on the healthcare sector as well. This decade saw a surging growth in the managed care sector and the integrated delivery systems. Also, another significant step taken by the Institute of Medicine (IOM) was the call for "computer-based" patient and the EMR [9]. The EMR or electronic medical record began as an idea or concept of recording and later storing patient data and information in electronic form, instead of the conventional method which involved a paper. Though the roots of this concept date back to 1972 when the Regenstreif Institute in Indianapolis developed this idea and labeled it as a huge advancement in healthcare and medical practices. Despite of being developed in the early 70, this concept was not used widely because of the associated high costs. This thwarted its mass use and limited the scope of operation to certain specific government hospitals [8]. Later, in 90, this concept became the backbone of healthcare analytics and increased its scope and usage. Also, 1990 marked the advent of the World Wide

Web. It was an era where efforts of certain IT giants led to considerable drop in cost of hardware. The proliferation of Internet has a positive impact on the healthcare industry. The use of Internet allowed the clinic and hospitals to skip the humongous costs of setting up a specific network for some of their intra-domain practices. The healthcare organizations took advantage of this Internet boom and a study by the Institute of Medicine predicted that by the end of this decade, there would be at least one computer at every physician's office which would be dedicated to use of computers for improving the patient care [8, 9].

The 2000 led to strong reforms which established both the necessity and presence of information system and analytics in healthcare industry. During this decade, the international organization gave a report on patient safety and medical error which demanded policies which could address the same issues. This decade marked a turmoil in the all of the major industries and healthcare industry was not spared, as the world faced spiraling health cost. It was also during this decade when the Technology Informatics Guiding Education Reform (TIGER) worked to advance the integration of informatics in the healthcare industry with the augmented application of information technology to improve the patient care while keeping an environment of learning health systems. Founded in the year 2004, the TIGER initiative was a revolutionary step in marking the goals and requirements toward allowing the various facets of the healthcare sector to use informatics tools, theories, principals and practices. These technologies were infused and interweaved together into pedagogy, research and practices which aimed at effective outcome of patient cases, their safely and cost reduction and optimization. In the USA, this was a time of economic upheaval and a growing number of citizens were uninsured. In order to mitigate the damage done by the economic recession of 2008, the 111th US congress signed a law called the American Recovery and Reinvestment Act of 2009 which also had provisions to boost up the healthcare sector [10]. As part of the American Recovery and Reinvestment Act, another program was enacted, namely the Health Information Technology for Economic and Clinical Health (HITECH) Act. This act ensured that proper measures are being taken to promote the adoption of information technology into the healthcare sector. In the following year, the landmark bill of the Affordable Care Act, formally known as Patient Protection and Affordable Care Act, and popularly called as the Obamacare was signed by the 111th US congress [9, 11]. This was consecutively signed into law by then President Barack Obama and was part of his vision to transform the healthcare sector. The various provisions present as part of Obamacare ensured that it allowed surplus funding to promote information technology into healthcare domain which resulted into the structured storing of clinical and medical data [8–11].

The last decade has witnessed a massive digital growth and has engendered something we call as the big data. While big data can be defined as the structured and non-structured set of enormous volume of data, "big data in healthcare" refers to the colossal health data amassed from innumerable wellsprings or sources. The data includs electronic health records (EHRs), genomic sequencing medical imaging, wearables, payer records, pharmaceutical research, medical equipment and so on. There are three specific characteristics which differentiate this segment of data from conventional electronic medical and human health data used for decision-making. This first characteristic is the availability in remarkably huge volume. The second characteristic is that it moves at high velocity and covers the healthcare sector's mammoth digital universe. The final characteristic lies in its diversity. Since it gets fetched from innumerable sources, it is highly variable in structure and nature. The above three characteristics are known as the 3Vs of big data in healthcare sector. Since it brings with it a profound diversity in format, type and context, it is onerous to merge big healthcare data into conventional databases such as the ones used in the previous generations to store medical and clinical records which were DB2 and IMS DB. This makes the task to process this data very challenging. Also, industry leaders find it arduous to utilize or mobilize its considerable potential to transform the healthcare industry. Despite these challenges, there are several new technological means and methodologies which allow the conversion of big data in healthcare sector utilizable and actionable commodity [12].

Once the appropriate set of big data is secured, the next steps involved its convergence which make it ready to be processed and used as a solution [12]. If one has to define the process of convergence in a hackneyed terminology, it can considered as the amalgamation of two very different entities. In the backdrop of analytics, it is the integration of two or more than two distinct technologies, commodities, data or systems into a single unit, device or system. One example of this process is emergence of a cellphone with a camera functionality. The process of convergence here has amalgamated a traditional camera and communication device to create a new piece technology or a device. One prime example of the use of convergence is the one currently take into task by Deloitte where one of its particular divisions is working on the convergence of healthcare trends [13]. This kind of convergence is transforming the traditional US healthcare industry using four major convergence trends. This is creating opportunities for innovation in the healthcare analytics sector. The four major trends are everywhere care; wellness and preventive care; aging, chronic and end of life care and personalized care. The convergence methodology is being used in the everywhere care trend to transpose the gamut of care from hospitals to comparatively lower-cost sites. The next trend wellness and preventive care focuses on repositioning the disease management from reactive to preventive. The third trends focused on utilizing big data to personalize and hence manage chronic conditions to provide support in aging, chronic and end-of-life care. The final trend of personalized care focuses on transmuting the healthcare services from mass generalization to mass customization which would add an extra layer of precision to it. A lot of organizations are considering to develop similar innovation strategies across the healthcare domain to remain afloat in this ever-changing scenario and landscape, by capitalizing on the emanating opportunities [13].

5 Results

After passing through the convergence technique, various algorithms can be applied onto the datasets to garner insights about the information, which a certain dataset wants to convey. For the datasets related to pandemics, a lot of information can be gathered by applying these algorithms on the cleansed and structured data which would provide trends and information about the worst-affected areas, countries or continents. They would also shed a light on the mortality rate and its metastasize across different places. The information could be generated by using a single piece of technology for data analysis or making histograms such as R or Python. To check the information provided by these histograms, data analysis was performed on several sample datasets of early twentieth century's Spanish Flu and COVID-19. The following set of graphs were developed using Seaborn library which is a tool for making statistical histograms in python. This library builds on top of matplotlib and consolidates intimately with pandas data structure. The plotting is carried out by firstly performing the semantic mapping and statistical aggregation on the dataframes and arrays which contain the whole dataset. This process results in the formation of informative plots. This technology rather than focusing on the details on how to draw a plot, prioritizes on defining what the different parts of the plot refer to. This technology is dataset-oriented can be considered as a declarative application programming interface [13, 14] (Fig. 1).

The above set of statements import the necessary libraries required to perform the analysis on the given dataset. Once the importing is done, the visuals would be created using another set of statements (Figs. 2 and 3) [14].

The above set of statement set the columns representing districts, dates and mortality as pivots. The dataset represents district-wise information about the mortality during the Spanish Flu phase. The data indicates the number of deaths per week due to fever in the districts of Assam, India, from 1916 to 1921. The numbers

Fig. 1 Code snippet of the import statements

```
import pandas as pd
import matplotlib.pyplot as plt
import seaborn as sns
import numpy as np
```

Fig. 2 Code snippet of setting the pivots

```
[ ] df = pd.read_csv('sp_1.tab', delimiter='\t')
    df.Date = pd.to_datetime(df['Date'])

    df = df[['District', 'Date', 'Mortality']]

    df['Year'] = df.Date.apply(lambda x : x.year)
    df['month'] = df.Date.apply(lambda x : x.month)
    df['day'] = df.Date.apply(lambda x : x.day)
```

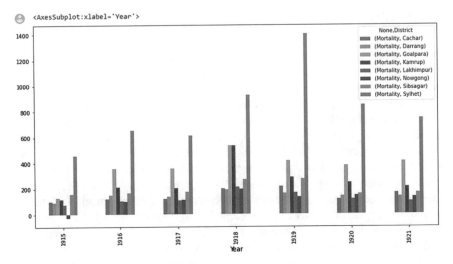

Fig. 3 Plot for average mortality for certain districts

are seasonally adjusted to show excess mortality due to the influenza pandemic of 1918–19 [15, 16].

The data here is a multi-level group by in order of Year and District to calculate the average mortality for those levels. The visual created is done by unstacking the pivots created before. This is done to convert the inner most labels as new column labels in the dataset. After that we have on the x-axis the years with each year representing the average mortality for each district. The following is not so different from a bar graph except that it contains multi-label for each first-level attribute upon which the data is grouped.

The above plot represents the average mortality of the districts of Cachar, Darrang, Goalpara, Kamrup, Lakhmipur, Nowgong, Sibagar and Sylhet from 1915 to 1921 during the trepidation period of Spanish Flu (Fig. 4) [16].

The above plot represents the month-wise mortality of the districts of Cachar, Darrang, Goalpara, Kamrup, Lakhmipur, Nowgong, Sibagar and Sylhet for a calendar year (Fig. 5) [16].

The above plot represents another view for the month-wise mortality of the districts of Cachar, Darrang, Goalpara, Kamrup, Lakhmipur, Nowgong, Sibagar and Sylhet for a calendar year.

Next, the data is separated for each mortality by each district and stored separately. These datasets are individually used to make uni-variate plots. The following plots are modeled upon the Gaussian distribution to express the normalness of the data. Each of the district's mortality is plotted in the same pyplot figure to give a comparative understanding of the mean and skewness of the rates in each district (Figs. 6, 7, 8 and 9) [16].

The next set of operations were performed on a dataset of COVID-19, where the total cases, recoveries and deaths were delineated across the top ten worst-affected

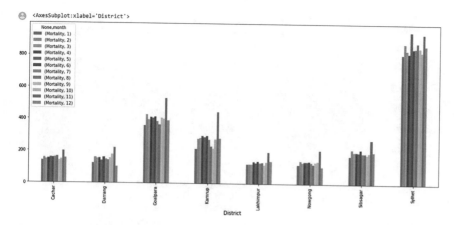

Fig. 4 Plot for month wise mortality for certain districts. Dataset: Mortality during the Spanish Flu

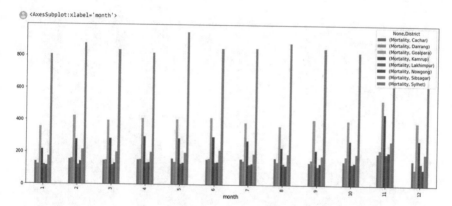

Fig. 5 Plot for month wise mortality for certain districts. Dataset: Mortality during the Spanish Flu

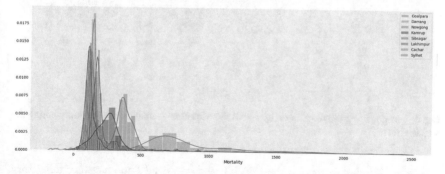

Fig. 6 Distribution of mortality across districts. Dataset: mortality during the Spanish Flu

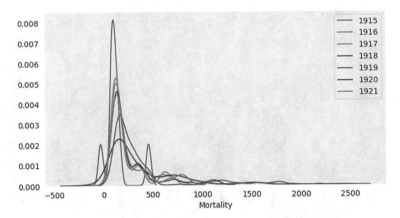

Fig. 7 Distribution of mortality across years. Dataset: mortality during the Spanish Flu

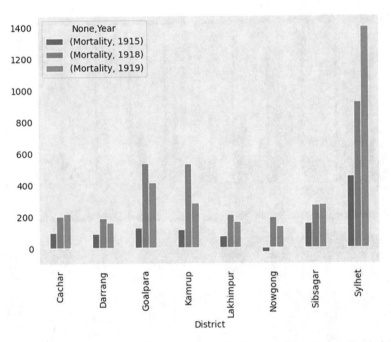

Fig. 8 Bar graph representing mortality across three specific years. Dataset: mortality during the Spanish Flu

countries. The dataset used is the COVID-19 cases, deaths and recovery repository from Worldometer. The dataset itself has been captured from a number of sources and also doubles up as a leaderboard to provide an updated count of the cases, recoveries and other information about COVID-19. The list of countries present in the dataset

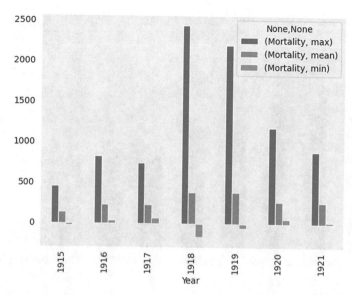

Fig. 9 Bar graph representing the max, min and mean mortality from 1915 to 1921. Dataset: mortality during the Spanish Flu

and territorial boundaries including continental regional classification has been done in accordance to the United Nations Geoscheme [17] (Figs. 10 and 11).

Once all the necessary data is accumulated, converged and processed, it has to be converted to an applicative software solution which contains multidimensional functionalities to support the operational performance, clinical outcomes and optimizing overall efficiency of healthcare as a service. A lot of operational and business factors need to be taken into account for developing a solution which gets a wide range of insights such as hospital management, costs, diagnosis and patient records. The majority of the healthcare big data has to be scrutinized and handled with utmost care in the most discreet way which would respect the patient's privacy and shield the sensitive information. During the case of a pandemic, all of these actions have to be carried out in a very short period of time, and hence, a proper business suite is required for the healthcare analytics solution to fight a pandemic. There are a number of ways in which a business intelligence software can transform the way a healthcare analytics solution is developed and used. These difference methodologies include financial planning, evaluating performance, taking care of a patient's satisfaction, coordinating communication, managing reputation, predicting the future, data visualization supporting and improving decision-making and so on. The solution should be able to manage the healthcare outcome in such a fashion that it get optimized to achieve the optimum use of analytics (Fig. 12) [15].

The above diagram explicates the optimum use of analytics as the amalgamation of the clinical outcomes and operational outcome. Unless these two sectors function together as an entity, the optimized use of analytics cannot be achieved. The first facet is the clinical outcome and there are a number of factors which have to be taken into

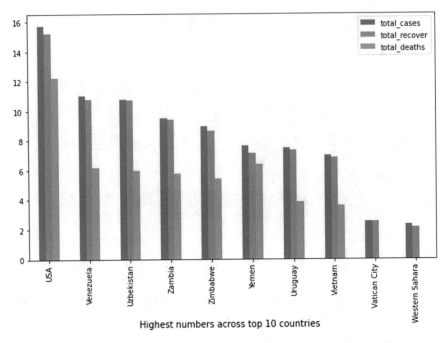

Fig. 10 Bar graph representing total cases, recoveries and deaths. Dataset: COVID-19 Worldometer

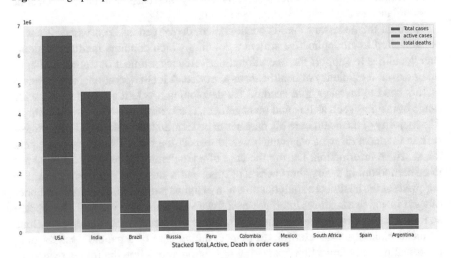

Fig. 11 Bar graph representing stacked total of cases, recoveries and deaths. Dataset: COVID-19 Worldometer

Fig. 12 Optimum use of analytics

consideration. The first facet is the diagnostic assistance. Analytical solution should be able to help into providing a diagnostic assistance for any aliment or symptoms which generally appear during an outbreak of a size of epidemic or pandemic. Second comes the clinical treatment effectiveness which can be monitored using this solution. The solution can also provide a grading mechanism using which the effectiveness of various clinical treatment can be evaluated and compared. The next factor on the side of clinical outcomes is the critical care intervention. The optimized use of analytics would help in predicting the requirement for a critical care among innumerable cases. The use of an intelligent analytics solution can also help in the research for improved disease management. The operational outcomes also lie on a number of factors which need to be taken care of before declaring the analytics solution as intelligent or optimized. The analytical solution can help in the prevention of readmission of cases. In case of a pandemic, the worst hit areas get multiple waves and a lot of readmissions which need to be mitigated. Next is the claims management, which is something that analytical solutions have been taking care of since a long time in term of providing multiple DRG, Medicare or Medicaid solutions and so on. The solution should also be able to detect the cases of frauds and incorrect entries. The solution can also contain a feedback mechanism to be the voice of patients which can provide trends on which hospitals or clinic is performing better than the rest. At last, the solution should look over the patient discharge and follow-up care [14, 18].

6 Illustration

In order to achieve the business transformation as mentioned in the previous section, a business solution must adopt an end-to-end approach toward analytics and not just follow one aspect. This end-to-end approach can be achieved by incorporating four specific layers which are business context and planning, analytics modeling, data layer, technology layer. The below diagram shows sample architectural diagram of a

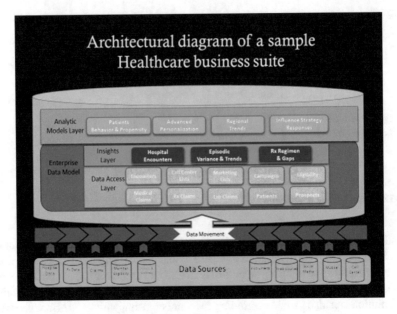

Fig. 13 Architectural diagram of a sample healthcare business suite

healthcare business suite which contains the required layer for fulfilling the business transformation [19] (Fig. 13).

The raw data as we see in the diagram comes from various sources such as hospitals, prescriptions, claims, member eligibility repository, disease and wellness repository, medical instrument records, web sources, social media, mobile, call centers and so on. In the diagram, the medical prescriptions have been denoted by Rx claims. The symbol "Rx" found its origin form a Latin word "recipe" which meant "to take." The Rx claims data is mostly formed by the customary part of a superscription which is the heading of a prescription. Next is the enterprise data model layer which has been further segregated into two parts—data access layer and insights layer. The enterprise data model layer contains an amalgamated view of both the data provided and used across the organization. This layer is responsible for incorporating the business standards, SOPs and an appropriate perspective of the healthcare industry. It also responsible for representing an unbiased view of a single integrated definition of data. This layer formulizes and brings together all the entities which are important to the sector or organization and the rules which govern them. This layer does not taken into account how a single unit of data or a chunk of data is physically stored, fetched, accessed or processed. The framework of these layers revolves around integration. It facilitates the identification of data within and outside the organizational boundaries which are sharable and the ones which are redundant. While looking upon all the layers in tandem, this layer can be label as the starting point of all system designs. It is more of a blueprint which contains maps of all the processes and provide a whole visualization of planning, development and implementation of the analytical solution. In the above sample architectural diagram, it is evident that all the raw data such as encounters, call center list, marketing list, campaigns, eligibility, medical claims, Rx or prescription claims, lab claims, patients and prospects have been clubbed together in the data access layer. While the more processed information such as hospital encounters as a whole entity, episodic variance and trends, Rx regimen and gaps have been clubbed together in the insights layer. The next layer is the analytic model layer which is a key element and is indispensable in comprehending the business data. This layer helps in making precise data-based predictions, and to extract the high-end insights which in turn, helps in making the correct business decisions. In a scenario where an analytical business solution is responsible for combating a pandemic, all of its internal layers should work in sync with each other to provide an optimized solution [14, 18, 19].

7 Conclusion

The methodical use of healthcare analytics has the potential to prepare mankind for the next pandemic. It also has the capacity to fight the current pandemic and become an effective tool of both defense and resurrection. However, the crux of the usage of this approach depends on a number of factors. It is essential for all those factors to be aligned in the correct form for this proposition to provide optimum results. To start with, organization and government policies should be favoring the approach of healthcare analytics. There should be adequate laws to facilitate and envisage the surfeit funding for the health information systems. Next, the raw data should be precise, voluminous and diversified to take in account a majority of the world population facing the pandemic. The gigantic amount of data has to be converged to be used as an input methodology for any analytical business suite. The data would have to be processed in accordance with the needs delineated in the enterprise data model of the solution. At last, a model solution keeps all the facets in sync which are essential in providing the optimum result. If the previous mentioned steps are carried out correctly, then the healthcare analytics software solution can fulfill applications such as diagnosis, preventive medicine, precision medicine, medical research, reduction of adverse medication events, cost reduction, population health monitoring and so on and so forth. The diagnosis process can be achieved by employing the means of data mining and analysis to identify the cause of illness. Significant success can be achieved in the course of discovering or detecting a preventive medicine for pandemics by using predict analytics and data analysis of lifestyle, genetic and social situations and circumstances to avert an outbreak. The aggregate data can be leveraged to drive a hyper-personalized care and provides precision medicine to masses. Healthcare analytics can bolster the medical research by employing a methodology of data-driven medical and pharmacological research to cure diseases and unearth novel treatments and medicines. The healthcare analytics can also contribute in the reduction of adverse medication events as the present of big data repositories can be channelized to find medication erratum and highlight potential adverse reactions. Identification of courses, actions and methodology which has the potential to be a cost effect optimized solution can drive toward long-term savings.

The statement that big data is transforming the healthcare industry is indisputable to the extent that the statement itself is almost considered canonical. But, similar to any other technology, healthcare analytics has also got its fair share of limitations. The first limitations comes in the form of weighted average approach which a lot of data models follow. Now, there are a number of sectors such insurance sector which rely on the actuarial models for risk-management. The mechanism used in the healthcare analytics system can only refine actuarial models till a certain juncture. Since a sizable amount of healthcare and clinical data is unstructured, we a get a resulting dataset which is not normally distributed, and hence, the weighted averages concept cannot be used in the resulting models. Instead, one needs to find datasets for different subgroups and in order to look at a minuscule level. It is a quite regrettable fact that till today, a number of big data extraction tools are not adept enough to analyze data on such a granular level which makes this scenario a limitation of healthcare analytics. The second limitation comes while predicting or accessing a doctor's performance using the patient data. A lot of doctors refer a high-risk or vulnerable patient to their colleagues to improve their track record. This practice hampers the accuracy of assessing a doctor's performance since in this case, the records have been compromised by the human action. This is also a limitation of healthcare analytics as we do not have a riveted mechanism to mitigate or appease this action [12, 14].

Despite all these limitations, the healthcare analytics domain has got the potential to monitor the health of masses to identify disease trends and can provided enhanced health strategies based on demographics, geography and socio-economics. Hence, healthcare analytics is indeed an advent to counter the imminent threats of outbreaks such as epidemics and pandemics, provided it gets the necessary path of action to fulfill its grail.

References

1. 20 of the worst epidemics and pandemics in history. Retrieved from: https://www.livescience. com/worst-epidemics-and-pandemics-in-history.html.
2. Gori, D., E. Boetto, and M. P. Fantini. 2020. The early scientific literature response to the novel Coronavirus outbreak: who published what? https://doi.org/10.1101/2020.03.25.20043315v1
3. Kagan, D., J. Moran-Gilad, and M. Fire. 2020. Scientometric trends for coronaviruses and other emerging viral infections. https://doi.org/10.1101/2020.03.17.995795v2
4. WHO. 2020. WHO Timeline-COVID-19. https://www.who.int/news-room/detail/27-04-2020-who-timeline—covid-19.
5. Centers for Disease Control and Prevention (CDC). 2009. Swine influenza A (H1N1) infection in two children–Southern California, March-April 2009. *Morbidity and Mortality Weekly Report* 58 (15): 400–402.
6. Centers for Disease Control and Prevention (CDC). 2010. The 2009 H1N1 Pandemic: Summary Highlights, April 2009–April 2010. https://www.cdc.gov/h1n1flu/cdcresponse.htm .

7. Andreasen, V., C. Viboud, and L. Simonsen. 2008. Epidemiologic characterization of the 1918 influenza pandemic summer wave in Copenhagen: Implications for pandemic control strategies. *Journal of Infectious Diseases* 197: 270–278.
8. Griffin, J. 2017. The History of Medicine and Organized Healthcare in America. Retrieved from: https://www.griffinbenefits.com/blog/history-of-healthcare.
9. Burnham, J. C. 2016. Health Care in America: A History. Journal of American History, Vol. 103, Issue 1, June 2016, pp. 167–168. https://doi.org/10.1093/jahist/jaw175.
10. The Pros and Cons of Obamacare. Retrieved from: https://www.healthline.com/health/consumer-healthcare-guide/pros-and-cons-obamacare.
11. Oberlander, J. 2013. The Future of Obamacare. Obstetrical & Gynecological Survey: April 2013, Vol. 68, Issue 4, pp. 265–266.
12. Big Data: The Management Revolution. 2012. Harvard Business Review. Retrieved from: https://wiki.uib.no/info310/images/4/4c/McAfeeBrynjolfsson2012-BigData-TheManagementRevolution-HBR.pdf.
13. Convergence of Health Care Trends. Deloitte US. Retried From: https://www2.deloitte.com/us/en/pages/life-sciences-and-health-care/articles/convergence-health-care-trends.html.
14. An introduction to seaborn. Retried from: https://seaborn.pydata.org/introduction.html.
15. Koh, H.C., G. Tan. 2011. Data mining applications in healthcare. Journal Healthcare Information Management pp. 65–73.
16. Chandra, Siddharth. 2015. Weekly mortality data for Assam, India from 1916 to 1921, Harvard Dataverse, Retrieved from: https://doi.org/10.7910/DVN/29045.
17. COVID-19 Coronavirus Pandemic. Retried from: https://www.worldometers.info/coronavirus/.
18. Janke, A.T., D. L. Overbeek, K. E. Kocher, P. D. Levy. 2016. Exploring the potential of predictive analytics and big data in emergency care. Ann. Emerg. Med. pp. 227–236.
19. Andreu-Perez, J., C.C Poon, R. D. Merrifield, S.T. Wong, G.Z. Yang. 2015. Big data for health. *IEEE Journal Biomedical Health Information*, pp. 1193–1208.

Image Classification for Binary Classes Using Deep Convolutional Neural Network: An Experimental Study

Biswajit Jena, Amiya Kumar Dash, Gopal Krishna Nayak,
Puspanjali Mohapatra, and Sanjay Saxena

Abstract Convolutional neural networks (CNNs) have proved itself a well-built model for image recognition in these modern computing days. Inclined by CNN's successes, we present an elaborative experimental assessment of CNN on image classification using a newly fabricated dataset of high-resolution images belonging to two different classes. The dataset partitioned into two distinct categories of high-resolution images of cats and dogs. This chapter presents an extensive experimental study of training size on training and validation accuracy and loss. We designed a fine-tuned predictive two-class image classification model for a large training size, which achieved a training accuracy of 100%, with validation accuracy close to 99.13%.

Keywords Convolutional neural network (CNNs or ConvNet) · Deep learning · Image classification · Binary classification

1 Introduction

In the present day, images are becoming an essential part of our daily life, so that we should keep our technology alerts to express the language of images. Recently, image recognition, image segmentation, detection, retrieval, and understanding image contents [1–3] are the issues addressed by the convolutional neural network with state-of-the-art results. For interpreting visual imagery, deep learning frameworks of CNNs are the best option, and the models are truly induced by how we, human beings, see the thing nearby us. The analogous between the CNN model and our brain function in many pretty similar ways. When our eye catches an object, and then the brain starts identifying that, so in a similar fashion, the CNN model recognizes and classifies the objects. The convolutional neural networks guides how the three-dimensional convolutional neural network clones the simple and complex cells of the human brain, including the receptive fields that human beings feel through their senses.

B. Jena · A. K. Dash (✉) · G. K. Nayak · P. Mohapatra · S. Saxena
International Institute of Information Technology, Bhubaneswar, India
e-mail: dash.amiya.k@gmail.com

© The Author(s), under exclusive license to Springer Nature Singapore Pte Ltd. 2021   197
S. Rautaray et al. (eds.), *Trends of Data Science and Applications*,
Studies in Computational Intelligence 954,
https://doi.org/10.1007/978-981-33-6815-6_10

The success behind the results of deep learning models is factors like scaling up the networks to billions of parameters, features, and a huge labelled dataset that can enhance the training activity. Beneath these scenarios, CNNs have been considered to learn complex and interpretable image features [4]. Uplifted by the optimistic outcomes in the domain of images, we re-examine the achievement of CNNs in high-resolution image classification. The networks have approaches to not only the appearance information present in single, static images but also their complex temporal evolution. There are various confrontations and questions to augment and pertaining CNNs in this context [5].

With the growing demands of deep learning, industries nowadays are usually on the lookout for a convolutional neural network guide, which is especially focused on the applications of CNNs to enrich the lives of people. Some broad areas of applications of CNNs like computer vision and natural language processing, which we can see in everyday life, are obvious choices, like facial recognition, image classification, scene labelling, action recognition, document analysis, speech recognition, text classification, etc. [6].

The remaining sections of this chapter are organized as follows. Section 2 discussed the coveted dataset used in this work. Then, various related literature to proceed with the work further are covered in section 3. The architecture and methodology used in this work are given in Section 4. Finally, Section 5 marks the conclusion.

2 The Dataset

Kaggle's Dogs versus Cats is a dataset of 25 thousand labelled high-resolution images belonging to mainly two categories. The dataset contains images of dogs and cats provided as a subgroup of images from a considerably bigger dataset of 3 million manually annotated photos. This huge benchmark dataset was created and modelled as an association between Microsoft and Petfinder.com. It is a binary classifier image dataset classifying dogs versus cats. The original dataset contains 20,000 training images with 10,000 per class and 5,000 validation images with 2,500 per class.

Our newly fabricated dataset is a subset of the original data set comprising 3,000 high-resolution images from both the classes. These 3,000 images used for training our CNN model are to find the effect of training size on training and validation accuracy and loss. Here, 2,000 training images with 1,000 images per each training class used for this experiment. There are also 5,00 images per each class as a validation image.

Here, we use a subset of the full dataset to decrease training time for educational purposes and work with limited resources. A note that we catch often is that "deep learning is exclusively appropriate when we have a large volume of data". While it is not completely unacceptable, this is relatively giving the wrong impression. Undeniably, deep learning essentially has the potential to learn features instinctively from the data, which is commonly only possible when a large amount of training data is there, particularly for issues where the input data are very high dimensional, like images [7].

Now there are few lines clarifying training, validation, and testing dataset. The validation dataset imparts an impartial assessment of a model properly suitable on the training dataset while tuning the model's hyper-parameters (e.g. the total of hidden units in a neural network). So, the validation dataset is a subset of the training data set. But, the data we choose for validation should be different from training data, even if they belong to the same set of data. Eventually, the test dataset is a dataset used to give an impartial assessment of the final model fit on the training dataset.

3   Literature Review

Artificial intelligence and machine learning methodology are current trends for object recognition, especially image processing and machine vision. The performance in this aspect can be improved by using a larger dataset, training more robust models, and the use of efficient ways to abstain from overfitting. Recently, datasets of labelled images were comparatively smaller on an approximation of thousands of images for educational learning purposes (e.g. Kaggle's Cat vs. Dog [8], and CIFAR-10/100 [9]).

The related work in [3], of ten class image classification on Thai food image, which got an average accuracy on a separate test set, can be predicted at 88.33%. Many people have worked or are working on constructing machine learning classifiers to address this image classification problem. In [5], work on large-scale video classification with a convolutional neural network gives a research path in this direction. In [11], a classifier based on colour features got 56.9% accuracy on the Asirra dataset. In [12], 82.7% accuracy was achieved from an SVM classifier based on a combination of colour and texture features again on the same Asirra dataset. The research carried in [17] discusses various machines learning framework for image classification using benchmark dataset Caltech 101 on colour feature. The machine learning for image classification approach is not only confined to general image but it is well implemented on medical image for better diagnosis and early detection of disease. The work in [18] classifies medical image to detect age-related molecular generation in fundus image using support vector machine method as a machine learning technique.

With the advancement of artificial intelligence in neural network and trending of deep learning models for image classification and object detection, the machine learning techniques of image classification lag in several aspects to the deep learning models. The work in [13] visualizes deep models by finding an input image that maximizes the neuron activity of interest by carrying out optimization using gradient ascent in the image space. The method was used to visualize the hidden feature layers of unsupervised deep architectures. Recently, the problems of ConvNet visualization were also addressed in [14]. Work on the segmentation of neural membranes of electron microscopy images using deep neural networks in [15] uses almost the same model of CNNs as in our case. The authors in [19] cover the feasibility of deep learning models for binary classification problems. In this research paper, authors explain the capability of deep learning model for classification high-dimensional data such as voice, image, video, and natural language. Also cover various issues related to deep leaning model like training the model, data augmentation, dropout, learning rate of optimiser, and various parameters to calculate accuracy of models.

4 Architecture, Methodology, and Results

(A) Architecture of CNN

There are no silver bullets in terms of CNN architecture modelling. The ultimate method to discover a model that is appropriate for a specific type is to begin with, some fundamental design and iteratively upgrade it over time and as per the requirements. We can train, validate, and test the same dataset over different models of CNN, which will be created by improving and incrementing the number of layers and hyper-parameters in it.

A convolutional neural network(CNN) is basically a neural network(NN) that involves the input and output layer, as well as several hidden layers. Apart from that, it consists of convolution, ReLU, polling, and fully connected layers as parts of CNN layers. Hidden layers are kept in between a fully connected layer and the output layer [10].

An input layer is a matrix of pixel values in the shape of width, height, channels. The utmost intention of the convolution layer is to obtain the feature maps. Usually, we begin with a few numbers of filters for low-level features detection. Gradually, we go deep inside the CNN, then more number of filters (usually they are also smaller) are required to detect high-level features. The convolution procedure requires sliding a filter window through the image to calculate the sum of the product under the window and the underlying image before down sampling [10].

ReLU layers work as an activation function. We will use the ReLU activation function that gives 0 for each negative pixel value in the input image while the value remains the same for each positive value of the pixel. So the shape of the matrix will remain unchanged.

The pooling layer's job is to impart spatial variance, which clearly measures that the system will be competent to acknowledge an object as an object truly when its appearance differs in some aspects. The pooling layer will accomplish down sampling functions along the spatial dimensions (width, height), giving the output over a pooling size. So, this layer is used to reduce the image dimension size.

In the fully connected layer, we flatten the last convolution layer's output and connect each node of the current layer with every other node of the subsequent layer. Neurons in a fully connected layer have full connections to all activations in the

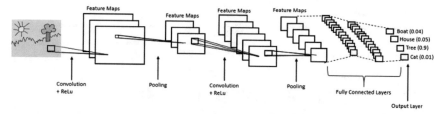

Fig. 1 Architectural design of CNN model. *Image source* medium.com

previous layer, as seen in regular neural networks, and it operates analogously. The probability scores are computed in the last layer of our CNN.

The details of the architectural model discussed so far are clearly depicted in Fig. 1 with various layers of deep learning models and input and output layers.

(B) **Methodology**

The guideline of our methodology for the classification of image for binary classes includes: input image, pre-processing, deep learning model, and result and classification [3].

The images that will go into our Convnet are 150 × 150 high-resolution colour images of dogs and cats. Our input feature map is 150 × 150 × 3, where 150 × 150 for the image pixels corresponds to the width and height of the image, and three for the three colour channels R, G, and B.

It is very uncommon to feed raw pixels into a Convnet. Hence, data pre-processing [16] is a very integral part, before feeding images into the Convnet. An image that goes into neural networks should usually be normalized in some way to make it more amenable to processing by the network. In this case, the pre-processing keeps the dataset image ready by minimizing the size from any high-resolution image into 150 × 150 pixels. This enhanced the proscessing hour while maintaining image details.

To build up the architecture for our methodology, we have a stack of three modules as convolution, relu, and maxpooling layers. Our convolutions operate on 3 × 3 windows, and our maxpooling layers operate on 2 × 2 windows. Our first convolution extracts 16 filters, the following one extracts 32 filters, and the last one extracts 64 filters. This is a configuration that is widely used and known to work well for image classification. Also, since we have relatively few training examples (2,000), using just three convolutional modules keeps the model small, which lowers the risk of overfitting. This configuration of the model follows the well-known LeNet [4] and AlexNet [5] models of CNNs. This is very similar to the architectures that Yann LeCun [4] advocated in the 1990s for image classification. Figure 2 represents the detailed summary of our model.

On top of our model, we stick two fully connected layers because we are facing a two-class classification problem, i.e. a *binary classification problem*; we will end our network with a sigmoid activation, so that the output of our network will be a single scalar between 0 and 1, encoding the probability that the current image is class 1, as

```
Model: "model"
```

Layer (type)	Output Shape	Param #
input_1 (InputLayer)	[(None, 150, 150, 3)]	0
conv2d (Conv2D)	(None, 148, 148, 16)	448
max_pooling2d (MaxPooling2D)	(None, 74, 74, 16)	0
conv2d_1 (Conv2D)	(None, 72, 72, 32)	4640
max_pooling2d_1 (MaxPooling2	(None, 36, 36, 32)	0
conv2d_2 (Conv2D)	(None, 34, 34, 64)	18496
max_pooling2d_2 (MaxPooling2	(None, 17, 17, 64)	0
flatten (Flatten)	(None, 18496)	0
dense (Dense)	(None, 512)	9470464
dense_1 (Dense)	(None, 1)	513

```
Total params: 9,494,561
Trainable params: 9,494,561
Non-trainable params: 0
```

Fig. 2 Model summary of our CNN model with different feature map in each layer

opposed to class 0. With these layers in our model, it uses 9,494,561 parameters, and all the parameters are trainable parameters as shown in Fig. 2 of model summery.

Next, we will configure the specifications for the training of our model. We will train our model with the *binary cross-entropy loss* because it is a binary classification problem, and our final activation is a *sigmoid*. We will use the *rmsprop* optimizer with a learning rate of 0.001. During the training and validation process, we want to monitor classification accuracy to measure the performance of the model. Now, we will define the performance metrics of the model, such as accuracy and binary cross-entropy loss as given bellow.

Accuracy: Accuracy is one metric for evaluating classification models. Informally, accuracy is the fraction of predictions our model got right. In easier terms, it means how much the resulting data is closer to the original data [20, 21].

$$Accuracy = Number\ of\ correct\ prediction\ /\ Total\ number\ of\ predictions$$

For binary classification, accuracy can also be calculated in terms of positives and negatives as follows:

$$Accuracy = (TP + TN) / (TP + TN + FP + FN) \tag{1}$$

where TP = True Positives, TN = True Negatives, FP = False Positives, and FN = False Negatives.

Binary Cross-Entropy: It is also known as log loss. It measures the performance of a classification model. Its value varies between 0 and 1. Cross-entropy loss increases when the predicted probability of the dataset starts to vary from the actual label of the dataset. The cross-entropy can be measured as [20, 21]:

$$-\sum_{c=1}^{M} y_{o,c}\log(P_{0,c}) \tag{2}$$

where

M = number of classes (i.e. 2), log = natural log.

y = binary indicator (values varies from 0 to1) if class label c is the correct classification for observation o.

p = predicted probability observation o is of class c.

(C) Result and Analysis

Let us train all 2,000 training images and validate on 1,000 test images for 20 and 100 epochs respectively. For both training and validation scenario, we consider batch size small as 10, i.e. 10 number of the sample (image) will be in a batch for training, and we have considered 20 epochs and 100 epochs to find out how the training and validation accuracy and losses changes with the epochs. Our training accuracy gets close to 100%, while our validation accuracy stalls near to 72% in both 20 and 100 cases of epochs. Our validation loss reaches its minimum after only five epochs.

Similarly, again we train our model with the whole dataset of 25,000 images, in which 20,000 are training images and 5,000 test images for 20 and 100 epochs respectively to know the effect of large training data on the training and validation accuracy and losses. We observed that our training accuracy gets close to 100%, while our validation accuracy stalls near to 99% in cases of 100 epochs.

To find the type of features that are learned by our convolutional network, we generate a figure for a random cat and dog image selected from the training sample as shown in Figs. 3 and 4. Each row in these images is the output of a layer in our CNN architecture and every single image in the row is a specific filter in that output feature map.

Another factor for consideration is the computation time, while GPU, TPU, and CPU execute our deep learning model. It takes on an average of 5.3 min while running under GPU for 20 epochs and nearly 13 min and 14.5 min under TPU and CPU, respectively, in case of newly fabricated dataset. The time gap between the execution times differs because of the architecture of the processor. A GPU is a processor in its own right. A TPU is a coprocessor; it cannot execute code in its own right; all code execution occurs on the CPU, which just feeds a stream of micro-operations to the TPU.

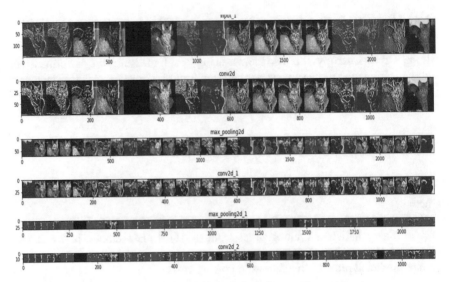

Fig. 3 Features learned by our CNN model in each layer for a random cat image

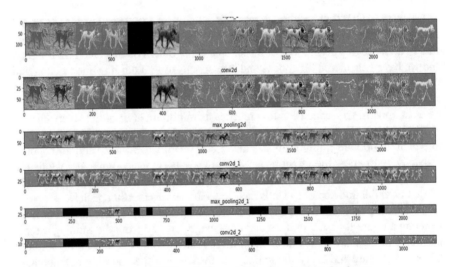

Fig. 4 Features learned by our CNN model in each layer for a random dog image

We plotted the training and validation accuracy and loss as collected during the training of the model with 3000 images, as shown in Figs. 5, 6, 7 and 8. Figures 5 and 6 represent the training and validation accuracy and loss of our model when run for 20 epochs. Similarly, Fig. 7 and 8 represent the training and validation accuracy and loss of our model when run for 100 epochs.

In Figs. 5 and 7, training and validation accuracy are represented by blue and red curve, respectively, where horizontal axis represents the number of epoch (for 20 and

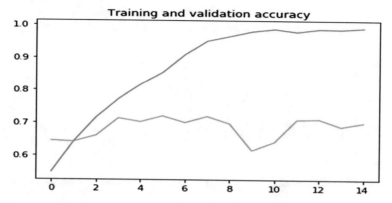

Fig. 5 Training and validation accuracy of our dataset with 3000 images for 20 epochs.

100 epochs) and vertical axis represents accuracy probabilities. Similarly, Figs. 6 and 9 represent both the training and validation loss in blue and red curve, respectively, where horizontal axis represents the number of epoch (for 20 and 100 epochs) and vertical axis represents loss probabilities.

We also plotted the training and validation accuracy and loss as collected during the training of the model with 25,000 images, as shown in Figs. 9, 10, 11 and 12. Figures 9 and 10 represent the training and validation accuracy and loss of our model when run for 20 epochs. Similarly, Figs. 11 and 12 represent the training and validation accuracy and loss of our model when run for 100 epochs.

From the above experiment, it has been observed that for a small training size of the data, the validation accuracy stalls at 72% even if we increased the epoch's number. From Fig. 5, 6, 7 and 8, it is cleared that our validation accuracy reached the minimum values after a few epochs, leading to overfitting the model. Overfitting occurred because our model was exposed to a small number of training examples. As

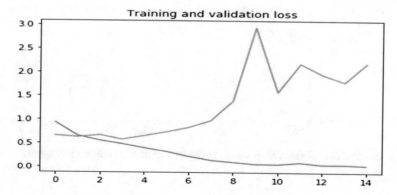

Fig. 6 Training and validation loss of our dataset with 3000 images for 20 epochs

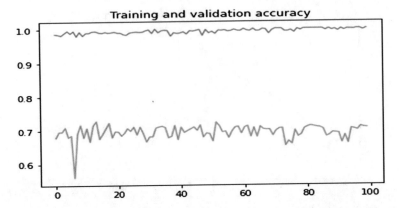

Fig. 7 Training and validation accuracy of our dataset with 3000 images for 100 epochs

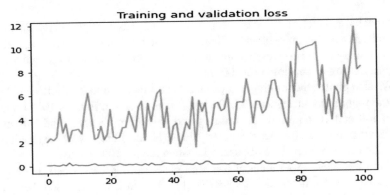

Fig. 8 Training and validation loss of our dataset with 3000 images for 100 epochs

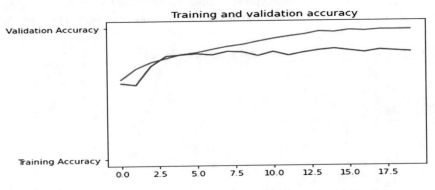

Fig. 9 Training and validation accuracy of our dataset with 25,000 images for 20 epochs

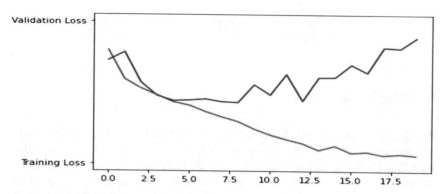

Fig. 10 Training and validation loss of our dataset with 25,000 images for 20 epochs

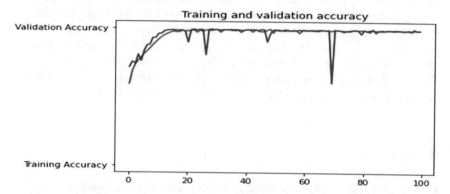

Fig. 11 Training and validation accuracy of our dataset with 25,000 images for 100 epochs

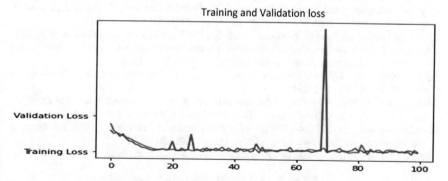

Fig. 12 Training and validation loss of our dataset with 25,000 images for 100 epochs

we increased the number of training samples, the training and validation accuracy increased even if for a small number of epochs, as shown in Fig. 9, 10, 11 and 12.

5 Conclusion

The extensive research goal related to image classification using deep learning models is stated here. Even if we train and test our deep learning network with fewer images, we still get 100% accuracy in training and 72% validation accuracy. Image classification using deep learning models requires an extensive training sample, as too few training samples learn patterns that do not generalize to new data. It leads to overfitting the model by learning insignificant features to make the prediction. Overfitting was a primary concern when we used a deep learning model for image classification. So, to improve the performance of the classifier, we need a large training sample for image classification. In our experiment, we achieved a validation accuracy of 99.13% for a large training sample. We also improve the state of the art on a plethora of standard image classification benchmarks. Further advanced research in this field may help to develop a mobile-based application for real-time image classification.

References

Krizhevsky, Alex, Ilya Sutskever, and Geoffrey E. Hinton. 2012. Imagenet classification with deep convolutional neural networks. *Advances in neural information processing systems.*

Yorozu, Y., M. Hirano, K. Oka, and Y. Tagawa. 1982. Electron spectroscopy studies on magneto-optical media and plastic substrate interface. *IEEE Translation Journal Magnetic* Japan, vol. 2, pp. 740–741, August 1987 Digests 9th Annual Conf. Magnetics Japan, p. 301.

Hnoohom, Narit, and Sumeth Yuenyong. 2018. Thai fast food image classification using deep learning. *2018 International ECTI Northern Section Conference on Electrical, Electronics, Computer and Telecommunications Engineering (ECTI-NCON). IEEE.*

LeCun, Yann, et al. 1998. Gradient-based learning applied to document recognition. *Proceedings of the IEEE* 86(11): 2278–2324.

Karpathy, Andrej, et al. 2014. Large-scale video classification with convolutional neural networks. *Proceedings of the IEEE conference on Computer Vision and Pattern Recognition*

Bhandare, Ashwin, et al. 2016. Applications of convolutional neural networks. *International Journal of Computer Science and Information Technologies* 7(5): 2206–2215.

Chollet, Francois. 2016 .Building powerful image classification models using very little data. *Keras Blog.*

Liu, Bang, Yan Liu, and Kai Zhou. 2014. *Image classification for dogs and cats.*

Krizhevsky, Alex, and Geoffrey Hinton. 2009. *Learning multiple layers of features from tiny images.* Vol. 1. No. 4. Technical report, University of Toronto.

Zeiler, Matthew D., and Rob Fergus. 2014. Visualizing and understanding convolutional networks. *European conference on computer vision* Springer, Cham.

Elson, Jeremy, et al. 2017. *Asirra: a CAPTCHA that exploits interest-aligned manual image categorization.*

Golle, Philippe. 2008. Machine learning attacks against the Asirra CAPTCHA. *Proceedings of the 15th ACM conference on Computer and communications security.* ACM.

Erhan, Dumitru, et al. 2009. Visualizing higher-layer features of a deep network. *University of Montreal* 1341(3):1.

Zeiler, M. D., and R. Fergus. 2013. Visualizing and understanding convolutional networks. CoRR, abs/1311.2901. arXiv preprint arXiv:1311.2901.

Ciresan, Dan, et al. 2012. Deep neural networks segment neuronal membranes in electron microscopy images. *Advances in neural information processing systems.*

Pal, Kuntal Kumar, and K. S. Sudeep. 2016.Preprocessing for image classification by convolutional neural networks. *2016 IEEE International Conference on Recent Trends in Electronics, Information & Communication Technology (RTEICT). IEEE.*

S. Loussaief and A. Abdelkrim. 2016. Machine learning framework for image classification. *7th International Conference on Sciences of Electronics, Technologies of Information and Telecommunications (SETIT)*, Hammamet, pp. 58-61. https://doi.org/10.1109/SETIT.2016.793 9841.

García-Floriano, Andrés, et al. 2019. A machine learning approach to medical image classification: Detecting age-related macular degeneration in fundus images. *Computers & Electrical Engineering* 75:218–229.

Kim, Kitae, Bomi Lee, and Jong Woo Kim. 2017. Feasibility of Deep Learning Algorithms for Binary Classification Problems. *Journal of intelligence and information systems* 23(1): 95–108.

Zou, Kelly H., et al. 2004. Statistical validation of image segmentation quality based on a spatial overlap index1: scientific reports. *Academic radiology* 11(2):178–189.

Tran, Giang Son, et al. 2019. Improving accuracy of lung nodule classification using deep learning with focal loss. *Journal of Healthcare Engineering.*

Leveraging Analytics for Supply Chain Optimization in Freight Industry

Kashyap Barua, Parikshit Barua, and Sandeep Agarwal

Abstract We live in a country whose logistics industry is slated to be worth $160 billion. Several start-ups have emerged in India trying to crack this globally yet unsolved problem statement. Moreover, the freight scene, being age old, has its segment of bottlenecks to deal with which are inclusive but not limited to problems like fragmentation, inflated costs, lack of visibility into lanes, limited digital capabilities and back hauling. Tech-driven start-ups provide solutions to overcome these challenges for the fleet owners in terms of optimized demand–supply matching, brokerage eliminations by connecting supply with relevant demand thereby reducing costs, expose truckers and fleet owners to unchartered lanes and also facilitate reverse loads to optimize costs for these businesses. This is where companies are leveraging data science and analytics to tackle these issues and help the businesses grow. Companies like Uber Freight, BlackBuck and Rivigo are using the best of technologies to monetize this industry. Data when logged in the right manner can help industries understand the intricacies of issues and help them overcome the same. A typical example of implementation would be using a simple regression technique to predict demand in a specific region so that supply can be exposed well within time in order to avoid idling period by these truckers. Tracking key metrics like supply turn around time (TAT), truck in transit duration (TiT), placement index and others can help organizations determine and optimize on these metrics to maximize revenue. Being convoluted of a system, this industry has been a tough nut to crack, especially in a country like India. This chapter discusses how companies set up the entire data platform and infrastructure, thereby facilitating the usage of data for advanced analytics techniques to solve some crucial supply chain problems in the freight industry. The

K. Barua (✉)
MiQ Digital India, Bengaluru, India
e-mail: kashyapbarua@gmail.com

P. Barua
KIIT University, Bhubaneswar, India
e-mail: parikshitb04@gmail.com

S. Agarwal
Axtria Inc, Bengaluru, India
e-mail: sandygarg65@gmail.com

© The Author(s), under exclusive license to Springer Nature Singapore Pte Ltd. 2021
S. Rautaray et al. (eds.), *Trends of Data Science and Applications*,
Studies in Computational Intelligence 954,
https://doi.org/10.1007/978-981-33-6815-6_11

chapter also talks about some of the use cases for analytics and machine learning to solve problems related to the freight industry. Firstly, we demonstrate some visualizations and representations as to how insights are drawn through analytics to solve these kinds of problems. We follow this up by discussing how data infrastructures are set up in organizations to collect freight data and then finally we showcase some ML techniques that are used in the freight sector of businesses. This in turn would help users understand the nuances of decision science and analytics with its capabilities in scaling businesses.

Keywords Freight industry · Analytics · Data science · Machine learning · Metrics · Supply chain

1   Introduction

Before we had a solid infrastructure to store and maintain data, the freight industry would be dependent on manual book entry in terms of data collection. Imagine keeping a record on a book about when a truck got booked, the truck arrived at a loading point, left the loading point and even fulfilled an order. This system of tracking is cumbersome, especially when you want to leverage analytics and data science techniques to upscale your business. With the advent of big data technologies and advanced data engineering techniques, the process of data collection and data audit has become even more simpler. Added to the above metrics/details, organizations have been collecting numerous data points to enhance their business output. They collect real-time GPS data about the trucks whereabouts, driver data through mobile applications suggesting each stop and at what location, user activity on the Load Board application and numerous other details. These organizations in turn provide shippers and carriers of every size, be it corporates or small and medium enterprises (SMEs), with actionable insights to build their supply chains even more efficiently than ever.

As shown in Fig. 1 [1], companies leverage data to track performance for their freight pickup and delivery performances. Here, it is observed that the overall freight ratings are the least in the northern region of India while on the contrary, the performances seem to be doing well in the Western region. The average ratings for pickups are 4.36 out of 5, and the same is 4.31 for the deliveries in the Western region encompassing states like Gujarat, Rajasthan, etc. Given that a vast majority of the trucking community come from the Western region of the country, it is evident that the same metrics would be well off for the same region. It is critical to track performance metrics across regions for any business case, to ensure smooth supply chain management.

From the analysis shown in Fig. 2, it is observed that FMC durables constitute almost 20% of all the products that have been moved in the month of June 2020. Other products that follow are paint, polyester, consumer durables, chemicals and oil. This is another instance wherein companies leverage data to keep track of

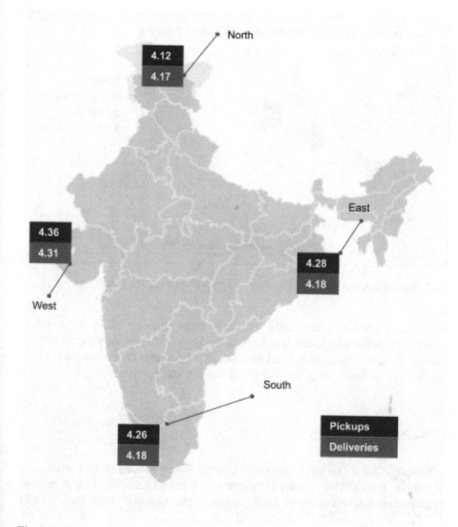

Fig. 1 Indian trucking report for pickup and delivery performance across regions

these nuances in their supply chain ecosystem. Demand and movement are about the type of product because the same type of trucks, be it half body, full body or standard containers, can be dispersed for the requirement with the clients. With the ever-increasing demands to ship products, the need to enhance this ecosystem will always increase. Hence, it becomes very crucial for the instrumentation of proper data warehousing infrastructures in organizations to draw insights and leverage data for optimized workflow.

Leveraging machine learning algorithms can help define a successful supply chain. For freight forwarders, implementation of these algorithms can be on a varied range of business fronts like prices, routes taken, volumes moved, customer requirements,

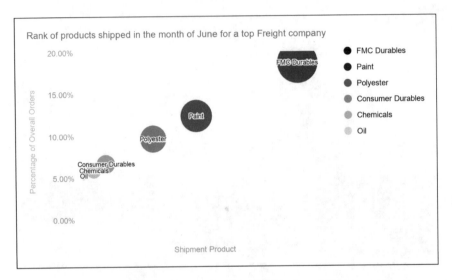

Fig. 2 Rank of products shipped in the month of June 2020 by a top freight company

etc. The data on these metrics can help optimize supply chains, yield business critical insights and processify operations. One of the key techniques used from machine learning is the technique of demand forecasting, which helps the demand side of the supply chain in anticipating demand across regions and businesses.

2 Literature Survey

In February 2020, Lóránt A.Tavasszy [2] emphasizes that the freight movement structure in the logistics industry is changing rapidly and the need to adapt to these developments has become even more important. He focuses on three basic scopes of the model improvement: the structural elements of the system that are modelled, the functional relations between these elements and the dynamic properties of these models. As per Lóránt, these innovation directions are independent of each other, and they help in reinforcing each other at various levels of the supply chain. At the end, he concludes that these areas need to be researched upon, to facilitate innovations at all levels which would facilitate in driving freight modelling research in the forward direction.

In October 2014, Paolo Ferrari [3], in his paper "The dynamics of modal split for freight transport", explained the dynamics of modal split in the transport system and the multimodality of the same. Also, the evolution of transport demand over the due course of time is also complimented by the evolutionary attributes of the modes of transport and the reaction of users to delay in cost variations.

In his chapter, Gregory Harris [4] focuses on improving the transportation and freight modelling by the implementation of a freight planning framework which leverages federal freight data and a variety of other tools and defines a model for statewide destination origin freight patterns, traffic models specific to freight movement and also system performance measures. His proposed FPF methodology considers data on freight flow or movements and structures the same into usable format for freight planning purposes at different levels. These types of model development and researches could be beneficial for the future supply chain industry.

Cruijssen Frans [5] discusses in his paper about the horizontal cooperation in the logistics and transportation sector. This method of horizontal cooperation has been an interesting topic of discussion from a theoretical standpoint of things, the reason being that it can be approached by personnel's coming from multiple disciplines like economics, operations and research. There is this collaboration between multiple businesses from different domains to ship products. A company like Flipkart or Amazon, who has their logistics department established can easily provide their services to SMEs and other businesses. This type of collaboration is quite evident in the country of India wherein these logistics dominators provide supply chain services to various other small-time businesses. This results in an ecosystem wherein each of these companies can benefit from the other in numerous ways. One of the reasons for the sustainability of this business model is because small businesses cannot start their own logistics services to transport goods, as oftentimes the system becomes complicated and expensive for them. This leads to the SMEs relying on the bigger players to help them ship their goods with sustained partnership and minimal costs, which benefits both the parties in their own ways.

3 Data Storage and Big Data Ecosystem

Leveraging analytics and data science techniques come with the responsibility of methodologies that an organization should adopt to collect and store data. This is an ever-increasing need, and companies have been on-boarding data engineers and data platform teams to upscale this aspect of technology. A data lake [6] is a centralized repository that allows one to store all their structured and unstructured data, at any scale. Any organization that is smart and proactively focuses on setting up these infrastructures will have a competitive advantage with their peers. These organizations with proper data lake implementations could do better predictions, customer segmentations and behaviour, understanding customer journey through clickstream in a much more efficient manner, thereby providing the competitive advantage in the business.

Companies have implemented data lakes wherein data from all sources and of all sizes are gathered in a single place. This facilitates processes such as reporting, visualizations, advanced analytics and machine learning to solve most business cases. A perfect example is a well-known tool called Qubole which is an open and secure data lake platform which enables machine learning, data streaming from different sources and advanced analytics. The advantages of implementing a data lake as a data single point of data storage unit can be numerous, ranging like

- Collect and obtain quality data
- Support for all format of data
- Schema flexibility
- Democratize data
- Advanced analytics and machine learning
- Scalability (Fig. 3).

To highlight some of the key features depicted in the diagram above are as follows:

- Machine Learning—Data lakes allow organizations to leverage historical data to draw extensive insights and perform forecasting techniques to determine likely outcomes, which depends on historical data. This in turn provides a range of prescriptive solutions to achieve the optimal outcomes
- Advanced Analytics—Data lake architecture allows various professionals like data scientists, data analysts, business analysts to collaborate and access the

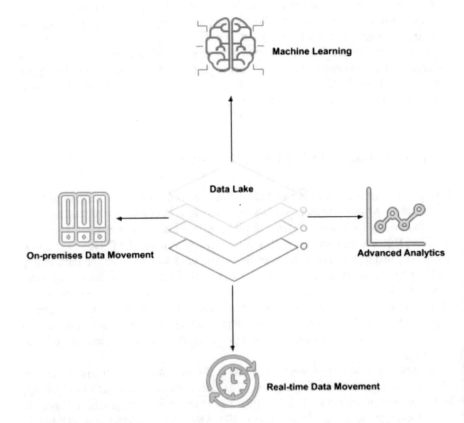

Fig. 3 Data lake key features

data in their choice of tools and framework to tackle a problem. Some of which includes frameworks such as Apache Hadoop, Presto, Apache Spark and other data warehousing tools.

- Real-time Data Movement—Data lakes facilitate the ability to stream/collect data in real time. This allows organizations in terms of saving time, scaling data to any size and saves the cumbersome task of defining schema, data structures and metadata information.
- On-Premises Data Movement—This allows to understand what data are present in the lake through methods of indexing, cataloguing and crawling.

4 Data Processing and Manipulation

One of the critical steps in a data science life cycle for any project would involve data processing [7] and manipulations (right after data collection and storage). This step involves shaping and cleaning the data as per business requirement or use case to solve problems.

The current trend for data capture and storage involves setting up data processing pipelines which pre-emptively cleans and shapes data even before ingesting them into storages. This enables analysts and data scientists to fast-track the analytics process as data cleaning/processing is a tedious task to perform (Fig. 4).

As per the diagram above, raw data can flow from multiple sources, be it real-time GPS data, survey data or salesforce data which is then pre-processed and streamed into the operational data storages. This is where the data quality assessment is done within the entire pipeline, and then the data are pushed to the data lakes. Organizations leverage the data lake for advanced analytics and insights to solve critical problems then.

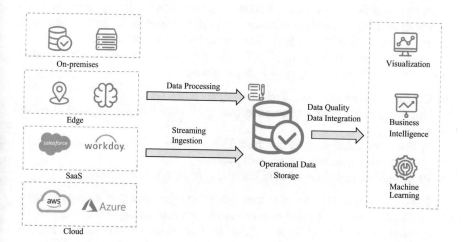

Fig. 4 Typical data processing framework

How can you create data processing pipeline architecture?

- Enterprise data preparation tools (EDPT) allow users to automatically convert data preparation steps into data pipelines that can be easily documented or modified. Traditional data preparation tools hinder this operationalized method of the data processing workflow.
- Traditional coding practices are using data processing technologies like Kafka, SQL, Spark and Pandas. These tools require understanding and expertise of the same so that they can be leveraged and used to streamline the data processing workflow

Data processing pipelines have been implemented for a lot of years now. Typical processes would be inclusive of reading data, transforming it in the required format and providing an output of a new data set for usage. Be it any use case from a wide array, a data processing pipeline must connect efficiently, collect effectively, integrate and prepare data as per requirement and deliver data at scale and also at the speed of the businesses at the same time.

5 Analytics and Insights

In a supply chain life cycle in the freight industry, a load is created when a demand wants to ship goods from point A to point B within x amount of time (where x can be any duration). Once the load is created, it gets exposed to the entire supply base or the fleet owners who own the trucks and are ready to provide trucks to ship the product. Once a supplier relates to the relevant demand, this is a successful matchmaking done. The supplier sends the truck to the loading point, and the products are loaded on the truck. In post loading, the truck is stated as truck in transit wherein the truck is moving from point A to point B. On reaching the unloading point, the products are delivered in the destination location, and the full and final settlement is done. This is how a supply chain looks like for an order in the freight businesses.

Now, these individual phases of the life cycle may be susceptible to a lot of issues and problems. Some of the typical issues might be

- Load expires even before the relevant supply could see it
- Document or truck-type mismatch resulting in order cancellations at the loading point
- Spillage or product damage discovered at unloading point leading to deductions
- Truck malfunction during the transit process
- Payment/fulfilment issues due to settlement document mismatches.

Hence, it becomes important to track each and every aspect of the supply chain life cycle, till the order is fulfilled to ensure smooth functioning of the same. This is where data capture comes into play. If data are tracked in each of these phases in the correct manner, advanced analytics methods and machine learning can be leveraged to solve these issues.

This is where companies use data to monitor for signals that are key to their businesses and solve for crucial problems.

6 Machine Learning Implementation

Once the data collection platforms and data analytics infrastructure are set up and ready to go, this is where organizations can leverage the rich data and implement and optimize supply–demand matchmaking, pricing and various other services KPIs [8]. Machine learning is a technique wherein data are used to understand the underlying pattern in the same. The algorithm learns the patterns from the data through iterations and helps in solving problems like forecasting, clustering, predictions and factor significance determination. Machine learning can broadly be classified into three categories, namely supervised, unsupervised and semi-supervised learning. In supervised learning, the input variables are known from the beginning wherein the algorithm is used to learn the mapping function between the input and the output variables. In unsupervised learning, we only have the input variables, and there is no understanding of the corresponding output variables. The goal of the algorithm is to understand the underlying structure and affinity of the data points with one another. Finally, semi-supervised are the types of problems with a large amount of input data out of which only some of the data are labelled. These algorithms/problems lie between supervised and unsupervised machine learning algorithms.

6.1 Demand–Supply Matchmaking

With the understanding of the regional requirements of fleet owners and customers in terms of supply and demand, the problems for deficit or surplus can be tackled for both parties. As shown in Fig. 6, the predicted demand has increased by the end of the year, whereas supply has remained constant even for the predicted duration. This seems to be a potential problem for the company as the suppliers remaining constant might lead to a deficit in supply base for the increasing demand. By understanding the predictions, organizations can ramp up on relevant requirements to avoid any form of issues. The models can account for unprecedented demand shocks and can help ramp up on requirements accordingly. Throughout the shipping life cycle, machine learning algorithms help organizations anticipate the complex dynamics of the system starting from tendering freight, matching trucks to load and hauling. This prediction of loads can be further split into other levels like truck-type mapping, regional demand–supply, cancellation affinity, trucker traits and many other factors (Fig. 5).

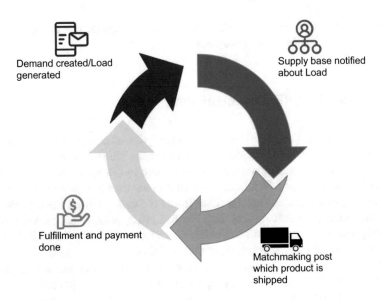

Fig. 5 Supply chain life cycle in freight

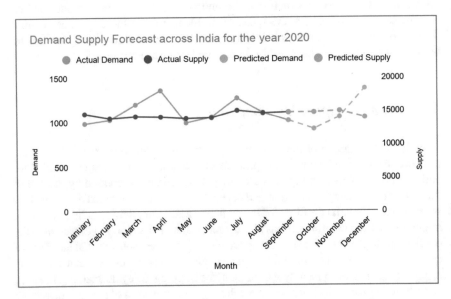

Fig. 6 Demand–supply predictions for fourth quarter of the year of trucking market

6.2 Pricing and Incentives

The freight market relies on the dynamic and complex nature of pricing to a big extent. Pricing especially in India is dependent on factors like market standard, seasonality, affinity for products movement and a lot of other factors [9]. Without the efficient pricing model under play, there would be an enormous amount of cancellations and inefficient matchmaking processes. A typical instance would be if there is a load available to be sent from point A to point B, and it has been priced at $x per ton. But it has been understood that the price changed due to some seasonality factor or other extrinsic factor (maybe due to brokers negotiating at a cheaper price for the same load). The current market price is evaluated at $(x-50) per ton, and this would result in inefficient matches made for that load. This is where efficient pricing models come into play, wherein these extrinsic roles are accounted for while listing the appropriate price for the load. An effective model could account the factors well while stating the price, and it could also account for the dynamic nature of pricing in real time, while monitoring the same.

As depicted in Fig. 7, brokers intervene in the entire shipping cycle in terms of pricing, and this results in a revenue loss for the freight company. Hence, it is quite important to fulfil the pricing expectations for truckers in the market which can easily be captured by broker networks, especially in a broker heavy market of a country like India. It becomes important that these metrics and factors are kept in check in real time to avoid such bottlenecks for the companies. With effective data storage capacity and machine learning models, these daunting issues can be catered to.

Other than that, there is the active competitiveness of having the best incentive programs for drivers, consumers or truckers in any sort of businesses, who want to give the best that they can to their employees. There are companies which rely on user behaviour and transaction data to roll out the best schemes possible and incentivize their efforts. This in turn leads to maximization of effort and results and leads to retention of these drivers/truckers in their businesses. A typical example would be that of incentivizing the cab drivers (Ola/Uber) who are paid some bonuses for completing milestones on a daily cadence. This data on driver completions and incentive earnings can help organizations understand how affine these drivers are to stick to their business and how well they are bound to perform in the long run.

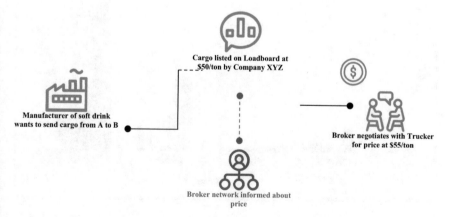

Fig. 7 Broker intervention for loads by gaming pricings

Added to that, these data sets could be fed into machine learning algorithms and the behaviour could be predicted, if their performances tend to go down. It is in the interest of the organizations that these drivers/truckers are incentivized well so that the businesses could be monetized to the maximum.

6.3 User Segmentations to Understand User Activities

Clustering is an unsupervised machine learning algorithm. In these types of problems, the user only has the input data, and the corresponding output variables are missing. The primary objective of the algorithm is to model the underlying structure of the data points in order to understand the data better. Unlike supervised machine learning algorithms, these algorithms do not rely on past data to understand the underlying pattern in the same. There is no correct answer as such, and hence, the algorithm is left on their own to discover the structure and underlying patterns in the data.

A lot of businesses use clustering techniques to understand their user base and their affinity towards the business. As shown in Fig. 8, a sample size of 5000 users were considered and various metrics like average number of trips, supply fleet size (or trucks owned) and average orders cancelled were used to define clusters of users. Cluster 1 seems to be the least engaging with the business while users belonging to Cluster 4 are the highly engaging users based on trips done, trucks owned and order cancelled. These overviews understanding of the user base can help roll out user-centric business plans, target users with custom notifications and other campaigns.

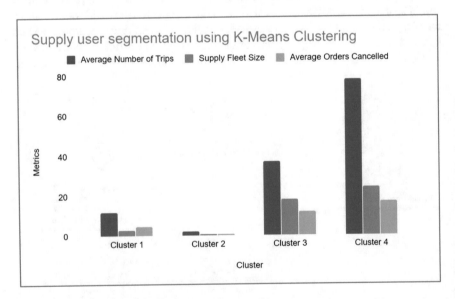

Fig. 8 Supply base segmentation based on user properties

K-means clustering was implemented for the activity with the number of clusters kept at 4. The number of clusters is identified using the elbow method for WSS calculation which provides the optimal number of clusters to be used while modelling.

Hence, clustering techniques can be very useful in a lot of industries ranging from freight to retail. Another instance of clustering implementation is in the retail sector wherein there is a need to understand customer buying patterns to be able to provide customer tailored offers and shelf placement optimization techniques. With proper clustering implementation, users can be segmented into different clusters like high value customers, returning customers, churned segments and highly active buckets. These buckets can then individually be targeted for marketing and other sorts of campaigns.

7 Comparative Study of Different Techniques

Even if we can categorize machine learning algorithms [10], the underlying assumptions and details that need to be considered while doing the same can be even more complex of a process. Below shown are some of the algorithms and their strengths and weaknesses pertaining to each of them. This section discusses which model to use and why you should use it, based on its pros and cons.

Algorithm	Strength	Weakness
Linear regression	Simple and the problem of overfitting can be avoided by regularization	Performs poorly when there are nonlinear relationships
Naive Bayes	They are easy to implement and can scale data over the time	Beaten by other models which are properly trained and tuned
K-means clustering	It is simple, fast and flexible if the data is pre-processed well, then significant features are engineered	The number of clusters must be determined before modelling, which is often not an easy task
Deep learning	Performs very well when classifying for text, audio and image data	It requires a large amount of data set to train due to which it is not considered as a general-purpose model
Classification trees	These algorithms are robust to outliers, and they can model nonlinear decision boundaries	Individual trees are very prone to overfitting
DBSCAN	This algorithm does not require each point to be assigned a cluster thereby reducing the noise of the clusters	The user must fine-tune the hyperparameters which define the density of the clusters

8 Chapter Takeaways and Significance

With the advancement of human requirements, the need to scale technology has also become the need of the hour. We have only discussed the freight industry in the chapter along with the implementation of machine learning and advanced analytics to solve problems. There are numerous other sectors which require the same level of business acumen and technological advancement to proceed ahead. Some of those industries being fintech, gaming, programmatic ads, health care and many more can leverage ML and advanced analytics to meet the consumer demands. This chapter focused on some of the key aspects of an analytics and machine learning infrastructure for a company dealing in the freight sector of the businesses.

We discussed data storage and big data ecosystems set-up in companies which are capable of processing terabytes of data in a couple of seconds. These types of processing power and infrastructure are required to solve problems and businesses in real time without having to suffer any down times. Even a single second of down time can incur huge losses on the consumer side of businesses, and hence, it becomes very important as to what type of data engineering or platforms you employ, who can set up this infrastructure up in the best optimized way. Another factor associated with data storage and data minimization efforts is the cost optimization aspect of it wherein organizations want to save dollar values worth of processing or storage systems. We also discussed what steps go into the data processing steps in the companies. Even before modelling or doing any sorts of predictions, it is quite important to understand the intricacies of the data and derive meaningful metrics out of them. These metrics in turn shape the overall outcome of the business along with conclusive predictions and model deployments. Finally, when the organization is all set with the data architecture and schemas set in place, along with the appropriate data pipelines, this is where the modelling phase starts and teams can start deploying real-time predictive or segmentation models for streamlined operations.

For further scope of research and improvements, these data storage technologies can be further researched upon to facilitate an even more streamlined process of capturing data and storing them in the right schema in the right manner. The optimal data processing tools need to be determined which can effectively result in cost and speed optimization for the teams. Other than that, deep learning can be implemented for incrementality in prediction accuracies.

9 Conclusion and Future Scope

With the current progression in the data science and AI domain and with the advent of efficient data collection techniques, there is wider scope to leveraging data to understand nuances in business problems. Real-time tracking mechanisms can be set up in the freight businesses to scale the supply chain. Machine learning and artificial intelligence tools can be used to predict demand in various geographical locations in real time so that supply is available to the audience at the right time. These signals will pre-emptively be detected by the advanced algorithms so as to keep the supply chain engaged, thereby avoiding staleness in any indents or loads.

For instance, one of the most common issues faced in the freight industry is the expiry of indents, without prior engagements. In a typical supply chain process, an indent is created by demand. An indent is a load requirement that is created by the relevant demand team, who wants to ship some product from location A to location B. If this indent is not exposed to the relevant demand well within time, it might expire, thereby impacting the supply chain cycle. Advanced prediction mechanisms and signal detection techniques can track these demand creations in real time and expose them to the relevant supply base which would in turn deter indent expiration/staleness.

References

1. Facility Insights Report. 2020. Uber Freight. https://info.uberfreight.com/UberFreight-Facility-Ratings-Insight-Report_2020.html.
2. Tavasszy, Lóránt A. 2020. Predicting the effects of logistics innovations on freight systems: Directions for research. *Transport Policy* 86: A1–A6.
3. Ferrari, Paolo. 2014. The dynamics of modal split for freight transport. *Transportation Research Part E: Logistics and Transportation Review* 70: 163–176.
4. Harris, Gregory. 2017. A Freight Planning Framework. Research Issues in Freight Transportation, Congestion and System Performances, Number E-C225 November 2017, pp. 45–48.
5. Cruijssen, Frans, et al. 2007. Horizontal Cooperation in Transport and Logistics: A Literature Review. *Transportation Journal JSTOR* 46 (3): 22–39. www.jstor.org/stable/20713677. Accessed 25 Sept. 2020.
6. What is Data Lake? Amazon. https://aws.amazon.com/big-data/datalakes-and-analytics/what-is-a-data-lake/.
7. Data Processing Pipeline patterns. https://blogs.informatica.com/2019/08/20/data-processing-pipeline-patterns/.
8. How Machine Learning Improves the Efficiency of Freight Operations. https://www.freightwaves.com/news/how-machine-learning-improves-the-efficiency-of-freight-operations.
9. Freight Costs: An Insider's Look on Freight Pricing Buyers Should Know. https://www.intekfreight-logistics.com/freight-logistics-cost-buyers-guide.
10. Arun Kumar, P., S. Agrawal, K. Barua, M. Pandey, P. Shrivastava, and H. Mishra. 2020. Dynamic Rule-Based Approach for Shelf Placement Optimization Using Apriori Algorithm. In *Frontiers in Intelligent Computing: Theory and Applications. Advances in Intelligent Systems and Computing*, edited by S. Satapathy, V. Bhateja, B. Nguyen, N. Nguyen, D.N. Le, vol. 1014. Singapore: Springer. https://doi.org/https://doi.org/10.1007/978-981-13-9920-6_23.

Trends and Application of Data Science in Bioinformatics

P. Supriya, Balakrishnan Marudamuthu, Sudhir Kumar Soam, and Cherukumalli Srinivasa Rao

Abstract Advancement of sequencing technologies, rapid advances in omics generated an extensive volume of biological data in recent years. It requires sophisticated analytical tools to analyze and draw conclusions from such massive amount of data. Bioinformatics is an inter-disciplinary science of analyzing and interpreting biological data by application of statistics, computational methodologies, and information technology. As huge volume of genomic, proteomic, and other data is generated, analysis and interpretation of such biological data sets involves use of data science and data mining tools. Hence, researchers are required to rely increasingly on data-science tools to store and analyze the data. Data science is an inter-disciplinary science that uses algorithms and scientific methods to derive information and insights from the big data. Data science extracts scientific work out of a wide variety of subjects viz., computer science, mathematics, statistics, databases, machine learning and optimization, etc. These strategies promote investigation and advancement of innovative methods to improve the incorporation of big data and data science into biological research. Advancements in computing and data science offers viable analytical techniques for processing huge biological data. Consequently, there is a huge possibility to enhance the interaction between bioinformatics and data science. Future applications of data science should concentrate on creating high-end integrated technologies for relatively low-cost processing of enormous biological data, greater efficiency, and reliable protection measures to advance bioinformatics research.

Keywords Big data · Bioinformatics · Data mining · Deep learning · Machine learning · Cloud computing

1 Introduction

Bioinformatics is the inter-disciplinary science which uses information technology and computer science to interpret biological data. As we continue to produce and

P. Supriya (✉) · B. Marudamuthu · S. K. Soam · C. S. Rao
ICAR-National Academy of Agricultural Research Management, Hyderabad 500030, India
e-mail: puramsupriya@gmail.com

© The Author(s), under exclusive license to Springer Nature Singapore Pte Ltd. 2021
S. Rautaray et al. (eds.), *Trends of Data Science and Applications*,
Studies in Computational Intelligence 954,
https://doi.org/10.1007/978-981-33-6815-6_12

incorporate large quantities of biological data, the value of this new field of inquiry will expand continuously. A specific active field of bioinformatics research is the use and advancement of techniques in data science to solve biological problems. The study of broad biological data sets allows the data to be pertinent by understanding the organization of the data. This type of study includes gene prediction/identification, functional classification, protein structure prediction, gene expression, protein–protein interaction studies, etc. Such big data enforces a range of technical challenges due to their diverse data sources, storage facilities, and need for flexible data analysis without sacrificing user privacy. Such problems include an iterative redesign of the current data management structures. So, there is a great opportunity to improve the relationship between bioinformatics and data science.

Data science is essentially about deriving useful insights from results. Data science is evolving to handle the complexities of extremely huge datasets with the continuous proliferation of different data produced from different sources viz., smart devices, internet, and social media. Data science is a collection of basic concepts which directs the abstraction of knowledge out of data. Perhaps data mining-the actual extraction of data knowledge by technology-is the most highly related field to data science. Data science tends to draw scientific progress from a wide range of topics such as computer science, statistics, cloud computing, machine learning, signal processing, information recovery, database development, and so on.

Big data analytics is an efficient method for high-speed and low-cost processing of large datasets. Big data architectures enable distributed computing at a stable pace within commodity hardware. It can be observed that analytics have a steady rate of influence on bioinformatics to assist realistic and dynamic applications.

2   Data Science

Data science is an inter-disciplinary science that uses algorithms and scientific methods to derive information and insights from the massive collection of big data. Data science at a greater extent is a set of principles that guide and support the retrieval of data and information. These challenges require an iterative restructuring of the existing data analytics. Big data analytics is an efficient method for low-cost and high-speed processing of large datasets.

Data mining applications to bioinformatics comprise data cleansing gene prediction, motif identification, detection of protein function domains, reconstruction of the network of protein, and gene interactions. But data science requires far more than simple algorithms for data mining. Data science extracts from many fields of research which are 'traditional.' There must be understanding of the basic concepts of causal analysis. A large amount of what was historically learned in the area of statistics is important to data science. Tools and methodologies are important for envisioning the data. There are unique fields in which intuition, imagination, and experience of a certain application have to appear. A data-science approach provides the researchers with framework and concepts that provide the data scientist with a basis for systematically resolving issues of extracting applicable data information. Biological data sets have increased exponentially in size over the last decade, largely due to advancements in data collection and storage technologies. Big data, therefore, has a substantial effect on the field of bioinformatics research. It is therefore important to advance procedures and tools to analyze big data so as to continue with our expertise to quickly obtain useful knowledge from the data, thus more advancing the decision-making process relating to various biological processes [1].

3 Application of Data Science in Bioinformatics

Bioinformatics can be described as using computer technology to manage biological knowledge. Bioinformatics is the science in which information from biological sequences and molecules is processed, extracted, organized, analyzed, interpreted, and utilized. Rapid advances in genomics and other molecular science technologies and IT innovations over the last few decades have combined to generate vast quantities of molecular biology related data. Bioinformatics' prime objective is to improve knowledge of biological processes. Bioinformatics is moving into a new age of 'big data' by generating massive data every day. One such example is that of research into omics. The completion of human genome project (HGP) laid the foundation for the recent developments in genome sequencing [2]. After HGP 's success, a range of projects were launched. The 1000 genomes project began on wide-scale sequencing of genomes, producing around 100 terabytes of massive genomic data [3]. With these advances, the quantity and quality of genomic data have been rapidly increased. Techniques such as next-generation sequencing (NGS) and high-performance screening (HTS) are widely used in genome sequencing. According to McAuley, by outpacing Moore 's rule, the cost of genome sequencing is falling at faster speed [4].

Health informatics also produces enormous biological data on similar lines of omics science. The data generated from a patient's clinical report is quite complex and sizeable [5]. Over the last few years, the growth of big data in bioinformatics has provided vast potential for researchers to mine the information hidden in raw data. Modeling multidimensional bioinformatics applications as combinations of workflow-driven web services is a method of developing to address the complexity and heterogeneity of applications in bioinformatics. A lot of workflows have been built in the last few years to solve these data through the implementation of multi-stage simulations and analytics on distributed networks. Many of the leading workflows for bioinformatics applications are discussed in the following sections. Galaxy is a web-based, open source bioinformatics platform. It provides access to numerous datasets and integrated visualization tools for biological data [6]. Taverna is a multithreaded model for scientific applications [7]. Although Tavaxy is a workflow focused on pattern recognition, used in bioinformatics, it combines current Galaxy and Taverna systems into one cloud-supporting environment [8]. Applications of data science and bioinformatics and some of the tools that are useful in different fields of omics are discussed below.

3.1 Genomics

Genomics is an inter-disciplinary field of biology that focuses on genome structure, evolution, function, mapping, and editing. A genome is the complete collection of DNA that an organism contains. Through the rising amount of sequencing data generated from several sequencing platforms, the existing analytical tools have indeed been considered to be insufficient. Developing a high performance and flexible architecture for the analysis of DNA data with a stronger emphasis on big data management is a high demand issue in recent days. The assembly of genomes is a very complex computational problem due to the huge amount of data to be put together. Preferably an assembly program must produce one contig for each sequenced chromosome of a genome. However, due to the complex existence of the genomes, the optimal conditions are actually never feasible, which contributes to genome gaps. Repeats are similar sequences that occur at various places in the genome and are also seen in varying lengths and multiple copies. Many forms of repeats exist such as tandem repeats or interspersed repetitions. The reads coming from various copies of the repeats tend to be identical with the assembler, triggering assembly errors. The data-science applications in the form of fast and accurate data analysis tools need to be built for de novo and reference genome assembly and short-read mapping which can produce longer reads leading to extremely high-quality genome [9]. Therefore, there is a need for algorithms and systems to study genomes, transcriptomes, proteomes and epigenomes at once. The list of tools that are useful in genomics data analysis are given in Table 1.

3.2 Transcriptomics

Transcriptomics focuses on analyzing the entire collection of RNA molecules in a specified study and is used exclusively for understanding gene expression. The three dominant techniques in transcriptomics viz., microarrays, qRTPCR, and RNA sequencing (RNA-seq) are these days applied to single cell applications [27, 28]. Due to more precise and quantitative measurements of gene expression, RNA-seq has evolved as a potential replacement to microarrays. RNA-seq data, however, contains some more information that is often ignored. Such data requires complex machine learning algorithms for more accurate analysis and results [29]. Using advanced machine learning techniques, big data technology can be utilized to find gene expressions, classify mutations, and exogenous RNA content from RNA sequence data. Hence, there is a need to improve the increased power of machine learning through the creation of innovative and reliable tools for big data analytics. Apache's Hadoop is the basis for utmost current RNA-seq big data applications [30], but equally exciting are spark and other upcoming frameworks. Big data provides biologists with a powerful source for enhancing experimental design based on their research work [31]. Several

Table 1 A list of bioinformatics tools/softwares for genomic data analysis

S. No	Name	Description	Web link/ Source	References
1	FastQC	Quality assessment of NGS data	https://www.bioinformatics.babraham.ac.uk/projects/fastqc/	Andrews [10]
2	Trimmomatic	A read trimming tool that removes low-quality reads and adapters	https://www.usadellab.org/cms/index.php?page=trimmomatic	Bolger et al. [11]
3	Seqtk	Tool for removal of adapters and low-quality reads	https://github.com/lh3/seqtk	Li [12]
4	FastX	Pre-processing of FASTQ and FASTA data	https://hannonlab.cshl.edu/fastx_toolkit/	Gordon and Hannon [13]
6	SPAdes	Genome assembly	https://cab.spbu.ru/software/spades/	Bankevich et al. [14]
7	Velvet	de novo short-read assembly	https://www.ebi.ac.uk/~zerbino/velvet/	Zerbino and Birney [15]
8	ABySS	Genome assembly of short reads	https://www.bcgsc.ca/resources/software/abyss	Simpson et al. [16]
9	ALLPATHS-LG	de novo assembly of small and large genomes	https://software.broadinstitute.org/allpaths-lg/blog/	NA
10	SOAPdenovo2	de novo assembler for short reads	https://github.com/aquaskyline/SOAPdenovo2	Luo et al. [17]
11	BLAST	Sequence searching and alignment tool	https://blast.ncbi.nlm.nih.gov/Blast.cgi	Altschul et al. [18]
12	MUMmer	Comparison of large genomes	https://mummer.sourceforge.net/	Delcher et al. [19]
13	MUSCLE	Multiple sequence alignment tool	https://www.drive5.com/muscle/	Edgar [20]
14	BWA	Short read alignment tool	https://bio-bwa.sourceforge.net/	Jo and Koh [21]
15	Bowtie 2	Mapping of reads to reference genome	https://bowtie-bio.sourceforge.net/bowtie2/index.shtml	Langmead and Salzberg [22]
16	SAMtools	Consensus generation and variant calling	https://samtools.sourceforge.net/	Li et al. [23]
17	GATK	Analyzing next-generation sequencing data	https://software.broadinstitute.org/gatk/	McKenna et al. [24]
18	MEGA	Phylogenetic analysis	https://www.megasoftware.net/	Kumar et al. [25]
19	BEDTools	Manipulation of genome data for genomic analysis	https://bedtools.readthedocs.io/en/latest/index.html	Quinlan and Hall [26]

Table 2 A list of bioinformatics tools/softwares for transcriptomic data analysis

S. No.	Name	Description	Web link/ Source	References
1	Trinity	Transcriptome assembly	https://github.com/trinityrnaseq/trinityrnaseq	Grabherr et al. [33]
2	Trans-ABySS	Assembly of RNA-seq data	https://www.bcgsc.ca/resources/software/trans-abyss	Robertson et al. [34]
3	RNA-SeQC	Quality control and optimization of RNA-seq data	https://software.broadinstitute.org/cancer/cga/rna-seqc	DeLuca et al. [35]
4	Rcorrector	Pre-processing of FASTQ and FASTA data	https://github.com/mourisl/Rcorrector	Song and Florea[36]
5	EdgeR	R package for the analysis of differential gene expression	https://bioconductor.org/packages/release/bioc/html/edgeR.html	Robinson et al. [37]
6	RNAeXpress	de novo short-read assembly for RNA-seq data	https://sapac.illumina.com/products/by-type/informatics-products/basespace-sequence-hub/apps/rna-express.html	Forster et al. [38]
7	rSeqNP	R package for differential expression and splicing of RNA-seq data	NA	Shi et al. [39]

large datasets are poorly analyzed and can be reprocessed to tackle many biological problems [32]. In conventional biology, this involves a certain rethinking that aims to generate far more new data. The tools for transcriptomic data analysis are given in Table 2. Since the initial steps of data analysis such as quality control, trimming, assembly are similar for both DNA and RNA data, the tools required for these analyses were already mentioned in Table 1.

3.3 Proteomics

Comparison of different samples from various physiological states is crucial in the identification of putative biomarkers, thus proteomics offers effective methods for this purpose, by quantifying proteins. Proteomics offers some advantages over transcriptomics, as it can also be used in cell-free biological fluids and provide additional information such as post-translational modifications [40]. The importance of this technology, however, depends on the quality of the methods of analysis used to process the produced data [41]. Advances in technology continue to give scientists more data to handle. Far better, a computational approach can be used to uncover other protein structures within that family until a process shows the structure of one member of a protein family. Computing plays a major part in advancing proteomics data. Solving a protein's exact structure is an extremely costly and time-consuming process. Researchers were required to manually carry out this prediction before machine learning techniques came into existence.

The latest state-of-the-art secondary structure prediction uses a method called DeepCNF (deep convolution neural fields) that depends on the machine learning model of artificial neural networks to obtain an accuracy of approximately 84% when dealing with classifying the amino acids of a protein into one of three structural groups (helix, sheet, or coil). Complexes and variations in protein—protein interactions impede high quality of knowledge about different diseases. PPI networks are being examined with the development of voluminous data in various domains of life sciences. The number, speed, and data variety make PPI complex analysis a true big data challenge.

Hence, there is an immediate demand to build fast and accurate big data tools for finding and rating PPI complex. The tools for proteomic data analysis are given in Table 3.

3.4 Metabolomics

Metabolomics includes recognizing and quantifying metabolites in order to compare their modifications with clinical conditions or by the influence of external factors for instance medications or contaminants [51]. Metabolomics, together with key 'omics' fields forms the building blocks of system biology. Due to its emphasis on small molecules and their interactions, it has recently extended extensive use in various fields such as toxicity evaluation, nutritional genomics, molecular epidemiology, discovery and recognition of biomarkers, and drug discovery [52]. The most useful technique should be able to monitor the transcript, protein and metabolite levels simultaneously and combine the data obtained to determine the structure as well as dynamics of biological networks underlined in the datasets of investigation undertaken. In order to analyze and incorporate this complex and abundant volume of data, many computational methods are necessary. Computational tools that can support data visualization and metabolite mapping significantly help to determine the related metabolites and the biological processes involved in the study environments. A list of tools and softwares for proteomic data analysis are given in Table 4.

Table 3 A list of bioinformatics tools/softwares for proteomic data analysis

S. No.	Name	Description	Web link/source	References
1	Mascot	Database searching algorithm for peptides	https://www.matrixscience.com/	NA
2	Protein prospector	Integrated framework of proteomic analysis tools	https://prospector.ucsf.edu/prospector/mshome.htm	[42]
3	TopPIC	Proteoforms identification	https://proteomics.informatics.iupui.edu/software/toppic/	[43]
4	DBParser	Shotgun proteomic data analysis	https://omictools.com/dbparser-tool	[44]
5	Sequest	Protein identification	https://fields.scripps.edu/sequest/	[45]
6	OpenMS	Analysis of mass spectrometry data	https://www.openms.de/	[46]
7	Skyline	Analysis of data from proteomic experimentation	https://skyline.ms/project/home/software/Skyline/begin.view	[47]
8	SOPMA	Protein secondary structure prediction	https://npsa-prabi.ibcp.fr/cgi-bin/npsa_automat.pl?page=npsa_sopma.html	[48]
9	SWISS-MODEL and Swiss-PdbViewer	Protein 3D structure analysis	https://spdbv.vital-it.ch/	[49]
10	Geno3D	Protein molecular modeling	https://geno3d-prabi.ibcp.fr/cgi-bin/geno3d_automat.pl?page=/GENO3D/geno3d_home.html	[50]

3.5 Epigenetics

Epigenetics includes studying the mitotically heritable mechanisms that control gene expression, irrespective of alterations in the DNA. Modern technological developments like whole genome bisulfite sequencing and inexpensive epigenomic array-based methods enable scientists to quantify large-scale cohort epigenetic profiles at genome level thus creating detailed, high-dimensional datasets that could comprise valuable facts for disease identification followed by treatment. The epigenomic expression for a certain condition is also a function of the dynamic interplay among various genetic and environmental influences that presents a huge challenge in the assessment and understanding of these results. Moreover, identification of causal relationships from the multiple associated associations is important due to the complex nature of the epigenome [61].

Table 4 A list of bioinformatics tools/softwares for metabolomic data analysis

S. No.	Name	Description	Web link/source	References
1	MetPa	Pathway analysis of metabolomics data	https://metpa.metabo lomics.ca	[53]
2	MetaboAnalyst	Integrative assessment of metabolomics data	https://www.metaboana lyst.ca/	[54]
3	Paintomics	Integrative visualization of metabolomic and transcriptomic data	https://www.painto mics.org/	[55]
4	IMPaLa	Integrated pathway analysis of omics data	https://bio.tools/impala	[56]
5	Metabominer	Identification of metabolites from NMR data	https://wishart.biology. ualberta.ca/metabo miner/	[57]
6	MeltDB	Metabolomic data analysis	https://meltdb.cebitec. uni-bielefeld.de	[58]
7	MetabolomeExpress	Web-based metabolomic data processing and analysis	https://www.metabo lome-express.org/	[59]
8	metaP-server	Data processing and analysis of metabolomics experiments	https://metap.helmho ltz-muenchen.de/met ap2/	[60]

Understanding the epigenetic mechanisms holds promising potential for medical treatments. Developments in machine learning are important if this expectation is to be realized so as to create an accurate feature selection mechanism and fix the issue of imbalances in genomic data. A combination of active learning (ACL) and imbalanced class learning (ICL) has been used in addressing earlier problems with machine learning. It indicates the strength of this innovative approach to machine learning and power to predict epigenetic processes related to disease/condition. The present methodology involves comprehensive computation of genome-specific characteristics.

An emerging new strategy is to implement deep learning (DL) to produce and simultaneously compute unique genomic features customized to the task of classification [62]. Various computational strategies for the processing of genome-wide profiling data were developed such as MethylCoder [63], Bismark [64], BRAT [65] and methylPipe [66]. Several computational methods were developed for prediction of DNA methylation [67, 68], CpG islands [69, 70] and nucleosome positioning [71, 72].

4 Techniques in Data Science that Can Be Used for Bioinformatics

4.1 Machine Learning and Deep Learning

The fundamental concept of machine learning is to consider the complications that mankind is solving in practice. Methodological descriptions, like unsupervised, semi-supervised and supervised learning, represent the bioinformatics dataset labeling element [73, 74]. Using the previous background of datasets, the goal is to acquire information and behave accordingly. Neural networks get into the picture in order to solve this difficulty. The challenge of analyzing huge datasets can be overcome by the use of deep neural networks (DNNs). Only when we have big datasets can the fundamental potential of the DNN be explored. The bigger the dataset used for preparation, the greater the precision of the test [75, 76]. Some of the deep learning applications include disease prediction, natural language processing, drug discovery, voice recognition, video processing, web content filtering, etc. [77] Since the spectrum for learning algorithms develops, the applications of deep learning are growing dramatically [78].

4.1.1 Deep Learning Applications in Bioinformatics

The science of analyzing biological data by means of computer science is bioinformatics or computational biology. Significant amounts of data are being produced due to the massive production of protein sequences, three-dimensional modeling of biomolecules, etc. [79] In order to draw inferences out of this enormous volume of biological data, complex analysis is needed [80]. Approaches to the study of bioinformatics data require strong knowledge of molecular biology and computer science. With the large-scale growth in the generation of genomic and proteomic data, the study of these data gains more attention [81].

To analyze these datasets, data mining methods serve as a foundation. The findings of broad data studies must be rational with respect to the structure proposed by the data [82]. Microarray data classification and gene classification are some of the applications under classification assignments. Prediction of the protein structure, protein–protein interaction statistical modeling, gene discovery, domain detection of protein function, inference of protein function, optimization of disease treatment, data cleaning, reconstruction of the gene interaction networks, etc., are a few additional applications [83]. The association of bioinformatics and deep learning is, therefore, on a road of superior growth and potential.

4.2   Parallel Computing

Parallel computing is essentially used to reduce the calculation time of a tedious process by breaking the large datasets into small parts in order to achieve appropriate results out of it. The challenge of analyzing terabytes of data every day is faced by medical imaging, social media, web services, bioinformatics, and several associated fields. There are several issues where, even with several processors, the run-time complexity cannot be improved [84]. Therefore, such operations must be performed in a sequential manner only and the evaluation of the propensity to make them parallel or serial is a separate problem domain.

4.2.1   Applications of Parallel Deep Learning in Bioinformatics

- Identifying the molecular cause for an illness;
- Quicker understanding and analysis of data through deep learning and machine learning;
- To increase the precision of outcome in detecting mutations with next generation sequencing;

4.2.2   Challenges in Using Parallel Deep Learning for Bioinformatics

- Ensuring the efficiency of parallelization;
- Reducing the overhead time;
- Developing algorithms and effective bioinformatics methodologies for detection and validation of targets.

4.3   Cloud Computing

The term cloud was previously used as a metaphor for the internet that represents the telephone and computer network. But it has become a buzz word today that alters the scenario and even its definition. Forrester defined the cloud as an abstract, extremely scalable computing infrastructure pool. The cloud now includes an application used to simulate raw data information, a database that stores all information, a physical storage system, and servers.

4.3.1   Genome Informatics and Biomedical Information

The trivial approach to genome informatics continued and worked smoothly until the 1980s. But the data flood forced the operating process to think of an alternative after the introduction of the next generation sequencing techniques. Users accessed data or information directly from the databases or the organization that keeps the websites in the trivial approach, downloaded them and kept local copies in their own resources and then analyzed them in their own environment. With the help of the file transfer protocol or hypertext transfer protocol, the entire transfer took place. But due to network conflicts, it was hectic to maintain the requirements with the exponential increase in data through new sequencing techniques. For researchers and database organizers, cloud computing has come as a ray of hope. This strategy has merely condensed cloud resources, data and all tools and users can access them.

In virtual mode, such information can work with that information in the cloud itself without downloading and maintaining a local copy in the personal system [85, 86].

4.3.2 Comparative Genomics

For comparative genomics studies in bioinformatics, the reciprocal shortest distance (RSD) algorithm is used, that further increases with the rise in the size of the genome to be analyzed. For comparative studies, RSD uses three bioinformatics applications: BLAST, ClustalW, and Codeml. The sequence is searched, aligned, and the shortest distance is calculated using maximum parsimony and is annotated based on the best score. Then, it is retained using the shortest distance phylogenetic analysis sequence and tested by reciprocal blast. This is a comprehensive and time-consuming process of computation and was impractical with the tremendously increasing data.

With the analysis of entire genome data, this algorithm was implemented in the cloud, and orthologs were calculated [87, 88]. The code and executables were installed on the master node and the input sequences were distributed into chunks using Amazon's elastic map reduce (EMR) algorithm, and the process was distributed over slave nodes, and the results from the slave nodes were reduced to a single file. One must fully rely on it and use it in an uninterrupted manner to harness the cloud in the beneficial and best possible way.

To accomplish this, in order to avoid termination and recreation of the cloud instances, one must first optimize the commands in a proper channel. This could be achieved by encoding the call to executables in a single program according to a workflow so that the cloud instances can be used uninterruptedly. Data security and backup recovery, however, are still the key issue in the adoption of the cloud for the research works, and it needs to be worked on to ensure that users move to the cloud. In addition, debugging becomes a challenge in the event of an error during execution because decreasing MapReduce 's program decreases the outcomes in a single file and the slave node used cannot be studied. The speed can vary as it implements a parallel computation and can give worse results due to load unbalancing and slow transmission of data on the slave nodes.

5 Future Perspectives

In recent decades, unrelenting technical developments rapidly modified the life science research practice. One of the best central role now played by bioinformatics is most noticeable transition. Big data incorporation is truly important to bioinformatics in this context. There are several open questions for the management and analysis of big data, especially in the field of computational biology. In computational terms, the novel concept is that roles are directly responsible for the management of input data, relevant procedures for the partitioning, arrangement, and integration of intermediate results.

Novel algorithms can comprise large sections of non-functional code, however, necessary for the general tasks to be exploited. Because of the practical impossibility of transporting all the data around scattered geographic areas, there is a need for computing resources capable of integrating massive storage facilities with high performance computing (HPC). Virtualization could be a solution to this problem, since it is used to build storage facilities that can accelerate data-intensive activities by using in-memory solution databases.

Academic research activities are the most significant projects for the use of big data techniques in bioinformatics. On the other hand, some commercial initiatives are also existing to cope with the enormous amount of data currently generated in the field of molecular biology, which utilizes high-throughput omics technologies. Such technologies are designed to assist computational biology researchers mainly using cloud infrastructures. Some of the examples are Era7 Bioinformatics [89], EagleGenomics [90], DNAnexus [91], and MaverixBio [92]. It is therefore clear that the big data environment has a lot to do with making these technologies efficient and easy to use, particularly given that they will be used in the near future by even small laboratories.

6 Conclusion

The rapid increase in biological data will continue to grow in the future due to the emergence of modern and inexpensive data collection methods. Biological data is huge, heterogeneous, and globally dispersed. Big data analytics are thus required to solve the bioinformatics challenges. The bioinformatics problems, data sources are of complex nature. Current big data technologies do not offer a robust solution that are large-scale, fast, distributed, fault tolerant, and optimized for complex computations. An integrated architecture of big data analytics which can meet the expectations of the bioinformatics challenges is an immediate need. Machine learning is one of the most used data analytics tools. Large-scale data, particularly in bioinformatics, existed long before the big data era. Machine learning algorithms are commonly used to evaluate both small and large-scale data using different approaches, such as sampling, distributed computations and feature selection. On the other hand, big data presents more challenges in the form of variety, size and incremental data on the conventional learning techniques. Big data analytics in bioinformatics needs to be thoroughly approached from the perspective of big data technology and realistic approaches to data analytics, including machine learning, due to the broad data growth in bioinformatics and the increasing research opportunities.

References

1. Kumari, D., and R. Kumar. 2014. Impact of biological big data in bioinformatics. *International Journal of Computer Applications* 101 (11).
2. Venter, J.C., M.D. Adams, E.W. Myers, P.W. Li, R.J. Mural, G.G. Sutton, and J.D. Gocayne. 2001. The sequence of the human genome. *Science* 291 (5507): 1304–1351.
3. Siva, N. 2008. 1000 genomes project. *Nature Biotechnology* 26 (3): 256.
4. Nagaraj, K., G.S. Sharvani, and A. Sridhar. 2018. Emerging trend of big data analytics in bioinformatics: A literature review. *International Journal of Bioinformatics Research and Applications* 14 (1–2): 144–205.
5. Burghard, C. 2012. Big data and analytics key to accountable care success. *IDC Health insights* 1–9.
6. Goecks, J., A. Nekrutenko, and J. Taylor. 2010. Galaxy: A comprehensive approach for supporting accessible, reproducible, and transparent computational research in the life sciences. *Genome Biology* 11 (8): R86.
7. Wolstencroft, K., R. Haines, D. Fellows, A. Williams, D. Withers, S. Owen, and J. Bhagat. 2013. The Taverna workflow suite: Designing and executing workflows of Web Services on the desktop, web or in the cloud. *Nucleic Acids Research* 41 (W1): W557–W561.
8. Abouelhoda, M., S. A. Issa, and Ghanem. 2012. MTavaxy: Integrating Taverna and galaxy workflows with cloud computing support. *BMC Bioinformatics* 13 (1): 77.
9. Berlin, K., S. Koren, C.S. Chin, J.P. Drake, J.M. Landolin, and A.M. Phillippy. 2015. Assembling large genomes with single-molecule sequencing and locality-sensitive hashing. *Nature Biotechnology* 33 (6): 623–630.
10. Andrews, S. 2010. FastQC: A quality control tool for high throughput sequence data.
11. Bolger, A.M., M. Lohse, and B. Usadel. 2014. Trimmomatic: A flexible trimmer for Illumina sequence data. *Bioinformatics* 30 (15): 2114–2120.
12. Li, H. 2012. seqtk Toolkit for processing sequences in FASTA/Q formats. *GitHub* 767: 69.
13. Gordon, A., and G. J. Hannon. 2010. Fastx-toolkit. FASTQ/A short reads preprocessing tools (unpublished). https://hannonlab.cshl.edu/fastx_toolkit, 5.
14. Bankevich, A., S. Nurk, D. Antipov, A.A. Gurevich, M. Dvorkin, A.S. Kulikov, and A.V. Pyshkin. 2012. SPAdes: A new genome assembly algorithm and its applications to single-cell sequencing. *Journal of Computational Biology* 19 (5): 455–477.
15. Zerbino, D.R., and E. Birney. 2008. Velvet: Algorithms for de novo short read assembly using de Bruijn graphs. *Genome Research* 18 (5): 821–829.
16. Simpson, J.T., K. Wong, S.D. Jackman, J.E. Schein, S.J. Jones, and I. Birol. 2009. ABySS a parallel assembler for short read sequence data. *Genome Research* 19 (6): 1117–1123.
17. Luo, R., B. Liu, Y. Xie, Z. Li, W. Huang, J. Yuan, and J. Tang. 2012. SOAPdenovo2: An empirically improved memory-efficient short-read de novo assembler. *Gigascience* 1 (1): 2047–2217.
18. Altschul, S.F., W. Gish, W. Miller, E.W. Myers, and D.J. Lipman. 1990. Basic local alignment search tool. *Journal of Molecular Biology* 215 (3): 403–410.
19. Delcher, A.L., A. Phillippy, J. Carlton, and S.L. Salzberg. 2002. Fast algorithms for large-scale genome alignment and comparison. *Nucleic Acids Research* 30 (11): 2478–2483.

20. Edgar, R.C. 2004. MUSCLE: A multiple sequence alignment method with reduced time and space complexity. *BMC Bioinformatics* 5 (1): 113.
21. Jo, H., and G. Koh. 2015. Faster single-end alignment generation utilizing multi-thread for BWA. *Bio-Medical Materials and Engineering* 26 (s1): S1791–S1796.
22. Langmead, B., and S.L. Salzberg. 2012. Fast gapped-read alignment with Bowtie 2. *Nature Methods* 9 (4): 357.
23. Li, H., B. Handsaker, A. Wysoker, T. Fennell, J. Ruan, N. Homer, and R. Durbin. 2009. The sequence alignment/map format and SAMtools. *Bioinformatics* 25 (16): 2078–2079.
24. McKenna, A., M. Hanna, E. Banks, A. Sivachenko, K. Cibulskis, A. Kernytsky, and M.A. DePristo. 2010. The Genome Analysis Toolkit: A MapReduce framework for analyzing next-generation DNA sequencing data. *Genome Research* 20 (9): 1297–1303.
25. Kumar, S., G. Stecher, and K. Tamura. 2016. MEGA7: Molecular evolutionary genetics analysis version 7.0 for bigger datasets. *Molecular Biology and Evolution* 33 (7): 1870–1874.
26. Quinlan, A.R., and I.M. Hall. 2010. BEDTools: A flexible suite of utilities for comparing genomic features. *Bioinformatics* 26 (6): 841–842.
27. Esumi, S., S.X. Wu, Y. Yanagawa, K. Obata, Y. Sugimoto, and N. Tamamaki. 2008. Method for single-cell microarray analysis and application to gene-expression profiling of GABAergic neuron progenitors. *Neuroscience Research* 60 (4): 439–451.
28. Tang, F., C. Barbacioru, Y. Wang, E. Nordman, C. Lee, N. Xu, and K. Lao. 2009. mRNA-Seq whole-transcriptome analysis of a single cell. *Nature Methods* 6 (5): 377–382.
29. Angerer, P., L. Simon, S. Tritschler, F.A. Wolf, D. Fischer, and F.J. Theis. 2017. Single cells make big data: New challenges and opportunities in transcriptomics. *Current Opinion in Systems Biology* 4: 85–91.
30. O'Driscoll, A., J. Daugelaite, and R.D. Sleator. 2013. 'Big data', Hadoop and cloud computing in genomics. *Journal of Biomedical Informatics* 46 (5): 774–781.
31. Dolinski, K., and O.G. Troyanskaya. 2015. Implications of big data for cell biology. *Molecular Biology of the Cell* 26 (14): 2575–2578.
32. Marx, V. 2013. Biology: The big challenges of big data.
33. Grabherr, M.G., B.J. Haas, M. Yassour, J.Z. Levin, D.A. Thompson, I. Amit, and Z. Chen. 2011. Trinity: Reconstructing a full-length transcriptome without a genome from RNA-Seq data. *Nature Biotechnology* 29 (7): 644.
34. Robertson, G., J. Schein, R. Chiu, R. Corbett, M. Field, S.D. Jackman, and M. Griffith. 2010. De novo assembly and analysis of RNA-seq data. *Nature Methods* 7 (11): 909–912.
35. DeLuca, D.S., J.Z. Levin, A. Sivachenko, T. Fennell, M.D. Nazaire, C. Williams, and G. Getz. 2012. RNA-SeQC: RNA-seq metrics for quality control and process optimization. *Bioinformatics* 28 (11): 1530–1532.
36. Song, L., and L. Florea. 2015. Rcorrector: Efficient and accurate error correction for Illumina RNA-seq reads. *GigaScience* 4 (1): s13742–s14015.
37. Robinson, M.D., D.J. McCarthy, and G.K. Smyth. 2010. edgeR: A Bioconductor package for differential expression analysis of digital gene expression data. *Bioinformatics* 26 (1): 139–140.
38. Forster, S.C., A.M. Finkel, J.A. Gould, and P.J. Hertzog. 2013. RNA-eXpress annotates novel transcript features in RNA-seq data. *Bioinformatics* 29 (6): 810–812.
39. Shi, Y., A.M. Chinnaiyan, and H. Jiang. 2015. rSeqNP: A non-parametric approach for detecting differential expression and splicing from RNA-Seq data. *Bioinformatics* 31 (13): 2222–2224.
40. Swan, A. L., A. Mobasheri, D. Allaway, S. Liddell, and J. Bacardit. 2013. Application of machine learning to proteomics data: classification and biomarker identification in postgenomics biology. *Omics: A Journal of Integrative Biology* 17 (12): 595–610.
41. Bantscheff, M., M. Schirle, G. Sweetman, J. Rick, and B. Kuster. 2007. Quantitative mass spectrometry in proteomics: A critical review. *Analytical and Bioanalytical Chemistry* 389 (4): 1017–1031.
42. Chalkley, R. J., P. R. Baker, L. Huang, K. C. Hansen, N. P. Allen, M. Rexach, and A. L. Burlingame. Comprehensive analysis of a multidimensional liquid chromatography mass spectrometry dataset acquired on a quadrupole selecting, quadrupole collision cell, time-of-flight mass spectrometer: II. New developments in Protein Prospector allow for reliable and

comprehensive automatic analysis of large datasets. *Molecular & Cellular Proteomics* 4 (8): 1194–1204.

43. Kou, Q., L. Xun, and X. Liu. 2016. TopPIC: A software tool for top-down mass spectrometry-based proteoform identification and characterization. *Bioinformatics* 32 (22): 3495–3497.

44. Yang, X., V. Dondeti, R. Dezube, D.M. Maynard, L.Y. Geer, J. Epstein, and J.A. Kowalak. 2004. DBParser: Web-based software for shotgun proteomic data analyses. *Journal of Proteome Research* 3 (5): 1002–1008.

45. Tabb, D. L., J. K. Eng, and J. R. Yates. 2001. Protein identification by SEQUEST. In *Proteome Research: Mass Spectrometry*, 125–142. Berlin, Heidelberg: Springer.

46. Sturm, M., A. Bertsch, C. Gröpl, A. Hildebrandt, R. Hussong, E. Lange, and O. Kohlbacher. 2008. OpenMS–an open-source software framework for mass spectrometry. *BMC Bioinformatics* 9 (1): 1–11.

47. MacLean, B., D.M. Tomazela, N. Shulman, M. Chambers, G.L. Finney, B. Frewen, and M.J. MacCoss. 2010. Skyline: An open source document editor for creating and analyzing targeted proteomics experiments. *Bioinformatics* 26 (7): 966–968.

48. Geourjon, C., and G. Deleage. 1995. SOPMA: Significant improvements in protein secondary structure prediction by consensus prediction from multiple alignments. *Bioinformatics* 11 (6): 681–684.

49. Guex, N., M.C. Peitsch, and T. Schwede. 2009. Automated comparative protein structure modeling with SWISS-MODEL and Swiss-PdbViewer: A historical perspective. *Electrophoresis* 30 (S1): S162–S173.

50. Combet, C., M. Jambon, G. Deleage, and C. Geourjon. 2002. Geno3D: Automatic comparative molecular modelling of protein. *Bioinformatics* 18 (1): 213–214.

51. Mehrotra, B., and P. Mendes. 2006. Bioinformatics approaches to integrate metabolomics and other systems biology data. In *Plant metabolomics*, 105–115. Berlin, Heidelberg: Springer.

52. Joyce, A.R., and B.O. Palsson. 2006. The model organism as a system: Integrating 'omics' data sets. *Nature Reviews Molecular Cell Biology* 7 (3): 198–210.

53. Xia, J., and D.S. Wishart. 2010. MetPA: A web-based metabolomics tool for pathway analysis and visualization. *Bioinformatics* 26 (18): 2342–2344.

54. Xia, J., and D. S. Wishart. 2016. Using MetaboAnalyst 3.0 for comprehensive metabolomics data analysis. *Current Protocols in Bioinformatics* 55 (1): 14–10.

55. García-Alcalde, F., F. García-López, J. Dopazo, and A. Conesa. 2011. Paintomics: A web based tool for the joint visualization of transcriptomics and metabolomics data. *Bioinformatics* 27 (1): 137–139.

56. Kamburov, A., R. Cavill, T.M. Ebbels, R. Herwig, and H.C. Keun. 2011. Integrated pathway-level analysis of transcriptomics and metabolomics data with IMPaLA. *Bioinformatics* 27 (20): 2917–2918.

57. Xia, J., T.C. Bjorndahl, P. Tang, and D.S. Wishart. 2008. MetaboMiner–semi-automated identification of metabolites from 2D NMR spectra of complex biofluids. *BMC Bioinformatics* 9 (1): 507.

58. Neuweger, H., S.P. Albaum, M. Dondrup, M. Persicke, T. Watt, K. Niehaus, and A. Goesmann. 2008. MeltDB: A software platform for the analysis and integration of metabolomics experiment data. *Bioinformatics* 24 (23): 2726–2732.

59. Carroll, A.J., M.R. Badger, and A.H. Millar. 2010. The MetabolomeExpress Project: Enabling web-based processing, analysis and transparent dissemination of GC/MS metabolomics datasets. *BMC Bioinformatics* 11 (1): 376.

60. Kastenmüller, G., W. Römisch-Margl, B. Wägele, E. Altmaier, and K. Suhre. metaP-server: A web-based metabolomics data analysis tool. *Journal of Biomedicine and Biotechnology*.

61. Cazaly, E., J. Saad, W. Wang, C. Heckman, M. Ollikainen, and J. Tang. 2019. Making sense of the epigenome using data integration approaches. *Frontiers in Pharmacology* 10: 126.

62. Holder, L.B., M.M. Haque, and M.K. Skinner. 2017. Machine learning for epigenetics and future medical applications. *Epigenetics* 12 (7): 505–514.

63. Pedersen, B., T.F. Hsieh, C. Ibarra, and R.L. Fischer. 2011. MethylCoder: Software pipeline for bisulfite-treated sequences. *Bioinformatics* 27 (17): 2435–2436.

64. Krueger, F., and S.R. Andrews. 2011. Bismark: A flexible aligner and methylation caller for Bisulfite-Seq applications. *Bioinformatics* 27 (11): 1571–1572.
65. Harris, E.Y., N. Ponts, K.G. Le Roch, and S. Lonardi. 2012. BRAT-BW: Efficient and accurate mapping of bisulfite-treated reads. *Bioinformatics* 28 (13): 1795–1796.
66. Kishore, K., S. de Pretis, R. Lister, M.J. Morelli, V. Bianchi, B. Amati, and M. Pelizzola. 2015. methylPipe and compEpiTools: A suite of R packages for the integrative analysis of epigenomics data. *BMC Bioinformatics* 16 (1): 313.
67. Fang, F., S. Fan, X. Zhang, and M.Q. Zhang. 2006. Predicting methylation status of CpG islands in the human brain. *Bioinformatics* 22 (18): 2204–2209.
68. Das, R., N. Dimitrova, Z. Xuan, R.A. Rollins, F. Haghighi, J.R. Edwards, and M.Q. Zhang. 2006. Computational prediction of methylation status in human genomic sequences. *Proceedings of the National Academy of Sciences* 103 (28): 10713–10716.
69. Feltus, F.A., E.K. Lee, J.F. Costello, C. Plass, and P.M. Vertino. 2003. Predicting aberrant CpG island methylation. *Proceedings of the National Academy of Sciences* 100 (21): 12253–12258.
70. Bock, C., J. Walter, M. Paulsen, and T. Lengauer. 2007. CpG island mapping by epigenome prediction. *PLoS Computational Biology* 3 (6): e110.
71. Segal, E., Y. Fondufe-Mittendorf, L. Chen, A. Thåström, Y. Field, I.K. Moore, and J. Widom. 2006. A genomic code for nucleosome positioning. *Nature* 442 (7104): 772–778.
72. Peckham, H.E., R.E. Thurman, Y. Fu, J.A. Stamatoyannopoulos, W.S. Noble, K. Struhl, and Z. Weng. 2007. Nucleosome positioning signals in genomic DNA. *Genome Research* 17 (8): 1170–1177.
73. Ravi, D., C. Wong, F. Deligianni, M. Berthelot, J. Andreu-Perez, B. Lo, and G. Z. Yang. 2016. Deep learning for health informatics. *IEEE Journal of Biomedical and Health Informatics* 21 (1): 4–21.
74. Akay, A., and H. Hess. 2019. Deep learning: Current and emerging applications in medicine and technology. *IEEE Journal of Biomedical and Health Informatics* 23 (3): 906–920.
75. Schmidhuber, J. 2015. Deep learning in neural networks: An overview. *Neural Networks* 61: 85–117.
76. Wei, L., R. Su, B. Wang, X. Li, Q. Zou, and X. Gao. 2019. Integration of deep feature representations and handcrafted features to improve the prediction of N6-methyladenosine sites. *Neurocomputing* 324: 3–9.
77. Luo, F., M. Wang, Y. Liu, X.M. Zhao, and A. Li. 2019. DeepPhos: Prediction of protein phosphorylation sites with deep learning. *Bioinformatics* 35 (16): 2766–2773.
78. Goh, G.B., N.O. Hodas, and A. Vishnu. 2017. Deep learning for computational chemistry. *Journal of Computational Chemistry* 38 (16): 1291–1307.
79. Fu, H., Y. Yang, X. Wang, H. Wang, and Y. Xu. 2019. DeepUbi: A deep learning framework for prediction of ubiquitination sites in proteins. *BMC Bioinformatics* 20 (1): 1–10.
80. Raza, K. 2012. Application of data mining in bioinformatics. arXiv preprint arXiv:1205.1125.
81. Jurtz, V. I., A. R. Johansen, M. Nielsen, J. J. Almagro Armenteros, H. Nielsen, C. K. Sønderby, and S. K. Sønderby. 2017. An introduction to deep learning on biological sequence data: Examples and solutions. *Bioinformatics* 33 (22): 3685–3690.
82. Rhee, S.Y., J. Dickerson, and D. Xu. 2006. Bioinformatics and its applications in plant biology. *Annual Review of Plant Biology* 57: 335–360.
83. Min, S., B. Lee, and S. Yoon. 2017. Deep learning in bioinformatics. *Briefings in Bioinformatics* 18 (5): 851–869.
84. Alipanahi, B., A. Delong, M.T. Weirauch, and B.J. Frey. 2015. Predicting the sequence specificities of DNA-and RNA-binding proteins by deep learning. *Nature Biotechnology* 33 (8): 831–838.
85. Stein, L.D. 2010. The case for cloud computing in genome informatics. *Genome Biology* 11 (5): 207.
86. Rosenthal, A., P. Mork, M.H. Li, J. Stanford, D. Koester, and P. Reynolds. 2010. Cloud computing: A new business paradigm for biomedical information sharing. *Journal of Biomedical Informatics* 43 (2): 342–353.

87. Wall, D.P., P. Kudtarkar, V.A. Fusaro, R. Pivovarov, P. Patil, and P.J. Tonellato. 2010. Cloud computing for comparative genomics. *BMC Bioinformatics* 11 (1): 259.
88. Kudtarkar, P., T. F. DeLuca, V. A. Fusaro, P. J. Tonellato, and D. P. Wall. 2010. Cost-effective cloud computing: a case study using the comparative genomics tool, roundup. *Evolutionary Bioinformatics* 6, EBO-S6259.
89. Era7 Bioinformatics, https://era7bioinformatics.com.
90. EagleGenomics, https://www.eaglegenomics.com.
91. DNAnexus, https://dnanexus.com/.
92. MaverixBio, https://www.maverixbio.com.

Mathematical and Algorithmic Aspects of Scalable Machine Learning

Gananath Bhuyan and Mainak Bandyopadhyay

Abstract Although a number of machine learning models have been proposed and successfully deployed in organizations, there is a new emerging challenge that the organizations are going to face in the upcoming years. As various organizations are relying on data for decision-making and optimization of processes, the volume of data is an important factor for developing precise models. The ever-increasing volume of data and its storage is contributed by the advancement of communication technology and storage services like cloud computing. The large volume of data collected is usually stored in a distributed storage and computing environment to ensure fault tolerance and scalability. The development of machine learning models is quite inefficient in a distributed environment using traditional machine learning algorithms. The inefficiency is attributed to the distributed nature of the dataset and computing. The development of the models needs to be carried out in a distributed manner. Thus, additional challenges related to distributed computing need to be addressed by the machine learning algorithms. Scalable machine learning is an updation of traditional machine learning in a distributed environment. As the nature of computing changes, the mathematical formulas and equations need to be revisited along with the algorithms to make it suitable for a distributed environment. This chapter discusses the challenges faced by the traditional machine learning algorithms in distributed environments, the various mathematical backgrounds of scalable machine learning models, and the state-of-the-art distributed algorithms for scalable machine learning models.

Keywords Scalable machine learning · Predictive modeling · Analytics; distributed computing

G. Bhuyan · M. Bandyopadhyay (✉)
Deemed To Be University, KIIT, Bhubaneswar, India
e-mail: mainak.bandyopadhyayfcs@kiit.ac.in

G. Bhuyan
e-mail: gananatha.bhuyanfcs@kiit.ac.in

© The Author(s), under exclusive license to Springer Nature Singapore Pte Ltd. 2021   245
S. Rautaray et al. (eds.), *Trends of Data Science and Applications*,
Studies in Computational Intelligence 954,
https://doi.org/10.1007/978-981-33-6815-6_13

1   Introduction

Machine learning is a branch of artificial intelligence (AI) that helps in understanding the structure and nature of existing data resulting in a model developed out of it that can be used to analyze real-world data. Machine learning is different from the traditional computation methodologies of computer science even though the former is a field of the later. In the traditional approach of computing, algorithms are designed by explicit programming. On the other hand, ML algorithms let computers be trained from data inputs so that statistical analysis can be used on real-world inputs resulting in an automatic decision-making system. ML is a continuously developing field, due to which some consideration has to be kept in mind while working with machine learning methodologies.

Data science integrates data and statistical machine learning algorithms. As the volume of data grows, there is a need for machine learning algorithms that can be scaled according to the increase and decrease in the volume of data.

Previously, the amount of data was not that big. Hence, keeping the entire data in one computer and applying machine learning algorithms to it was not a very challenging task. As the size of data keeps increasing exponentially, it is not feasible to follow the traditional methods to get the job done. Not only the rapidly increasing size of data but also the variety of data as well the different sources of data is also a matter of concern here. One of the best solutions that can be thought of is to use a distributed environment and process huge data concurrently.

Distributed and parallel processing of large datasets have been employed for decades now in many fields. Recent years have brought huge changes in terms of parallel computation as their demand increases in data analysis and machine learning tasks. One undeniable fact is that with the new settings that we are talking about now, the resource requirement is also going to be increased. Although this fact was well known to experts since the very beginning, nothing could really be done due to the lack of hardware support. The current rise in interest in scaling up machine learning is due to the improved hardware support and the evolution in programming frameworks which make things much more realizable. The requirement of scaling up is also due to the spread of large datasets in various applications. Such big datasets are normally distributed over multiple nodes, motivating the development of learning algorithms that can be distributed appropriately.

1.1   Challenges in Scalable Machine Learning

As of now, we have understood that scalability in machine learning is very much essential so as to cope with the increasing size and the scatteredness of data. But the volume or the variety of data is not the only concern while talking about scalability in machine learning. There are various challenges that need to be addressed before a scalable machine learning which can be designed, while designing and implementing a scalable system a few things, such as efficiency, distributedness, terminatability, parallelism, correctness, etc., have to be taken care of. Below, a few of the challenges are mentioned.

Efficient parallelization: With the increase in the amount of data, the number of machines on which the data needs to be distributed also increases. To analyze such a huge amount of data, an appropriate scale is required so that independent tasks can be assigned to different machines. Executing such an operation in a single machine is not that complex but doing the same in a distributed environment could really be challenging. As the computation is costly in the cloud environment, it is very much essential to design schedulers in a way to reduce I/O waste and the network overhead while utilizing CPU resources to the optimal.

Termination: Termination condition is one of the most important parts of an ML algorithm that needs to be implemented effectively. In the case of a distributed setting, the implementation of a terminating condition is quite challenging so as to maintain a balance between accuracy and performance. Normally, in ML techniques, the standard termination condition assessment requires knowledge of the global state.

Correctness: The next challenge is to ensure consistency and convergence of results. As in distributed settings, computation is performed at different nodes; the ML algorithm will have a different convergence behavior than the sequential version. Hence, choosing the right model and algorithm, based on the nature of the data and application, has a huge impact on the correctness of the result.

Usability: In order to achieve greater performance in parallel computing, many experts use very high-level tools. But those kinds of advanced setups may not be attainable by non-experts. Hence, the design must have a high level of abstraction in order to simplify the implementation and insulate users from the complexity.

Keeping the amount and diversity of data, it is clear that conventional machine learning techniques are not suitable enough to process these data. Thus, a definite scale-up is required for these machine learning techniques.

1.2 Reasons for Scaling up Machine Learning

There are many instances a machine learning user would like to scale it up. Below mentioned are a few of the instances.

Large dataset: Nowadays, it is a very common thing that several domains are generating thousands of terabytes of data per day. And due to the emergence of IoT, the data collected by various sensors each moment is also huge. Storing and processing this scale of data are a big challenge. The preferred way of processing such a huge amount of data is distribution and parallel processing. Many computational frameworks such as MapReduce and DryadLINQ have made this process easy which we will be seeing in this chapter.

Input dimension: Machine learning tasks that involve images, videos, natural languages have a large input dimension. Handling such a large dimension of data using conventional machine learning techniques is way too inefficient. Thus, parallelizing computation across features could really be helpful to handle such data.

Complexity: Data in some domains are inherently non-linear with respect to the basic features. Although feature engineering can achieve high accuracies with computationally cheap and simpler linear models in these domains, the interest of learning automatically from the base representation is growing rapidly. Even though the training data can fit easily for these algorithms, learning could be way too time taking.

Time constraints: As mentioned in the above point, even though data can somehow be managed to fit on the conventional machine learning algorithms, the time they would take for the learning process could make the whole system inefficient.

As now, we know why scaling in machine learning is that important, we shall know the different types of scaling approaches. Scaling is basically of two types: Vertical scaling and horizontal scaling.

Vertical scaling means to increase the system performance by adding more resources (i.e., GPUs, storage, etc.) to it. This increases the cost of the system drastically but the performance doesn't get improved that much (Fig. 1).

Horizontal scaling means adding more systems of the same capability in order to make them work parallelly. And this type of scaling doesn't add up the cost a lot but due to parallelization, the performance gets improved marginally, which is exactly required to handle such a huge amount of data.

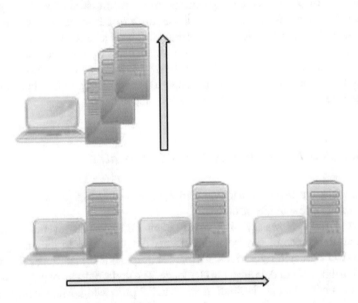

Fig. 1. Vertical and horizontal scaling [5]

2   The Infrastructure of Scalable Machine Learning

Day by day the amount of data is growing rapidly, and it continues to grow. As data is being collected from different sources and in different formats, it is getting difficult to handle such a huge amount of data by using traditional machine learning techniques. Even more important than this is the fact that the size of data will be increasing exponentially in near future as the areas like the Internet of Things and M2M are getting popular day by day. Machine learning has already emerged as a technique to handle such massive data flawlessly. However, when it comes to the diversity of type as well as the source of data, different more powerful software architecture is required. That is where the need for scalable machine learning comes into the picture which allows us to store and process such diversifies natured data efficiently. But developing a scalable infrastructure using machine learning is not an easy task as it has its own challenges. Here, in this section, we shall discuss a few possible architectures for scalable machine learning.

As we talked about storing and processing the data, the first thing that needs to be taken care of is the file system above which the whole architecture is going to be developed. As the size of the data is huge, it is very difficult to keep the whole data in a centralized manner. Accessing information from such a big amount of data, placed centrally, will also be a tedious task. Hence, a centralized file system can not be chosen; instead, a decentralized or distributed file system will be required.

2.1   Distributed File System

A distributed file system is a file system, where the entire data is not kept at a single location in a centralized manner; rather, the entire data is placed in different locations in a distributed manner. Having the entire data, not in the same location, improves parallelization and fault tolerance. Parallelization improves as data can be retrieved from different locations at the same time. Similarly, fault tolerance is improved because if one data center goes down due to some reason, then others can easily work it out. At the same time, keeping data in a distributed manner increases communication cost and the data security-related concern. Because when data is placed in different locations, they have to be connected through a network and can easily be affected by malicious attacks. Hence, a distributed system has its own pros and cons but the pros are way too superior to the cons, and with the help of active research and development, the cons are constantly being diminished.

As we can see in figure given below, in the case of centralized design, different users are dependent on the common central database (HQ). But in the case of distributed design, each user has its own copy of data. Some of the topologies that employ a distributed file system and are widely used in distributed machine learning are discussed in Fig. 2.

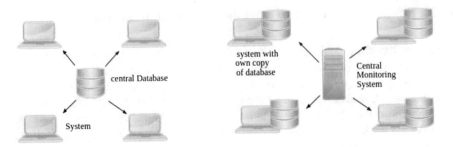

Fig. 2 Centralized and distributed architecture

2.2 Distributed Topology for Machine Learning

Here, in this section, we will be seeing some topologies that use a distributed file system and can be considered for scaling up the machine learning algorithms.

Ensembling: Here, the copy of the same data is placed in different places. A different machine learning algorithm is applied to each copy of the dataset. Then, the result of each algorithm is merged back to get the global result. This is not truly a distributed topology because data is not being stored in a distributed manner (Fig. 3).

Tree: Tree-like topologies are popular as they are easy to scale and manage. In tree-like topology, communication is only between parent and child nodes. For example, nodes can calculate the local gradient and pass it to its parent and so on in order to calculate the global gradient (Fig. 4).

Rings: In situations where the communication cost needs to be kept low and it doesn't give efficient support for broadcasting, a ring-like topology is used which requires communicating only to its neighboring nodes (Fig. 5).

Fig. 3 Centralized (Ensembling) architecture (Reproduced from A Survey on Distributed Machine Learning [1])

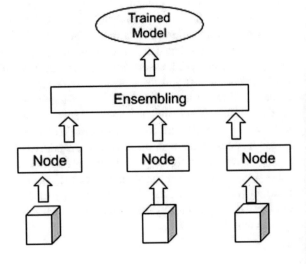

Fig. 4 Distributed tree topology (Reproduced from Survey on Distributed Machine Learning)

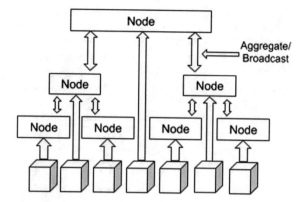

Fig. 5 Distributed ring topology (Reproduced from A Survey on Distributed Machine Learning [1])

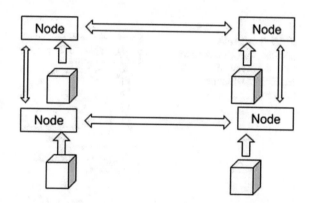

Parameter server: This topology uses a decentralized set of nodes monitored by a centralized master. The nodes are allowed to read from and write to a shared space from which the client can take the information. One of the advantages of this topology is the whole system which can be monitored centrally. A disadvantage of the topology is that the parameter servers can form a bottleneck because they are handling all communication (Fig. 6).

Peer-to-Peer: In this type of topology, each node has its own set of parameters and can communicate directly with each other. The advantages of this topology overcentralized system are higher scalability which can be achieved, and single-point failure can be eliminated. The communication cost mostly is the drawback here (Fig. 7).

3 MapReduce

Traditional systems use a centralized server for storing data. But the rapid growth of data in almost all fields makes it quite impossible to have a centralized system to

Fig. 6 Distributed
parameter server (
Reproduced from A Survey
on Distributed Machine
Learning)

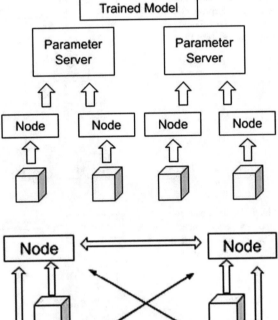

Fig. 7 Distributed
peer-to-peer (Reproduced
from A Survey on
Distributed Machine
Learning)

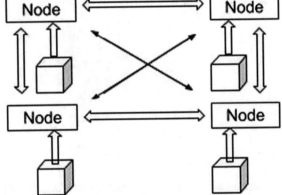

store and retrieve such a huge amount of data. Processing multiple files concurrently
in a centralized system is also a headache. To perform the distributed computation
for large-scale datasets, MapReduce, developed by Google, provides a framework
that overcomes such bottlenecks. In this section, we will be discussing MapReduce,
and how it is useful in parallel processing of large-scale data.

"*MapReduce is a computational component of the Hadoop framework for easily
writing applications that process large amounts of data in parallel and stored on large
clusters of cheap commodity machines in a reliable and fault-tolerant manner.*"

How does it work?

The whole framework is basically performed in two phases: First, Map phase; second,
Reduce phase. The entire dataset is split into disjoint sets and is distributed among
multiple systems or worker processes. Each system performs its task defined by
the user, and then, the output of each system is combined to get the output of the

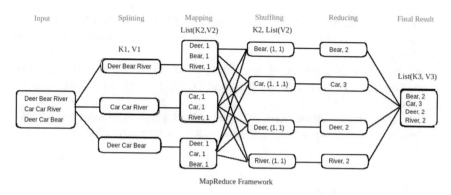

Fig. 8 MapReduce framework

MapReduce. This way a large dataset can be processed parallelly resulting in the reduction in processing time (Fig. 8).

Map phase: This is the phase, where the big dataset is divided and distributed among a number of worker processes called mappers. Each mapper performs the user-defined task/function defined by the user on the sub-dataset assigned to it concurrently with other mappers. The result of this phase is a set of key-value pairs. For example, if you want to group the songs in your playlist on the basis of the singer name, then this phase will analyze the entire playlist and result in a key-value pair list, where the key will be the singer's name, and the value will be the song itself. So basically mapper converts a single data into a number of small data chunks with respect to key.

Reduce phase: The combination of *shuffle* and *reduce* is considered as a single phase, i.e., the reduce phase. The result of the map phase, which is basically key-value pairs that are collected, shuffled, and redistributed among a set of worker processes called as reducers by the master process. Each reducer performs a user-defined reduce function on the corresponding assigned keys to get its output. The collection of outputs of all the reducers is the final output of the MapReduce.

3.1 Benefits of MapReduce

Fault tolerance: During the MapReduce process, if one of the systems goes down due to some failure, then the architecture takes care of it by considering replicated copies of the data in other machines to continue the process.

Resilience: Each node sends an update to the master node in a certain time interval. If a node doesn't do so, then the master node chooses another node in the cluster for the assigned task of the current node.

Quick: MapReduce is quick enough to handle terabytes of unstructured data in minutes as it works on HDFS.

Parallel processing: As the entire dataset is divided into multiple chunks of data and is processed parallelly, the total time required to process them reduces significantly.

Scalability: MapReduce used the HDFS and not the traditional RDBMS. Due to this reason, it is comparatively easy and effective to scale the system as per requirement.

Cost-effective: Using the horizontal scaling approach, a significant performance can be gained without spending a lot more, making it very cost-effective.

4   Linear Regression

Linear regression is a prediction technique developed in the field of statistics and is studied as a model in understanding the relationship between input and output in order to predict the output for a new set of inputs. It is included under the supervised learning techniques of machine learning. The prediction is done by following a simple equation of a line.

$$y = mx + c \tag{4.1}$$

where x: the feature or input and y: the prediction or output.

How Linear regression works?
As mentioned earlier, linear regression is a supervised learning technique. The model training process is carried out by feeding a set of inputs and their corresponding outputs. The machine tries to figure out in what way the input is related to the output. Then, on the basis of the relationship, the model can predict the output, when an unknown input is fed to it.

This technique is efficient when all the training data points are present on one computer. But considering the rapid increase in data size, it is quite impossible to have all the data in one place. A parallel approach that can perform linear regression concurrently on multiple machines can definitely help in dealing with this situation. In this chapter, a parallel linear regression using MapReduce on the top of HDFS has been discussed.

In regression, the goal is to predict a target value that represents a continuous-valued outcome such as the conditional probability of a discrete outcome, a conditional mean, or a conditional quantile, and so forth depending on the context. The regression application can be formulated based on the following optimization problem:

Given a training dataset $\{x_i, y_i\}$ $N_i = 1$, where the x is the vector of input features, y is the corresponding target feature, and N is the number of training data examples, along with a loss function L(y, ŷ) that measures the discrepancy between the target value y and a predicted value given by $ŷ = f(x)$, find the regression function $\hat{f}(x)$ that minimizes the average loss on the training data,

$$\hat{f}(x) = Min\emptyset(f) \tag{4.2}$$

where

$$\emptyset(f) = \frac{1}{N} \sum_{i=1}^{N} L(y_i, f(x_i)) \tag{4.3}$$

The optimization formulation Eq. (4.1) can be regularized by controlling the class of functions for $f(x)$ as the regression function $\hat{f}(x)$ may overfit the training data.

The loss function can be represented as a squared-error loss function as follows,

$$L(y, f(x)) = \frac{1}{2}(y - f(x))^2 \tag{4.4}$$

assuming y is Gaussian with mean $f(x)$ and constant variance.

4.1 Parallel Version of Linear Regression

In order to perform linear regression, we try to minimize $\|X\beta - Y\|^2$ and the derivative $2X^T(X\beta - Y)$. In small data settings, we can set the derivative to 0 and can solve it directly. But in the case of large data settings X would be too big to fit into memory and may be hard to solve directly.

Normally, the cost of gradient descent is calculated by taking the average of all the data points. This approach could be useful when the number of data points is less. But when there are a huge number of data points (let's say in millions), this approach will take a lot of time. Instead of calculating the average, the gradient can be calculated by approximation which will save a lot of time and cost. Stochastic Gradient Descent (SGD) approximates the gradient using only one data point in a given iteration. By performing several such iterations (let's say in hundreds or thousands), SGD may get a reasonable solution compared to summing millions of data points.

The cost function would be as follows if the regression on squared loss (as assumed above),

$$J(\theta) = \frac{1}{2m} \sum_{i=1}^{m} (h_\theta x_i - y_i)^2 \tag{4.5}$$

and the gradient is

$$\frac{dJ(\theta)}{d\theta} = \frac{1}{m} \sum_{i=1}^{m} (h_\theta x_i - y_i)x_i \tag{4.6}$$

for gradient descent, we update the parameter by,

$$\theta_{\text{new}} = \theta_{\text{old}} - \alpha \frac{1}{m} \sum_{i=1}^{m} (h_\theta x_i - y_i) x_i \tag{4.7}$$

For SGD, we don't have to find the summation, and the $1/m$ also can be eliminated. Only the gradient for the current data point (x_i, y_i) is to be calculated, by saving time.

$$\theta_{\text{new}} = \theta_{\text{old}} - \alpha \frac{1}{m} \sum_{i=1}^{m} (h_\theta x_i - y_i) x_i \tag{4.8}$$

If we have let's say 1 billion data points. In SGD, we can think of it as trying to get an approximated gradient instead of an exact gradient. The approximation can be deduced from a single data point or from a small subset of the entire dataset resulting in a faster updation of the parameter in the case of SGD. The trick is that in SGD, we do not need to have 1 billion iterations/updates, but much fewer iterations/updates, let's say 1 million, and we will have a "good enough" model to use.

One of the best ways to parallelize linear regression is to use an iterative method with the help of stochastic gradient descent, where the gradient can be approximated as explained above.

The iteration can be performed on a single data point or on a small subset of the entire dataset at a time. If x_s, y_s is a subset of the data point, then the gradient can be approximated by,

$$2X_s^T(X_s\beta - Y_s) \tag{4.9}$$

and we can update β with the approximated gradient.

In addition, for the R^2 statistic, we can compute R^2 for all data in parallel or approximate it by using a subset of the data.

To achieve parallelization, each subset of the dataset can be placed in different computers (e.g., if we have 1 million data points, then it can be divided into 100 subsets of 10,000 data points each. Each subset can be placed in different computers requiring 100 computers.), and the gradient of each computer can be calculated concurrently (Map), and then, the average of the output of the 100 computers (Reduce) can be considered as the final gradient value.

The vector of weights "β" in linear regression can be evaluated as follows:

$$\beta = (X'X)^{-1} X'Y. \tag{4.10}$$

The product $X'X$ can be evaluated in p^2n operations and then inverting it takes p^3 operations.

Time complexity: Assuming n being the no. of training samples and p being the no. of features, the upper bound of linear regression is: training the model: $O(p^2n + p^3)$, Prediction: $O(p)$.

5 Clustering

Clustering is one of the most important subfields of machine learning and data mining. Clustering is the process of dividing a dataset in such a way that data points present in one group are more similar to the data points in that group than that of other groups. It is basically the collections of data points on the basis of their similarities and dissimilarities with respect to certain characteristics. So there is no good clustering or bad clustering. It completely depends on what characteristic is being chosen by the user. A single dataset may be clustered in various ways by applying the clustering algorithm to different attributes. Clustering is a technique that basically falls under unsupervised learning (Fig. 9).

Clustering is performed to group similar elements together. The task seems not to be that complex when the entire dataset is present at one location. There are so many clustering techniques such as density-based methods, hierarchical methods, partitioning methods, grid-based methods, etc., that can perform this task on a centralized dataset efficiently. One of the most popular and easiest clustering algorithms is the K-mean clustering algorithm, which is discussed next.

5.1 K-Mean Clustering

K-mean clustering is an unsupervised learning technique that is used on unlabeled data (i.e., data points without having defined groups or categories). The aim of this algorithm is to categorize the data in k groups. For each group, one centroid will be considered at random, and the algorithm iteratively assigns each data point to its closest centroid. The data points are grouped on the basis of their feature similarity. Each centroid is nothing but a collection of features to which all the similar featured data points are attached.

Algorithm

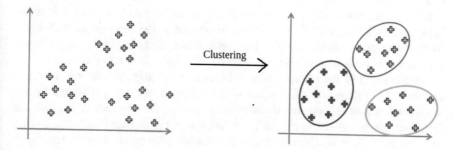

Fig. 9 Clustering

1. Initial data assignment: The initial data assignment is done by finding out the minimum of Euclidean distance of the point from all the centroids. If S is the set of initially chosen centroids and s_i be one element in it, then the data point x will be assigned to s_i using the following formula.

$$\min Ecd\_Dist(1D460_i, x), \quad \forall s_i \in S \tag{5.1}$$

where Ecd_Dist() gives the Euclidean distance between s_i and x.

2. *Centroid update*: The centroid is now updated by computing the mean of the distance between each data point (x) and the associated centroid as follows:

$$s_i = \frac{1}{|N_i|} \sum_{x_i \in N_i} x_i \tag{5.2}$$

where N_i is the set of points associated with the centroid s_i. x_i is a point that belongs to N_i.

Step 2 of the algorithm iterates until a stopping criterion has been achieved. A stopping criterion is achieved when the sum of the distances is minimum, no point changes cluster or the maximum number of iteration is reached.

How to choose K?

The above algorithm is applied for a pre-chosen value of k. So the user has to apply the algorithm on a range of k values and find out for what value of k, the sum of the distances remains the minimum.

5.2 Parallel K-mean for a Scalable Environment

The keep on growing volumes of information, being generated by the advancement of technology, makes clustering a challenging task. And the large scale of data is stored in a distributed manner. In such a scenario, the above algorithm will not produce the desired output, as data points stored at different locations may have similar features. Hence, a parallel K-mean clustering that can operate on the data present at different locations simultaneously can solve the problem [6]. Here, in this section, we will discuss a parallel k-mean clustering based on the MapReduce framework.

MapReduce is a combination of two functions: one map function and a reduce function. The assignment of samples to it's closest centroid is performed by the map function, and the updation of the centroid value is done by reduce function. As data is present at different locations, there has to be a certain kind of communication between them. There will be a combiner function that takes care of the cost of network communication. Now, we shall see how each function works.

Map function

The HDFS file system is being used to store the input dataset as a sequence of <key, value> pairs. Each pair denotes a record in the dataset. The key is the offset to the start point of the record, and the value is a string of contents of the record. There is a mapper at each location, and each sub-dataset is being broadcasted to all rest of the locations. The mapper is capable of computing the nearest center for each sample. The result of the map function, which is also the intermediate result of the whole process, is a <key, value> pair, where key is the index of the closest center, and value is the sample information.

Combine function

After each map task, a combiner function is applied to combine the intermediate data at each node. As the intermediate data is stored in the local machine, there won't be any communication cost. In this function, the partial sum of values of each point in one cluster is being calculated. The number of samples has to be recorded so as to find the mean value of the objects in the cluster.

Algorithm 5.1 Map function (Parallel K-mean by Weizhong Zhao et al.) [6].

```
map(key, value)

Input: Global variable centers, the offset key, the sample value
Output: <key', value'> pair
1. Construct the sample instance from value;
2. minDis = Double.MAX VALUE ;
3. index = -1;
4. For i=0 to centers.length do
          dis = ComputeDist(instance, centers[i]);
          If dis < minDis {
                    minDis = dis;
                    index = i;
          }
5. End For
6. Take index as key';
7. Construct value' as a string comprise of the values of different dimensions;
8. output < key, value > pair;
9. End
```

In step 2 and step 3, the auxiliary variables, minDis, and index are being initialized. In step 4, using the function ComputeDist, the minimum distance between the instance and the centroid is calculated which in turn gives the closed centroid of the data point.

Algorithm 5.2 Combine function (Parallel K-mean by Weizhong Zhao et al.) [6].

<u>*combine(key, V)*</u>

Input: *key* (index of the cluster), *V* (list of the samples assigned to the same cluster)
Output: < *key'* , *value'* > pair, where the key' is the index of the cluster, value' is a string
comprised of sum of the samples in the same cluster and the sample number
1. Initialize one array to record the sum of value of each dimensions of the samples contained
   in the same cluster, i.e. the samples in the list *V* ;
2. Initialize a counter *num* as *0* to record the sum of sample number in the same cluster;
3. *while(V.Next()){*
          Construct the sample instance from *V.next()*;
          Add the values of different dimensions of instance to the array
          num++;
4. *End while*
5. Take *key* as *key'*;
6. Construct *value'* as a string comprised of the sum values of different dimensions and *num*;
7. *output < key , value > pair;*
8. *End*

Reduce function

Algorithm 5.3 Reduce function (Parallel K-mean by Weizhong Zhao et al.) [6]

<u>*reduce(key, V)*</u>

Input: *key* is the index of the cluster, *V* is the list of the partial sums from different host
Output: < *key'* , *value'* > pair, where the *key'* is the index of the cluster, *value'* is a string
          representing the new center
1. Initialize one array record the sum of value of each dimensions of the samples contained
   in the same cluster, e.g. the samples in the list *V* ;
2. Initialize a counter *NUM* as *0* to record the sum of sample number in the same cluster;
3. *while(V.Next()){*
          Construct the sample instance from *V.next()*;
          Add the values of different dimensions of instance to the array
          NUM += num;
4. *End while*
5. Divide the entries of the array by *NUM* to get the new center's coordinates;
6. Take *key* as *key'*;
7. Construct *value'* as a string comprise of the center's coordinates;
8. *output < key , value > pair;*
9. *End*

The result of the combine function is given as input to the reduce function. The
data in the *combine* function includes the partial sum of the samples in that particular
cluster and the sample number. The reduce function sums all the samples in that
cluster and finds out how many samples are assigned to that cluster. Therefore, the
new center can be figured out for the next iteration.

5.3 DBSCAN

Density-Based Spatial Clustering of Applications with Noise (DBSCAN) is a popular
unsupervised learning technique used in model building and ML algorithms. It is a

Fig. 10 Different types of points in DBSCAN

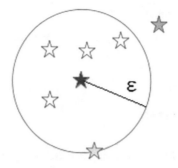

Red: Core Point

Yellow: Border Point

green: outlier

MinPts: 4

clustering technique that is useful in creating groups out of a given dataset on the basis of density. The portions having higher density are grouped as a single cluster. Basically, the algorithm seeks for the areas having higher density to form the cluster and leaves the areas having lower density for the next iteration to be included in other clusters. Another advantage of this algorithm is that it can process data of varying shapes efficiently.

Epsilon (eps), which is the radius of the neighborhood around a point x called as the ε-neighborhood of x; and minimum-points (MinPts), which is the minimum number of neighbors within eps radius of point x are two very important parameters associated with this algorithm. On the basis of eps and MinPts, the entire dataset can be categorized in three ways.

Core point: That point in the dataset whose neighbor count is greater than or equal to MinPts.

Border point: That point in the dataset whose neighbor count is less than MinPts.

Outliers: Those points who do not come under the ε-neighborhood of any core point (Fig. 10).

In the given figure, if we consider MinPts = 4 and eps = ε, then the red dot is a core point, the blue dot is a border point, and the orange dot is an outlier.

The following are some terms that are defined to understand the DBSCAN algorithm.

Direct density reachable: A point x_1 is called as directly density reachable from a point x_2 if x_2 is a core point and x_1 comes under the ε-neighborhood of x_2.

Density reachable: A point x_1 is density reachable from a point x_2 if there is a set of core points that leads from point x_2 to point x_1.

Density connected: Two points x_1 and x_2 are density connected if a core point x exists between them, such that both points x_1 and point x_2 are density reachable from point x.

The algorithm of density-based clustering works as follow:

1. For each point $x_i \in X$ [where X is the entire dataset]

Find the distance between x_i and other points.

if the number of points, having distance less than or equal to *eps,* is greater than or equal to MinPts, then declare x_i as a core point.

2 For each point $x_i \in X$

If x_i is not a core point, then

a. if it comes under the ε-neighborhood of any core point, then assign the point x_i to the cluster of that core point.
b. Else declare it as an outlier.

5.4 Parallel DBSCAN

In this section, we will be discussing a parallel approach of DBSCAN clustering algorithm, where multiple computers are interconnected through a network. For the parallel DBSCAN algorithm to work efficiently, the data placement strategy plays the most important role.

Considering a dataset of n-dimensional points $D = \{D_1, D_2, ..., D_N\}$, a density-based clustering is explained here. Along with the dataset D, a set of computers $C = \{C_1, C_2, ..., C_N\}$ connected through a network and a minimal density of clusters defined by *eps*, and *MinPts* are also considered [2]. Sub-dataset D_i is assumed to be present in computer C_i.

The method consists of three basic steps: **First**, the whole dataset is going to be partitioned and placed in different computers such that $\overset{N}{\underset{i=1}{\cup}} D_i = D$, and the sub-dataset D_i is placed in computer C_i, for $i = 1$ to N; **Second**, the data in different computers need to be clustered using DBSCAN concurrently as,

$$Cl_i = \text{DBSCAN}(D_i, \text{eps}, \text{MinPts}) \tag{5.3}$$

where Cl_i is the cluster formed in computer C_i for $i = 1$ to N; **Third**, the clustered data need to be merged back to find a global clustering for the given dataset such as, $Cl_1 Cl_2 ... Cl_N = Cl$ (final global cluster), where $\bar{U}$ is the merging function.

The most crucial part is the data placement in the case of a parallel DBSCAN algorithm. The following things need to be taken care of while placing the data in a distributed manner.

Load balancing: The data should be divided among the nodes equally so that the DBSCAN procedure that is going to be performed concurrently will be finished at the same time assuming that all computers have the same processing ability.

Minimized communication cost: Communication cost must be minimized, and to do so, it has to be made sure that inter-node communication during DBSCAN procedure is as less as possible.

Distributed data access: Inter-node communication must be made possible so that nodes can access data from other computers when it is required.

There are many data structures that can be used to store spatial data. One such data structure is an R-tree.

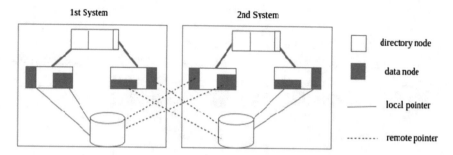

Fig. 11 Distributed R-tree

R-Tree: It is a data structure that is commonly used to store spatial data. If the root node is not the only node present in the tree, then the root must contain at least two children. The intermediate nodes have a pointer (ptr) that points to the child node and a minimal bounding rectangle (MBR) that covers the entire child node. But the child node structure is different. It contains a pointer (obj) to the object description and a minimal bounding rectangle (MBR) of the object. As all the leaf nodes are at the same distance from the root, it is a balanced tree.

In order to perform parallel DBSCAN, a distributed R-tree (dR-tree) is used which is a bit different from the traditional R-tree. Unlike R-tree, dR-tree has the entire data stored in multiple computers, and the indices of the data in each computer are replicated in each and every other computer. The intermediate nodes in dR-tree contain another field, i.e., an identifier(cptr) to locate the computer along with ptr and MBR to locate.

In Fig. 11, there are 4 data pages in total; out of which, 2 are present in first system, and other two are present in second system. The dotted line indicates the accessing of data present in the remote system.

The parallel DBSCAN is performed on the basis of a master–slave model, where the master is like the manager that controls the slaves and assigns works to them and collects output from them. But the slave does the actual computation. A slave is spawned in each computer by the master and is responsible for clustering the data in that computer. Then, it sends the output to the master. By performing data placement strategy, the dataset is partitioned into smaller data units called as load(S). load$_i$(S) is placed in computer C_i (Fig. 12).

The part DBSCAN is meant to find clusters in a partition, load$_i$(S) of the database D. As part DBSCAN performs only on a partition load$_i$(S) and not on the entire dataset like DBSCAN, a space constraint has to be adapted.

Algorithm 5.4 partDBSCAN (Clustering of spatial dataset by JOCHEN JÄGER et al. 1999).

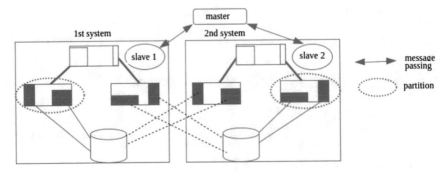

Fig. 12 Master–slave architecture for parallel DBSCAN

Algorithm: **PartDBSCAN** (S, dR-tree, Eps, MinPts)
//S is the workload and unclassified
//L is the list of merging candidates
1. Initialize *L* to be empty;
2. *for all* objects o in S *do*,
      if o is unclassified
            //construct a cluster w.r.t *S, Eps, MinPts* containing o
            if ExpandCluster(S, dR-tree, o, Clustered, Eps, MinPts, L)
                 increase ClusterId;
            end if
      end if
3. *End for*
4. *if L NOT* empty
      send L to the master;
5. *end if*

A DBSCAN with respect to space constraint can be understood by defining the terms directly density reachable, density reachable, and density connected in terms of space constraint.

Directly density reachable with respect to the space constraint S: A point *a* can be called as directly density reachable from another point *b* with respect to the space constraint S, Eps, and MinPts if

1. $b \in S$,
2. $a \in N_{eps}(b)$, and
3. $\|N_{eps}(b)\| \geq MinPts$.

In Fig. 13, point a and b are two core points. a forms a cluster with respect to the space constraint $S1$, and b forms a cluster with respect to the space constraint $S2$. Point *p* and *q* are the points falling in the intersection part and will be considered while taking care of the merging candidates which will be discussed shortly.

Likewise, density reachable and density connected can be defined with respect to space constraint and can be compared the same with respect to the entire database.

If C is a cluster and all its members are inside S, then the job is done but if some members of C are present outside C, then C may need to be merged to another

Fig. 13 Neighbor constraint space having merging candidate

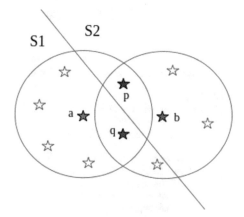

cluster formed with respect to S. Such a cluster is called a merging candidate (MC). All the merging candidates are collected together in a list L and sent to the master at the end. A merging candidate is represented by MC (C, S), where C is the cluster whose members are present outside C, and S is the constraint space. Two adjacent merging candidates are merged by the master if there is a core point located at their intersection. The merging candidate MC (C, S) is not exactly the original cluster C but a subset of it. The reduction of size can be explained as follows:

We know that two merging candidates will be merged if they have a common core point. From the definition of cluster, we know that each point present in the cluster has to be density reachable from a core point in that cluster in the space constraint. This implies no point of C, outside of S, has a distance more than eps from the boundary of S. Similarly, no point from another cluster can be present in C that has a distance greater than eps from the boundary of S. Hence, the intersection between the two clusters is a common region around the boundary of S with a distance maximum eps from both the sides. Thus, the merging candidate MC (C, S) can be concluded to be a subset of C.

Algorithm 5.5 Merging (Clustering of spatial dataset by JOCHEN JÄGER et al. 1999).

```
merging(LL)

For i from 1 to LL.size DO
    L1 = LL.get(i);
    for j from 1 to L1.size DO|
        C1 = L1.get(j);
        for k from i+1 to LL.size DO
            L2 = LL.get(k);
            for m from 1 to L2.size DO
                C2 = L2.get(m);
                if C1 ∩ C2 ≠ φ then
                    if C1 = C1 U C2;
                        L2.remove(C2);
```

Algorithm 5.6. ExpandCluster (Clustering of spatial dataset by JOCHEN JÄGER et al. 1999).

```
ExpandCluster(S, dR-tree, o, ClusterId, Eps, MinPts, L):

1. MC.init();                                      //initialize the merging candidate set;
2. retrieve Eps-neighborhood N_eps (o)
3. if |N_eps(o)| < MinPts                          //i.e. o is not a core point;
       mark o as noise and RETURN false;
4. Else                                            //i.e. o is core point
       select a new clusterId and mark all objects in N_eps (o) with clusterId;
       push all objects from N_eps (o)\{0} onto the stack seeds;
       While not seeds.empty() do
          currentObject = seeds.top();
          seeds.pop();
          if currentObject ∈ S
              retrieve Eps-neighborhood N_eps (currentObject);
              if |N_eps(currentObject)| ≥ MinPts
                  select all objects in N_eps(currentObject) not yet classified or marked as noise;
                  push the unclassified objects onto seeds and mark them with ClusterId;
                  if N_eps(currentObject) \ S Not Empty
                      insert {currentObject} U N_eps(currentObject)\S into the set MC
                  end if
              end if
          else                                     //currentObject is not element of S
              insert o and currentObject into set MC;
          end if
       end while
       if MC ≠ Empty THEN
           L.append(MC);
       end if
5. end if
6. Return True;
```

Another very important algorithm is the **ExpandCluster** algorithm. This algorithm basically expands the cluster to the very last point starting from the core point. The formation of clusters is explained in DBSCAN. Whenever a point outside S is encountered, while forming the cluster, the outside point with its core point is put in MC of the cluster. If MC is empty, then the cluster is a global cluster (a cluster with respect to DB) else it may need to be merged. Once the expansion process is over, MC will be appended to L.

There is another function: Merging, which needs to be executed in order to get the final result. Now, the master receives the list of merging candidates from each slave, if there. All the MC lists are collected from each slave and are added to another list LL. As long as the list LL is not empty, the function merging is executed. The function goes through each MC present in LL and merges those MCs whose intersection is not empty as they are the merging candidates of adjacent space constraints.

6 Parallelization of Support Vector Machine

Support Vector Machine (SVM) is a popular machine learning algorithm for supervised learning. The parallel version of SVM is based on solving the optimization problem for determining parameters in a distributed and parallel manner. It must be noted that the training of SVM is a computationally expensive process. The computational cost increases in case the size of the training dataset increases. The solution to the problem is proposed by many researchers as the parallelization of the SVM in various computation nodes. One of the key phases of the parallelization of SVM training is partitioning the training data into subsets and allocating them to various computing nodes in a distributed environment. The main challenge is, however, to train the SVM parallelly in the node in such a way that the weights or support vectors derived from training the subsets can be used to determine the weights or support vectors of the entire training dataset. The process must ensure that the accuracy of the classifier must match the accuracy of the traditional, non-parallelized SVM as much as possible.

Following are the brief description of some of the techniques used in parallelization of SVM:

1. Training SVM's in subsets of training data and combining the SVM classifiers into a global SVM classifier.
2. Training the SVM iteratively by sharing the training data subsets among various nodes. However, this may increase the overhead.
3. Parallelly, train multiple SVM by training using the support vectors collected from one SVM. Thus, eliminating non-support vectors and reducing overhead.
4. Using the MapReduce framework to train the SVM, the mappers generally produce partial weights which are combined or aggregated by the reducers.

Basically, support vector machine computes a discriminate linear classifier given a set of training data of the form

$$f(x) = w^T x + c. \tag{6.1}$$

So that predict $y = 1$ if $f(x) \geq 0$ and $y = 1$ if $f(x) < 0$.

This can be also expressed as Eq. (6.2), where n is the number of training samples.

$$f(x) = \left(\sum_{i=1}^{n} \varphi_i y_i x_i \right)^T + c$$

$$= \sum_{i=1}^{n} \varphi_i y_i (x_i, x) + c \tag{6.2}$$

The dual form of regularized support vector machine optimization for non-linearly separable datasets is derived from the optimal margin classifier given as

$$\min \frac{1}{2}||x||^2 + K \sum_{i=1}^{n} \xi_i \tag{6.3}$$

$$\text{s. t.} \quad y\left(w^{\mathrm{T}}x_i + c\right) \geq 1 - \xi_i, \quad i = 1, \ldots, n$$

$$\xi_i \geq 0, \quad i = 1, \ldots, n.$$

The Lagrangian of the above objective function can be defined as

$$\frac{1}{2}w^{\mathrm{T}}w + K \sum_{i=1}^{n} \xi_i - \sum_{i=1}^{n} \varphi_i \left[y_i \left(x^{\mathrm{T}}w + c\right) - 1 + \xi_i \right] - \sum_{i=1}^{n} m_i \xi_i \tag{6.4}$$

With φ_i and m_i are the Lagrange multipliers and are constrained to be greater than zero. After solving the Lagrangian by setting the derivatives with respect to w and c to zero and substituting, the final dual form obtained is

$$\max \sum_{i=1}^{n} \varphi_i - \frac{1}{2} \sum_{i,j}^{n} y_i y_j \varphi_i \varphi_j \left(x_i, x_j\right) \tag{6.5}$$

$$\text{s. t.} \quad 0 \leq \varphi_i \leq K, \quad i = 1, \ldots, n$$

and, $\sum_{i=1}^{n} \varphi_i y_i = 0$

With $w = \sum_{i=1}^{n} \varphi_i y_i x_i$, thus w is expressed in terms φ_i.

Based on Karush–Kuhn–Tucker (KKT) conditions, the sufficient condition for a solution to our objective that all φ_i satisfy are the following:

$$\varphi_i = 0 \quad \text{if,} \quad y_i \left(x^{\mathrm{T}}w + c\right) \geq 1$$

$$\varphi_i = C \quad \text{if,} \quad y_i \left(x^{\mathrm{T}}w + c\right) \leq 1$$

$$0 < \varphi_i < K \quad \text{if,} \quad y_i \left(x^{\mathrm{T}}w + c\right) = 1$$

The parallelization of SVM is basically to; solve the optimization problem in a distributed environment. Conventionally, to solve the optimization problem in a fast and efficient manner is through sequentially minimal optimization (SMO).

Sequential minimal optimization is based on coordinate ascent algorithm to gradually update the parameter φ_i keeping other $\varphi_{i+1} \ldots \varphi_n$ fixed to maximize the objective function. However, it can be shown that the constraint cannot be satisfied by only updating one parameter at a time keeping all the other parameters fixed. Platt [3] provided the selection of a set of only two points as the working set due to the following condition:

$$W(\varphi_i) = \sum_{i=1}^{n} \varphi_i y_i = 0 \tag{6.6}$$

The SMO algorithm is provided as follows:-

Repeat until convergence {

1. Select a pair (φ_i, φ_j) using a heuristic which will help to move toward the global maximum and update in the next iteration.

2. Re-evaluate $W(\varphi_i)$ with respect to φ_i and φ_j, while keeping the values of all the other φ_k fixed.

}.

The Karush–Kuhn–Tucker (KKT) conditions can be checked satisfied or not to ensure whether the algorithm is converging or not.

For parallelization of the SVM using the MapReduce framework, various algorithms are provided [4, 7]. MapReduce-based SMO provided in [4] uses a caching scheme to store the output temporarily and update the value after each successful step. The entire dataset is partitioned into subsets of data, and each partition is provided to a single map task or mapper to be optimized in parallel using the SMO algorithm. The number of mapping tasks is the same as the number of partitions created. The mapping functions output the partial weights w_partial for the partitions and bias c for the partition. The reducer function sums up the partial weights w_partial of each partition and for the bias c which varies among each partition is taken as average to output an overall w_global and bias c. In case of non-linear SVM [4], an array is created to output the values of φ_i for each partition. The reduce phase sums the values in the array and takes an average of the individual bias c of each partition (Fig. 14).

7 Decision Tree

Decision tree classification is organized as a tree with nodes defined using the features or attributes of the dataset. The leaf nodes are the predictive attribute or features. The tree is constructed using a top-down approach in a recursive way. The main challenge in this classification algorithm is the selection of attributes as the nodes at each level. The decision tree supervised training can be classification or regression based on the nature of the values of the predictive attribute. To create a tree, the attributes are selected at each node based on entropy and information gain. The main objective of splitting the node is to determine the partitions in which the tuples are as pure as possible means the tuples belong to a particular class.

$$\text{Entropy}(X) = \sum_{i=1}^{m} -p_i \log_2 p_i \tag{7.1}$$

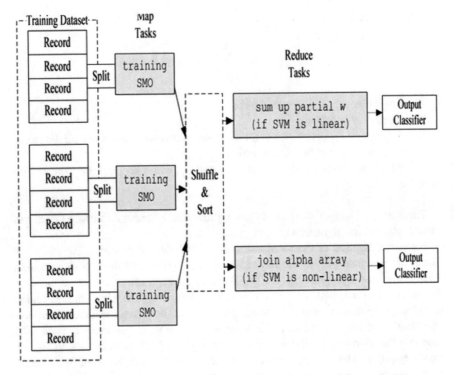

Fig. 14 Illustration of MapReduce SMO algorithm using map tasks and reduce tasks [4]

The information gain is used as a criterion for the selection of attributes at each node. The attribute having maximum information gain is selected.

$$\text{Gain}(D, A) = \text{Entropy}(D) - \sum_{j=1}^{v} \frac{|D_j|}{|D|} \text{Entropy}(D_j) \qquad (7.2)$$

where D is a partition of data, A is an attribute from attribute set, and v is the distinct values in tuple partition D on some attribute A.

The MapReduce framework is popular for parallelizing the decision tree construction [8, 9]. As the size and dimension of the data increase, the computational expense of selection criteria, i.e., information gain also increases.

The mapping function first splits the whole datasets into subsets of the dataset and stores them in the mapping nodes in the Hadoop framework. The information gain for each of the attributes are calculated in each mapping node with the subset of datasets and stored in the form of <Attribute, Gain> <key, value> pair. The reducer function calculates the sum of gain for all the attributes, and the attribute having maximum gain is selected as a splitting attribute.

8 Conclusion

Given that the size of the enterprise data is increasing day by day, the modeling and learning processes are becoming computational expensive using traditional computational environments. This chapter discussed this issue and the solution provided to it by a scalable, distributed environment of computation. The various algorithms of traditional machine learning, i.e., regression, clustering, and classification are discussed in the parallel environment. One of the popular frameworks for scaling up both the data and computation is the MapReduce framework. The chapter discussed in detail various ways reported in the literature that is used to transform popular machine learning algorithms into Map and Reduce functions. More specifically the mathematical equations related to various algorithms are placed in various Map and Reduce functions to distribute the computational expense to different computational nodes.

References

1. Verbraeken, J., M. Wolting, J. Katzy, J. Kloppenburg, T. Verbelen, and J. S. Rellermeyer. 2019. A survey on distributed machine learning. arXiv preprint arXiv:1912.09789.
2. Xu, X., J. Jäger, and H.-P. Kriegel. 1999. A fast parallel clustering algorithm for large spatial databases. In *High Performance Data Mining*, 263–290. Boston, MA: Springer, Boston.
3. Platt, J. 1998. Sequential minimal optimization: A fast algorithm for training support vector machines.
4. Alham, N. K., M. Li, Y. Liu, and S. Hammoud. 2011. A MapReduce-based distributed SVM algorithm for automatic image annotation. *Computers & Mathematics with Applications* 62 (7): 2801–2811.
5. Bekkerman, R., M. Bilenko, and J. Langford, (eds.). 2011. *Scaling up machine learning: Parallel and distributed approaches*. Cambridge: Cambridge University Press.
6. Zhao, W., H. Ma, and Q. He. 2009. Parallel k-means clustering based on mapreduce. In *IEEE International Conference on Cloud Computing*, 674–679. Berlin, Heidelberg: Springer.
7. Çatak, F. Ö., and M. E. Balaban. 2016. A MapReduce-based distributed SVM algorithm for binary classification. *Turkish Journal of Electrical Engineering & Computer Sciences* 24 (3): 863–873.
8. Mu, Y., X. Liu, Z. Yang, and X. Liu. 2017. A parallel C4. 5 decision tree algorithm based on MapReduce. *Concurrency and Computation: Practice and Experience* 29 (8): e4015.
9. Ben-Haim, Y., and E. Tom-Tov. 2010. A streaming parallel decision tree algorithm. *Journal of Machine Learning Research* 11 (2).

An Implementation of Text Mining Decision Feedback Model Using Hadoop MapReduce

Swagat Khatai, Siddharth Swarup Rautaray, Swetaleena Sahoo, and Manjusha Pandey

Abstract A very large amount of unstructured text data is generated everyday on the Internet as well as in real life. Text mining has dramatically lifted the commercial value of these data by pulling out the unknown comprehensive potential patterns from these data. Text mining uses the algorithms of data mining, statistics, machine learning, and natural language processing for hidden knowledge discovery from the unstructured text data. This paper hosts the extensive research done on text mining in recent years. Then, the overall process of text mining is discussed with some high-end applications. The entire process is classified into different modules which are test parsing, text filtering, transformation, clustering, and predictive analytics. A more efficient and more sophisticated text mining model is also proposed with a decision feedback perception in which it is a way advanced than the conventional models providing a better accuracy and attending broader objectives. The text filtering module is discussed in detail with the implementation of word stemming algorithms like Lovins stemmer and Porter stemmer using MapReduce. The implementation set up has been done on a single node Hadoop cluster operating in pseudo-distributed mode. An enhanced implementation technique has been also proposed which is Porter stemmer with partitioner (PSP). Then, a comparative analysis using MapReduce has been done considering above three algorithms where the PSP provides a better stemming performance than Lovins stemmer and Porter stemmer. Experimental result shows that PSP provides 20–25% more stemming capacity than Lovins stemmer and 3–15% more stemming capacity then Porter stemmer algorithm.

S. Khatai (✉) · S. S. Rautaray · S. Sahoo · M. Pandey
KIIT Deemed to be Univerisity, Bhuneshwar, Odisha, India
e-mail: k.swagat1391@gmail.com

S. S. Rautaray
e-mail: siddharthfcs@kiit.ac.in

S. Sahoo
e-mail: swetaleenafet@kiit.ac.in

M. Pandey
e-mail: manjushafcs@kiit.ac.in

© The Author(s), under exclusive license to Springer Nature Singapore Pte Ltd. 2021
S. Rautaray et al. (eds.), *Trends of Data Science and Applications*,
Studies in Computational Intelligence 954,
https://doi.org/10.1007/978-981-33-6815-6_14

273

Keywords Big data · Hadoop · MapReduce · Word stemming · PSP · Decision feedback

1   Introduction

In today's world, the amount of unstructured data is growing in an enormous way that the existing relational systems are incompetent in handling them. The form of data can be audio-video clips, textual data, software program logs, flight records, etc. The information hidden inside those data leads to a complete new world of opportunity and insight. This is the reason for why every organization and individual is demanding to explore these huge amount of data, which constructs the foundation of text mining. It is also called as a practice of textual form of data to discover the key conceptions, themes, hidden trends, and relationships without prior knowledge of exact terms that has been used by author to express the concept [1]. As part of text mining algorithms of data mining, text analytics, machine learning, natural language processing, and statistics are used to extract high quality, useful information from unstructured formats. Test mining is also popular as "text analytics" is a means by which unstructured data is processed for machine use. For example, if a Twitter comment "I don't find the app useful: it's really slow and constantly crashing." is taken into consideration then text mining of the contextual information is really important to help us understand why the tone might be negative and what may be the cause of such customer disappointment as shown in Fig. 1. These analyses may lead to the answer of questions like "Is the person replying to another negative tweet? or is this the original composition? or what is the application name? or is this the only problem with the app or there are other problems too?, etc.

1.1   Conventional Process Flow of Text Mining

Textual data are in the form of unstructured data are normally available in readable document formats. These formats can be user comments, e-mails, corporate reports,

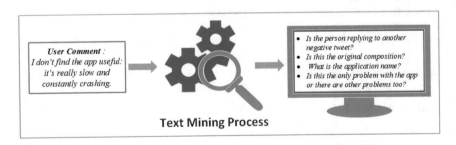

Fig. 1 Text mining

web pages, news articles, etc. According to conventional text mining process, the documents are first derived into a quantitative representation. Once the textual data is transformed into a set of numbers which precisely capture the hidden pattern in it, then any data mining algorithm or statistical forecasting model is applied on the numbers for generating insights or for discovering noble facts [2, 3].

A typical text mining process generally have the following sub-tasks to complete the process.

Data Collection Collection of textual data is the first step in any text mining research [3].

Text Parsing and Transformation The next step is to parse the words from the documents. Therefore, sentences, parts of speech, and stemming words [3] are identified from the document. Document variables associates with author, category, gender, etc., are also extracted with the parsed words.

Text Filtering After the parsing of words, there may be some irrelevant words which are not required in the analysis, and those words are removed from the document. This is done manually by browsing through the terms or words. This is the most time-consuming and subjective tasks in all of the text mining steps. A fair amount of subject knowledge and domain knowledge is required to perform this task. In case of document filtering [3], the selected keywords are searched in all the selected documents. If any document does not contain any of the keywords, then it is removed from the list of analysis.

Text Transformation In this step, the document is presented in a numerical form of matrix [3]. This matrix generally contains the occurrences of the words is also called as term frequency. Numerical presentation of the document is mandatorily required to perform any kind of analytics on the document. Therefore, this step converts the unstructured text to a workable analytical document.

Text Mining In this step, hidden patterns and knowledge are extracted using mining algorithms such as classification, clustering, association analysis, and regression analysis. As shown in Fig. 2, text mining is an iterative process where the process of filtering to mining is repeated based on the feedback received from this step [4].

1.2 Applications of Text Mining

Text mining process is being used to provide answers to industrial queries and to optimize daily operations efficiently. It is also used to develop business strategic decisions in finance, automobile, marketing, health care, etc. Hidden patterns, trends, and perceptions are discovered from a huge volume of unstructured data using techniques like data analytics, categorization, and sentiment analysis. In this research, we have discussed below applications of text mining.

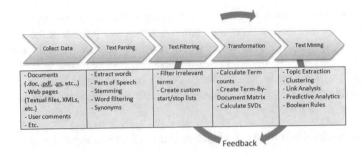

Fig. 2 Conventional text mining process

Risk Management Inadequate risk estimation is accounted for biggest reason of failures in any industry. In these cases, text mining is used to estimate the proper risk in business and also to identify the most adequate way to mitigate the risk [3]. Therefore, the application of text mining software has drastically increased the capacity of risk mitigation in industries.

Knowledge Management Managing huge volume of data containing the historical information creates many problems like huge storage space, latency in finding specific information, etc. The healthcare industries are a classic example for the above problems where the information of historical patients' data can be potentially used for medical analysis and product development [3]. Therefore, text mining is used to filter the useful informations by discarding the irrelevant ones. Then, many analytic algorithms are run on the filtered data to find and store the extracted unknown facts only, which reduces the storage issue, latency issue, etc.

Cybercrime Prevention Random availability of information over Internet can bear the brunt of cybercrimes. Text mining is used to trace the cybercrime activities and also helps to identify the source of intruders [3]. Therefore, text mining is used by law enforcement and intelligence agencies.

Customer Care Service Customer care services are better operated using text mining and natural language processing. Text analytics software improves customer experiences. These analytics use many valuable information sources such as survey and customer call notes which help effectiveness and speedy resolution of customer problems [3]. Test mining is also used for automated faster responses to customer queries.

Contextual Advertising Digital advertising has got a new height of safety and user's privacy by applying text mining as core engine of contextual retargeting [3]. It also provides better accuracy in contextual advertising.

Business Intelligence Text mining is uses to support faster decision making by taking consideration of valuable enterprise data [3]. It helps to find future insights for improving the business by monitoring huge number of data sources.

Social Media Social media is a potential source of huge amount of unstructured data inside which a lot of hidden patterns related to business, sentiment [5], and intelligence are there. Many organizations predict the future customer needs using text analytics. This information help organizations to extract the customer opinions, to understand their emotions, and also to predict their requirements. Text mining has made revolutionary modifications in social media.

2 Literature Survey

As text mining is our focus of research, therefore, some recent research artifacts are studied. All the related studies and analysis points that application of big data technologies like Hadoop MapReduce, k-means, particle swarm optimization (PSO), and cloud computing provides better result, reduced execution time and better solution for big data problems. Large data sets can be analyzed using Hadoop cluster and parallelization of clustering algorithms and using parallel k-means clustering provides a drastic reduction in execution time [1]. Document clustering, parallel k-means, and distributed computing [6] are the techniques that have been used with Hadoop MapReduce in the study. After selecting centroids randomly, every document is fed to one mapper. The mapper calculates the new centroids based on the Euclidean distance. The result of all mappers is sent to a reducer to calculate a resulting centroid which then compared with the assumed centroid [7]. If there is a difference in centroid value, then the process is iterated, otherwise, the centroid is considered as final output as shown in Figs. 3 and 4.

To settle the number of cluster and initial centroid, the parallel k-mean algorithm is modified which can be optimized using fuzzy logic, gravitational intelligence, and swarm optimization. Big data has its own challenge in terms of storing the data and retrieving it fast. Manual grouping of files is very complex when there is a huge amount of document. A new working k-means non-negative matrix factorization (KNMF) with modified guideline of non-negative matrix factorization [8, 9] is used for document clustering. Comparison study of iterated Lovins algorithm, Lovins algorithm, and Porter algorithm of text mining shows that maximum words are stemmed in iterated Lovins algorithm. Therefore, the characteristics of k-means non-negative matrix factorization help in clustering the documents with parallel implementation of MapReduce on large sized documents. This results in quick and easy clustering as well as less time consumption.

In order to shrink the computational time, HDFS, MapReduce, and clustering algorithms are used by distributing the clustering jobs on multiple nodes which means multiple clustering tasks run parallel on different nodes. A comparative review of components of Hadoop and MapReduce has been studied to compare result with the traditional partition-based algorithms with their implementation in MapReduce paradigm to achieve various clustering objectives on different size of data sets [7]. Introduction of combiner programs between the map and reduce function helps in

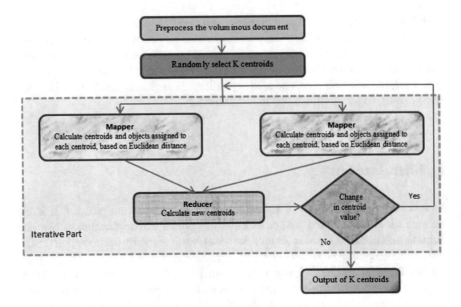

Fig. 3 Stages of document clustering using parallel k-means

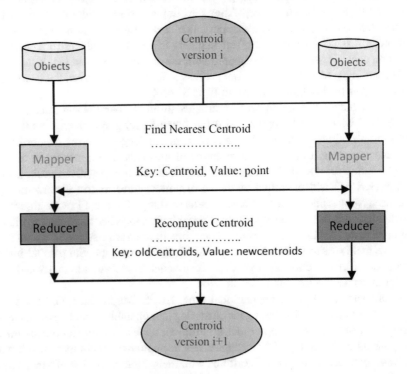

Fig. 4 Parallel k-means algorithm with MapReduce

reduction of volume of data to be written by mapper and volume of data to be read by reducers that decrease the overall operation time. The time reduction is highly realizable when the number of document is huge rather than smaller data sets [1]. The model for implementation of parallel k-means clustering in MapReduce without a combiner is shown in Fig. 4.

With above working methods, the advantage of the ability of global search in particle swarm optimization (PSO) is used for optimal generation of centroids. The power of parallel processing with global search supports data intensive distributed application with improved accuracy in generating compact clusters [4]. Some more literatures are studied in context of text mining and a comparative study is presented in Table 1 in terms of their objective, findings, and methods used. From the study, it is clear that MapReduce is the most popular technology to handle text mining problems. Therefore, in our context of research, we have implemented the proposed text mining model using Hadoop MapReduce with partitioner.

3   Proposed Decision Feedback-Based Text Mining Model

Textual content is typically available in comprehensive document format. These formats can be e-mails, text file lettering, user feedback, sentimental comments, corporate reports, sentimental comments, news reporting, Web pages, etc. The proposed text mining model tries to first instigate a quantitative representation of document and then transfer the document into a set of numbers where the numbers adequately capture the patterns of textual data. Any traditional statistic model, forecasting model, and analytical algorithm can be used on these numbers for generating insights or to produce a predictive modeling. Statistical-based systems count the word frequency of each word and calculate their statistical proximity toward related conceptual indexes. These systems may produce inappropriate concepts and miss the required words which in turn reduces the prediction model accuracy. Iterative text mining decision feedback model is the advanced form of a text mining where the process is repeated till the result is acceptable without getting completely out of process. In this model, the feedback block is the controller of number of iterations. Feature selection, data analytics, and evaluation phase constitutes the feedback block. Also, this process minimizes the interference of irrelevant words to increase the model accuracy. The iterative text mining model which we have proposed has the design as shown in Fig. 5. The steps involved in this model are:

Data Collection Collecting an unstructured data set for analysis is always the first step of any text mining process.

Text Parsing and Transformation In this step, the data set is cleaned and a dictionary of words is created from the document using NLP. This includes identification of sentence, word, parts of speech, and stemming words [1]. The extraction of each word from document is associated with a variable for further reference in the process.

Table 1 Comparative study of recent works in text mining

Title	Year	Objective	Technology used	Finding
An analysis of MapReduce efficiency in document clustering using parallel K-means algorithm [1]	2018	To design and experiment a parallel k-means algorithm using MapReduce programming model and compared the result with sequential k-means for clustering varying size of document data set. The result demonstrates that the proposed k-means obtained higher performance and outperformed sequential k-means while clustering	Hadoop and MapReduce	MapReduce programming model for Hadoop cluster is a recent and popular trend in analyzing large data sets in short span of time. It is important to parallelize clustering algorithms using MapReduce for efficiency in clustering result in terms of execution time. This work proposed a parallel k-means algorithm using MapReduce for document clustering
Research trends on big data in marketing: a text mining and topic modeling based literature analysis [2]	2018	Given the research interest on big data in marketing, we present a research literature analysis based on a text mining semi-automated approach with the goal of identifying the main trends in this domain	Big data	This research literature analysis focused on the application of big data in marketing, in an attempt to identify the trends in these applied domains through different dimensions
Partition based clustering of large datasets using MapReduce framework: an analysis of recent themes and directions [7]	2018	To provide a comprehensive review of Hadoop and MapReduce and their components to compare recent research works on partition-based clustering algorithms which use MapReduce as their programming paradigm	Hadoop and MapReduce	This paper focuses on recent technologies for partition-based document clustering

(continued)

Table 1 (continued)

Title	Year	Objective	Technology used	Finding
Text mining with Lucene and Hadoop: document clustering with updated rules of NMF non-negative matrix factorization [8]	2018	Proper alignment of document files is to be labeled, when large number of files increases characterizing the files are needed, therefore, here comes the clustering of data, i.e., document clustering	MapReduce	A new processing techniques mainly in stemming algorithms that is iterated Lovins stemmer algorithm have given better results when compared to Porter stemmer and Lovins stemmer algorithm, and a new algorithm KNMF which is furthurly used and application named as "text mining lead"
MapReduce based analysis of sample applications using Hadoop [10]	2018	The rate of increase of structured, semi-structured, and unstructured data is very high. To discover hidden Information from different types of data is a big challenge. The two techniques, word frequency count and string matching, are applied on a single node and multi-node cluster with an input data set	Cloud computing, Hadoop, HDFS, MapReduce	This paper shows how to operate, manage, process, and analyze structured, semi-structured, and unstructured data by exploiting word count and string matching applications on Hadoop by varying MapReduce configuration

(continued)

Table 1 (continued)

Title	Year	Objective	Technology used	Finding
Distributed document clustering analysis based on a hybrid method [4]	2017	PSO is used to take advantage of its global search ability to provide optimal centroids which aids in generating more compact clusters with improved accuracy. This proposed methodology utilizes Hadoop and MapReduce framework which provides distributed storage and analysis to support data intensive distributed applications	Hadoop, MapReduce	MR-P k-means was proposed to overcome the inefficiency of PK means (or big data sets). The proposed method can efficiently be parallelized with MapReduce to process very large data sets. In MR-PK means, the clustering task that is formulated by k-means algorithm utilizes the best centroids generated by PSO
Performance evaluation of word frequency count in Hadoop environment [11]	2017	The research objective of this study is to measure the execution time on different sizes of text files by performing a simple MapReduce simulation on the word count program which is very popular in the big data arena. Also, an improved version of the word count program has been designed and simulated using the same set of text files	Big data analytics, HDFS, MapReduce parallel processing	The revolution of big data analytics has made a remarkable impact on all spheres of our modern living and also in the IT sector. The Hadoop MapReduce framework has found many important applications in big data analytics. In this present study, a simple and an improved version of word count programs have been simulated in the Hadoop MapReduce environment using different file sizes

(continued)

Table 1 (continued)

Title	Year	Objective	Technology used	Finding
Text document tokenization for word frequency count using rapid miner (taking resume as an example) [12]	2015	RapidMiner is unquestionably the world leading open-source system for data mining. It is available as a standalone application for data analysis and as a data mining engine for the integration into own products. Tokenization is the process of breaking a stream of text up into words, phrases, symbols, or other meaningful elements called tokens	RapidMiner	In this paper, the word frequency count of text document is done using RapidMiner tool—transform case and tokenize operators, with their interconnection. In order to find the frequency of occurrence of particular word, the user has to scroll the scrollbar
A distributed data mining system framework for mobile internet access log based on Hadoop [13]	2015	Single node-based data mining platform has been unable to store and analysis the massive data. According to cloud computing technology, we preset a distributed data mining framework based on Hadoop. Then, we present the implementation of this system framework and process mobile Internet access los on the Hadoop cluster	Cloud computing, Hadoop, Hive	We can write complex MapReduce of logic query language efficiently and quickly using Hive. Compared to traditional databases, Hive supports only a small set of primitive data types. Some of Hive's extensions are in view

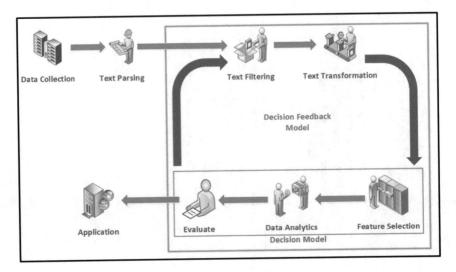

Fig. 5 Proposed model of text mining

Text Filtering In the parsed document, there will be some words which are not relevant to the mining process and those words need to be filtered out from the document called as word stopping and word stemming [14, 15]. This process requires an in-depth knowledge of the domain. Number of word stemmed are denoted by "S". The word stemming process has been discussed in more detail in further sections.

Text Transformation After text filtering, the document is presented by the occurrences words contained in it. After transformation, a document can be represented in two ways such as

- A simplified representation used in information retrieval and natural language processing where it contains the multiset of words irrespective of grammar is known as a bag of word. It is a JSON object representation.
  Bow1 = {"John": "3", "is": "1", "Good": "5"}
- Vector space model is an algebraic representation of text involving two steps. First, the document is represented in a vector of words and then the vector is transferred into a numerical format where the techniques of text mining can be applied. In this research, the documents have been represented in a vector space mode.

Feature Selection It is also known as variable selection in which we select a subset of more important features to be considered in the model creation. Irrelevant and redundant features are not to be used in model creation to improve the model accuracy.

Data Mining At this stage, the traditional data mining process is merged with text mining. Classical data mining techniques are used for clustering of the data that obtained from the quantitative representation of document to be associated in further evaluation steps. K-mean clustering [9] or a parallel k-mean clustering [13] technique is taken into consideration in this phase.

Evaluate In this step, we evaluate the mining result. After evaluation, if the result is not acceptable, then we discard the result and continue the process as an iterative model to get the best results. Once the result is acceptable, we proceed to next step.

In this step, word stem factor (WSF) is calculated to decide the result acceptance. Word stem factor (WSF) is defined as the percentage of number of word stemmed to total number of distinct word. Word Stem Factor (WSF) = (S/T) * 100

Application The evaluated model is now have a broader area of application in the different text mining process. This model is ready as a product to be deployed in real-life problems. The model can be applied in web mining, E-consultation in medical, Twitter data analysis, and resume filtering.

4 Big Data Technologies

In hope of using data in future organizations collect and store by organizations store enormous amount of data. A number of significant global challenges have been notified as revolution in big data technologies [16]. The way organizations are collecting, using, managing, and leveraging data using big data technologies is ways beyond of imagination. In this research, we have focused on the most popular big data technology—Hadoop. It is one of the most sophisticated and ever growing ecosystems in the era of big data. Different technologies of Hadoop ecosystem have been briefly discussed.

4.1 Hadoop Distributed File System

To store huge amount of data in cluster of computers and to channel them to the required applications at a high bandwidth Hadoop distributed file system (HDFS), it is used inside Hadoop ecosystem. Large cluster constituting hundreds and thousands of server nodes built of commodity hardware to execute user application tasks [16, 17]. Storage and computation are distributed across servers and the system provides a technique of parallel processing and the required resource for each node have the capability to grow with demand while cost remains economical at every size. Data is stored in files and files are placed on nodes providing replication for fault tolerance. Some unique features of HDFS are highlighted below.

- Physical location of node is considered in rack awareness for storage allocation and task scheduling.

- Minimal data motion that process is moved to data rather than moving data to process. This technique reduces bandwidth.
- The previous versions of storage are restored using standby name node and secondary name node in case of human or system errors.

4.2 MapReduce

As a parallel processing framework, Hadoop MapReduce is used for processing huge amount of data in very less time. Large amount of data are processed clusters containing thousands of node built from commodity hardware. The cluster is highly reliable and fault tolerant. Job tracker is the single master nose and multiple task trackers acting as slave node constitutes the initial architecture of Hadoop framework. Whereas yet another resource negotiator (YARN) is the advanced Hadoop architecture [10, 13]. Resource manager is responsible for job scheduling on slave nodes, monitoring the task execution, and re-executing the failed tasks. Some more advantages of MapReduce are mentioned below.

- Commodity hardware is added to the existing server to increase the capacity is also known as scale-out architecture or horizontal scaling.
- Failed tasks are automatically recovered proving the fault tolerance of cluster.
- Flexibility for amount of file systems and facility of serialization in multiple open frameworks.
- Intelligent data placing technique to maintain the load balancing with maximum utilization and efficiency.

A MapReduce process is shown in Fig. 6. The data split files are executed in mapper parallely. After mapper phase is completed, the interim results are sorted and shuffled. Then, the results are merged and fed to the reducer. The no of reducer is defined in the MapReduce program determines the number of output files.

4.3 Pig

Pig is a data flow tool used for analyzing large data sets. It is not specific to Hadoop only rather it can be used with any parallel data processing. Though it can support all types of data that is structured, semi-structured, quasi-structured, and unstructured data but very frequently used for structured and semi-structured data. It uses Pig Latin language [18]. Each line in Pig code is converted to a logical plan and series of MapReduce tasks. It creates a directed acyclic graph (DAG) for each job. Features:

- Pig provides ease of programming where developers have to write less number of coded than MapReduce for a particular requirement.
- In case built-in functions are not available, users can create custom programming which can be easily integrated with Pig.

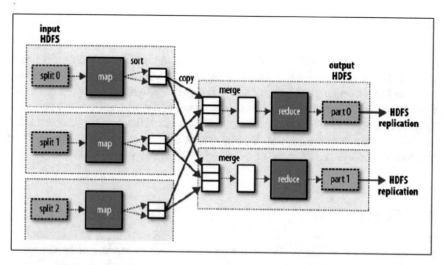

Fig. 6 MapReduce processing

4.4 Hive

As a data warehouse software Apache Hive inside Hadoop ecosystem helps to query, analyze, and manage large data sets stored distributed storage (HDFS). It provides the facility of HIVEQL, an SQL-like language for querying and retrieving data. All the Hive queries are converted into MapReduce job by Hive engine automatically and implicitly. When it is difficult to express logic in HIVEQL, it allows MapReduce programmers to be plugged in with Hive using custom mappers and reducers [17].

Hive allows indexing to provide acceleration in data search. Compaction and Bitmap indexes are also applicable in Hive. It supports different file types like plain text, RCFile, and ORC. It can operate on compressed data storages using GZIP, BZIP2 an SNAPPY. User-defined functions (UDFs) are supported by Hive when built-in functions are not available.

4.5 Sqoop

The facility to transfer data between HDFS and RDBMS (MySQL and ORACLE) is provided by Sqoop inside the Hadoop ecosystem. It imports data from RDBMS to HDFS to process it and again export the data to RDBMS [18]. It facilitates the connection of different database servers, controlling of import and export process. It can import data to Hive and HBase.

4.6 Oozie

Hadoop ecosystem provides the facility of a Web application based on Java used for scheduling Hadoop jobs is known as Apache Oozie. It sequentially combines multiple jobs to one logical unity of work. It supports MapReduce jobs, Pig scripts, Hive query, and Sqoop import exports. Jobs of a specific system like Java or a shell script program can also be scheduled in Oozie. Oozie workflow and Oozie coordinator are two categories of Oozie jobs. Multiple workflow and coordinators are bundled in Oozie to manage the lifecycle of running jobs. It is scalable and reliable.

4.7 Flume

Flume is used for efficient collection, aggregation, and movement of large amount streaming data like record logs. It has failover and recovery mechanism and it is used for online analytic application. Flume has a new data set sink Kiite API that is used to write data to HDFS and HBase.

4.8 ZooKeeper

Zookeeper is a centralized configuration and synchronization service in Hadoop ecosystem [17]. Every time a service is scheduled a lot of configuration need to be changed and resources are synchronized and this makes the service more fragile. Zookeeper is very fast with workloads with the ideal read-write ratio of 10:1. It can be replicated over multiple servers to avoid single point of failure.

5 Word Stemming

In context of information retrieval and linguistic morphology stemming, it is the process of tumbling any transformed word to its original stem word. Stem word is the base or morphological root form of any word. Stemming is a process that maps all related words to its stem [14]. Word stemming is an essential part of natural language processing and it is done by removing any suffix or prefix attached to the stem word. This conversion is also required in text clustering, categorization, and summarization as part of pre-processing in text mining.

5.1 Pre-requisites for Stemming

Word stemming requires tokenization and filtering from the document first. These two processes bring the document into the granular level required for word stemming.

Tokenization In tokenization, a document is split into a set of word based on some tokenizer or separator [12]. The separators can be a blank space or any special character. An example is illustrated as below.

Text = "Science brings the society to the next level."

The output of tokenization assuming blank space (" ") as a separator: ["Science", "brings", "the", "society", "to", "the", "next", "level"].

The punctuation marks and non-text characters are removed from the document in tokenization. Hence, the words are finally converted to nouns, verbs, etc. Another approach of word tokenization is focused on the statistical distribution of the words inside the document instead of following the occurrences of words. In the statistical analysis, it is important to index the texts into vectors. In this research, as the bag of word (BOW) approach has been adopted part of statistical representation of document.

Filtering This process removes the words which are not important for text mining process or which may degrade the result of analysis and it is also called as stop word filtering. Stop words [19] are the words which are not required in the text mining process. This filtering is controlled as per the requirement, i.e., a strong stop word list will create the best result in text mining process. The stop word lists are available in World Wide Web. One of the resource available in http://www.lextek.com has been considered in this research [20].

5.2 Classification of Stemming

Stemming algorithms are broadly classified into three groups. The classification of stemming algorithm is shown in Fig. 7.

Truncating Method This method removes the prefix and suffix of a word. Truncate (n) is the most basic stemming algorithm in which each nth position word is truncated and words existing in positions less than n are not truncated as well as no stemming rule is applied on them. Therefore, the chance of over stemming is increased. Another stemming algorithm where the plural words are transferred to singular form by removing the suffix 'S' [14]. There are four types of algorithm in truncating method as highlighted below.

Lovins Stemmer This algorithm contains 294 ending rules, 35 transformation rules, and 29 conditions. The longest suffix from any word can be removed by this stemmer.

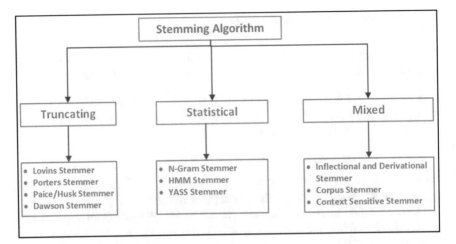

Fig. 7 Classification of stemming

After removing the suffix from the word, the word is referenced with different tables to convert it to a valid stem or root word after making some adjustments [15]. As a single pass algorithm maximum of one suffix is removed from a word. This algorithm can transfer the double letter words like "setting" to its original stem words very fast, i.e., "set" and also handles many asymmetrical plural forms to their singular transformations, for instance, "feet to foot," "men" to "man," etc. Lovins stemmer algorithm consumes more data and many suffixes are not available in the ending rules. Sometimes, it is very unreliable as it cannot match the stems of similar meaning.

Porter Stemmer Porter stemmer algorithm was proposed in 1980. Many modifications have been suggested and done on the elementary algorithm. There are 1200 suffix rules in the algorithm having five steps in each rule. The algorithm is iterated through the rules until one of them is accepted. Once a rule is satisfied, the suffix from the word is removed, then the resultant stem word is returned and next step is performer [15]. Also, there are 60 comprehensive conditions in this algorithm in the form of <Conditional Rules> with <Suffix> constitutes a <New Suffix>. For example, if a word ends with "EED" and has at least one consonant and vowel then the suffix can be changed to "EE." For instance, "Emceed" will be changed to "Emcee" but "Speed" will remain as it is. Porter stemmer algorithm is designed as a detail stemming framework where the key intension of the framework is that the programmers can develop new stemming rules for different sets of suffix.

Paice/Husk Stemmer It contains 120 rules indexed by suffixes and is iterative in nature. In each iteration, algorithm tries to find a match with the suffix and then either the suffix is deleted or it is replaced. Advantage of this algorithm is that it takes care of both deletion and replacement. But this is a very heavy algorithm which may create over stemming error [new paper].

Dawson Stemmer This is an extension of Lovins stemmer algorithm which have a 1200 extensive list of suffix transformation [15]. It is also a single pass algorithm therefore it is fast. The suffix is stored in reverse order indexed by their length and last letter.

Statistical Method These types of stemming algorithm remove the affixes (suffix and prefix) after applying any statistical analysis and technique. N-Gram stemmer, HMM stemmer, and YASS stemmer are statistical stemming algorithms. N-Gram stemmer is language independent and is based on n-gram and string comparison [14]. HMM stemming algorithm is unsupervised and language-independent stemming and it is based on hidden Markov model. YASS stemming corpus based and can be implemented without knowing the morphology. It uses hierarchical clustering and distance measure approach.

Mixed Method This type of stemming algorithms are composition of inflectional and derivational morphological methods, corpus-based methods, and context-sensitive methods [15]. As part of inflectional methods, the algorithms are correlated to syntactic variations such as plural, cases, and genders of a specific language. Krovetz and Xerox stemmers are example of inflectional and derivational methods. Corpus-based methods use the occurrences of word variants. Some drawbacks of Porter stemmer algorithm have been taken care here like "Iteration" is not converted to "Iter" and "General" is not converted to "Gener".

6 Proposed Porter Stemmer with Partitioner Algorithm (PSP)

This algorithm has about 60 rules which can be coded using MapReduce. When "Partitioner" technique is applied with all the porter rules, it provides better result. In partitioner of MapReduce, multiple partitions [21] are created based on conditions for data before data goes to reducer. The simplest partition technique is a hashing partition, but based on the condition, we can create required number of partition. For example, if special characters are not required for text mining process, then we can separate them in one partition and other alphabets and numbers will be in another partition. For this technique, the number of reducers needs to be set in the MapReduce program. Figure 8 shows the model for the proposed algorithm Porter stemmer partitioner which combines the rules of Porter stemmer implemented in MapReduce partitioner.

Fig. 8 Proposed Porter
stemmer algorithm with
partitioner (PSP)

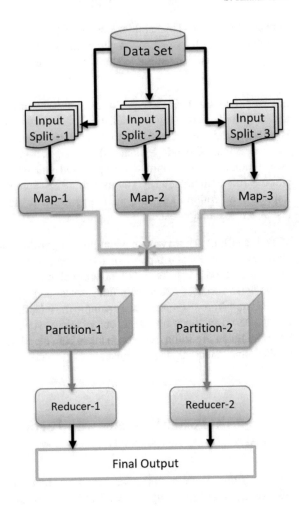

7 Hadoop Cluster Operation Modes

For this research, the selected documents have unstructured format of data. Therefore, Hadoop MapReduce and HDFS have been chosen for implementation. The selected documents are stored in HDFS and a MapReduce program is run on each document parallel [22]. For the purpose, a Hadoop cluster with Hadoop Archirtecture-2 has been set up. A Hadoop can run in three different modes as shown in Fig. 9.

Standalone Mode Standalone mode is the default operation mode of a Hadoop cluster also known as local mode. In this mode, none of the demons like name node, resource manager, secondary name node, data node, and node manager run inside the cluster. Therefore, it is mainly used for learning, debugging, and testing [23]. In this mode, the cluster runs faster than the two other modes. In this mode, HDFS storage architecture is not utilized, so it is like a system having the same kind of storage

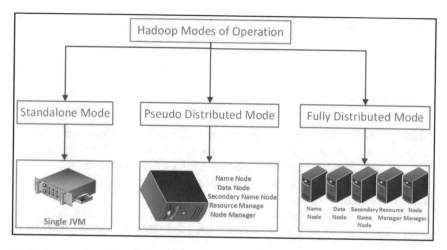

Fig. 9 Hadoop cluster operation modes

as in windows like an NTFS or FAT32 system. When this mode starts to run none of the configuration files like mapred-site.xml, hdfs-site.xml, core-site.xml, etc., are needed. All the processes run in a single JVM in this mode.

Pseudo-Distributed Mode In pseudo-distributed operation mode, all the demons run on a single node. This mode is a simulation of the cluster, therefore, all the processes run independently. Name node, resource manager, secondary name node, data node, and node manager run on separate Java virtual machines (JVMs) inside a single node. This mode mimics the operation of fully distributed mode on a single node [23].

The master-slave architecture of Hadoop cluster also exists in this mode is handled by a single system. Resource manage and name node are run as master, whereas data node and node manager run as slave. The secondary name node in this mode is used to handle the hourly back up of the name node. When this mode starts to run the configuration files (core-site.xml, mapred-site.xml, and hdfs-site.xml) need to be set up in the environment.

Fully Distributed Mode This is the production mode of Hadoop cluster where multiple nodes are used. Some of the nodes run master demons resource manager and name node, whereas rest of nodes in the cluster run slave demons node manager and data node. The HDFS storage architecture is fully followed here therefore the files are stored on multiple nodes [23]. The configuration parameters of the cluster environment need to be specified in this mode. This mode is highly scalable supporting both horizontal and vertical scaling. Also, this mode is completely reliable, fault tolerant and have the full capability of distributed computing.

Standalone mode has a very limited scope, whereas fully distributed mode is highly expensive and need a lot of configurations to be handled. Therefore, for this research, a pseudo-distributed cluster mode has been chosen. The chosen mode is a Horton works pseudo-distributed cluster running on Hadoop-2 architecture.

8 Environment Setup

A Hadoop cluster has been set up for implementation taking the Hortonworks Hadoop 2.2 version. It provides a command line interface to interact with the cluster and an easy accessible Web interface for displaying cluster-related informations.

Commands to make up the Hadoop cluster [24]. Figure 10 shows the Hadoop version installed on the cluster.

As the used Hadoop architecture is a second generation architecture, five demons always run on the cluster to make it operational [25, 26]. The running demons are shown in Fig. 11.

- Name node
- Data node

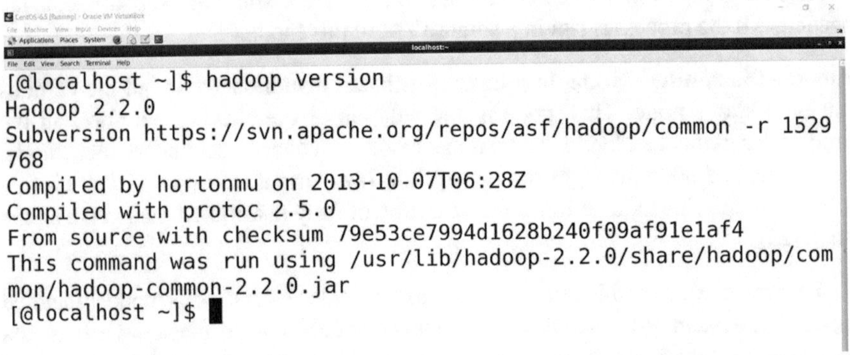

Fig. 10 Installed Hadoop version

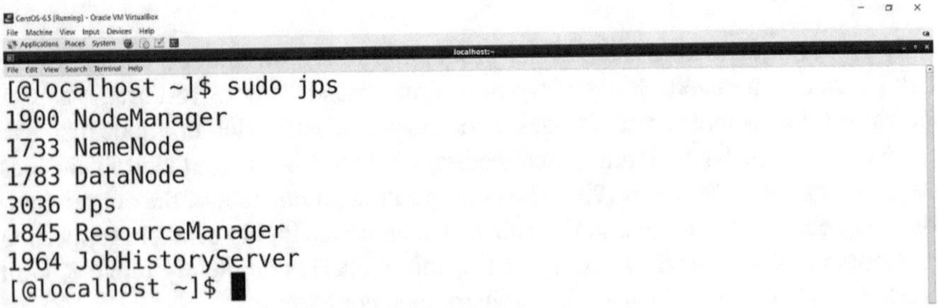

Fig. 11 Running demons on Hadoop cluster

- Node manager
- Resource manage
- Job history server.

Information about the name node are shown in Figs. 12 and 13. The name node runs on port 8020. There are total 38 blocks in the cluster. The cluster have 10.60 GB storage for Hadoop distributed file system out of total ~18 GB storage. Figure 14 shows internal storage structure of HDFS. This server has a block size of 128 MB and the files are stored as part files inside the blocks of HDFS. Part files are the logical partitioning of a bigger data set [24, 26]. The replication factor of cluster is

NameNode 'localhost:8020' (active)

Started:	Wed Dec 19 10:26:07 IST 2018
Version:	2.2.0, 1529768
Compiled:	2013-10-07T06:28Z by hortonmu from branch-2.2.0
Cluster ID:	CID-bef59e3e-aba0-4501-910b-d3685bbe282c
Block Pool ID:	BP-1756909416-127.0.0.1-1411538715533

Browse the filesystem
NameNode Logs

Cluster Summary

Security is OFF
90 files and directories, 38 blocks = 128 total.
Heap Memory used 44.42 MB is 91% of Commited Heap Memory 48.72 MB. Max Heap Memory is 966.69 MB.
Non Heap Memory used 18.05 MB is 99% of Commited Non Heap Memory 18.16 MB. Max Non Heap Memory is 96 MB.

Configured Capacity	:	17.23 GB			
DFS Used	:	1.09 MB			
Non DFS Used	:	7.73 GB			
DFS Remaining	:	9.50 GB			
DFS Used%	:	0.01%			
DFS Remaining%	:	55.12%			
Block Pool Used	:	1.09 MB			
Block Pool Used%	:	0.01%			
DataNodes usages	:	Min %	Median %	Max %	stdev %
		0.01%	0.01%	0.01%	0.00%
Live Nodes	:	1 (Decommissioned: 0)			
Dead Nodes	:	0 (Decommissioned: 0)			
Decommissioning Nodes	:	0			
Number of Under-Replicated Blocks	:	0			

Fig. 12 Name node information-1

NameNode 'localhost:8020'

Started:	Wed Dec 19 10:26:07 IST 2018
Version:	2.2.0, 1529768
Compiled:	2013-10-07T06:28Z by hortonmu from branch-2.2.0
Cluster ID:	CID-bef59e3e-aba0-4501-910b-d3685bbe282c
Block Pool ID:	BP-1756909416-127.0.0.1-1411538715533

Browse the filesystem
NameNode Logs
Go back to DFS home

Live Datanodes : 1

Node	Transferring Address	Last Contact	Admin State	Configured Capacity (GB)	Used (GB)	Non DFS Used (GB)	Remaining (GB)	Used (%)	Used (%)	Remaining (%)	Blocks	Block Pool Used (GB)	Block Pool Used (%)	Failed Volumes	Version
localhost	127.0.0.1:50010	1	In Service	17.23	0.00	7.68	9.55	0.01		55.43	38	0.00	0.01	0	2.2.0

Fig. 13 Name node information-2

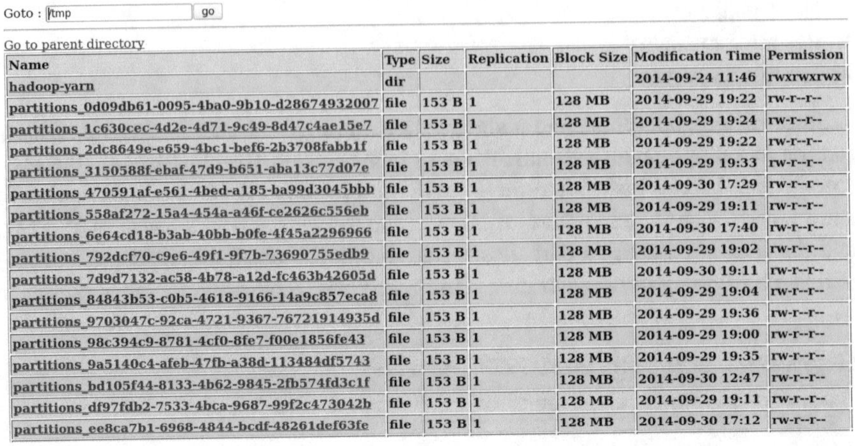

Goto : tmp   [go]

Go to parent directory

Name	Type	Size	Replication	Block Size	Modification Time	Permission
hadoop-yarn	dir				2014-09-24 11:46	rwxrwxrwx
partitions_0d09db61-0095-4ba0-9b10-d28674932007	file	153 B	1	128 MB	2014-09-29 19:22	rw-r--r--
partitions_1c630cec-4d2e-4d71-9c49-8d47c4ae15e7	file	153 B	1	128 MB	2014-09-29 19:24	rw-r--r--
partitions_2dc8649e-e659-4bc1-bef6-2b3708fabb1f	file	153 B	1	128 MB	2014-09-29 19:22	rw-r--r--
partitions_3150588f-ebaf-47d9-b651-aba13c77d07e	file	153 B	1	128 MB	2014-09-29 19:33	rw-r--r--
partitions_470591af-e561-4bed-a185-ba99d3045bbb	file	153 B	1	128 MB	2014-09-30 17:29	rw-r--r--
partitions_558af272-15a4-454a-a46f-ce2626c556eb	file	153 B	1	128 MB	2014-09-29 19:11	rw-r--r--
partitions_6e64cd18-b3ab-40bb-b0fe-4f45a2296966	file	153 B	1	128 MB	2014-09-30 17:40	rw-r--r--
partitions_792dcf70-c9e6-49f1-9f7b-73690755edb9	file	153 B	1	128 MB	2014-09-29 19:02	rw-r--r--
partitions_7d9d7132-ac58-4b78-a12d-fc463b42605d	file	153 B	1	128 MB	2014-09-30 19:11	rw-r--r--
partitions_84843b53-c0b5-4618-9166-14a9c857eca8	file	153 B	1	128 MB	2014-09-29 19:04	rw-r--r--
partitions_9703047c-92ca-4721-9367-76721914935d	file	153 B	1	128 MB	2014-09-29 19:36	rw-r--r--
partitions_98c394c9-8781-4cf0-8fe7-f00e1856fe43	file	153 B	1	128 MB	2014-09-29 19:00	rw-r--r--
partitions_9a5140c4-afeb-47fb-a38d-113484df5743	file	153 B	1	128 MB	2014-09-30 19:35	rw-r--r--
partitions_bd105f44-8133-4b62-9845-2fb574fd3c1f	file	153 B	1	128 MB	2014-09-30 12:47	rw-r--r--
partitions_df97fdb2-7533-4bca-9687-99f2c473042b	file	153 B	1	128 MB	2014-09-29 19:11	rw-r--r--
partitions_ee8ca7b1-6968-4844-bcdf-48261def63fe	file	153 B	1	128 MB	2014-09-30 17:12	rw-r--r--

Fig. 14 HDFS storage structure

set to 1, therefore, every file is present in a single rack only according to the rack awareness of Hadoop.

9 Implementation

Implementation of this research follows all the steps of proposed text mining model. Implementation of this research has compared the stemming performance of Lovins stemmer algorithm, Porter stemmer algorithm, and Porter stemmer with partitioner algorithm.

9.1 Data Collection

Three different data sets have been considered for the implementation of this research. All data sets are of different sizes and have different structures of data as described below.

Data Set-1 (CV Data Set) A CV structure has been considered as the first and smallest data set for this research. It has the text that is relevant to a CV like technologies, expertise, work experience, etc. This data set has been collected from an open source [27] of size 2 KB and total 260 words. The data set has the text data so it is unstructured in nature. A portion of the data set has been shown in Fig. 15.

Data Set-2 (Speech data set) Speeches have the most complex linguistic morphology. The second data set has been considered as a speech data set of PMO

CHRISTOPHER MORGAN
ADDRESS: 177 GREAT PORTLAND STREET, LONDON W5W 6PQ
PHONE: +44 (0)20 7666 8555
EMAIL: CHRISTOPER.MORGAN@GMAIL.COM
OBJECTIVE PROVIDE ANALYSIS DATA SUPPORT IN A COMPANY AS DATA ANALYST.
WORK
EXPERIENCE   04/2014 - 04/018
DATA ANALYST, GHT COMPANY, MADRID SPAIN
RESPONSIBILITIES:
ESTABLISH OPERATION STRATEGY IN A TEAM FOR IMPROVING SALES
PREPARE DATA AND INFORMATION FOR MAKING REGULAR REPORT DATA ANALYSIS
PERFORM DATA ANALYSIS FOR COMPLEX DATA AND FILES JAVA JAVA JAVA JAVA JAVA JAVA
03/2012 – 05/2014
DATA ANALYST, STARTUP CORPORATION, MADRID SPAIN
RESPONSIBILITIES:
COMPOSED JAVA PROGRAM FOR INTERFACING WITH ORACLE DATABASE
PERFORMED DATA ANALYSIS ESPECIALLY FINANCIAL DATA
PERFORMED STATISTICAL DATA ANALYSIS USING STATA
SHOWED DATA ANALYSIS IN REGULAR MEETINGS FOR CREATING NEW PROGRAM JAVA JAVA JAVA
JAVA JAVA
EDUCATION
2004 - 2008
BACHELOR DEGREE OF COMPUTER SCIENCE, TECHNICAL UNIVERSITY OF MADRID
2002 - 2004
CERTIFIED AS DATA ANALYST, DATA ANALYST CERTIFICATION, TECHNICAL UNIVERSITY OF MADRID

Fig. 15 Data set-1

India on 72nd Independence Day. Data has been collected from official site [28] of the India's Prime Minister. The data set has total 8000 words and unstructured in nature. A part of the data set is shown in Fig. 16.

Data Set-3 (Twitter data set) The third data set has been collected from the American microblogging site Twitter [29]. The data set contains social media comments and is the largest data set considered for this research. It have total ~52,00,000 words and is of 185 MB. A part of data set is shown in Fig. 17.

After the data sets are collected, they are transferred to HDFS, because data has to be present in HDFS for MapReduce processing. Command used to move data from local storage to HDF are given below.

- To check if file exists in local storage—"ls".
- To   move   file   from   local   storage   to   HDFS—"hdfs   dfs-copyFromLocal/home/local/textdata/textmining".

Figure 18 shows the data sets presence inside the HDFS.

9.2 Text Parsing

Text parsing is a technique to read the input data set and break it into granular levels which is a word. Text parsing is a logic that performs the above task inside

I hail from Gujarat. There is a saying in Gujarati 'Nishan Chuk Maaf Lekin Nahi Maaf Nichu Nishan' which means one should have big aims and dreams. However, for that, one has to strive hard and be answerable. But if the aims are not big, targets are not far-sighted, decisions are also not taken. Development comes to a halt. That is why my dear brothers and sisters, it is necessary for us to move ahead with big aims and resolve. When targets are vague, when the spirit is not strong, then important decisions in our social life also get held up for years. Take for instance the case of MSP – economists, farmer organizations, farmers as well as political parties had been demanding, that farmers should get an MSP which is one and a half times of their investment. The matter was debated for years, files moved to and fro, and but was stuck. Finally, we took the decision. We took a bold decision of giving the farmers the MSP which is one and half times their investment. There was unanimity on GST. Everyone wanted GST but they could not arrive at a decision because while on the subject, they were thinking in terms of their vested interests and whether this will translate into electoral gains. Today, with the help of small traders, their open mindedness and their attitude of accepting the new, the country has implemented GST. A new found confidence has been generated in the business community. The small entrepreneurs, small businessmen who faced teething difficulties in adopting GST, accepted the challenge and the country is now moving ahead. Today, we have enacted laws on Insolvency and Bankruptcy to strengthen the banking sector. Who opposed them earlier? Taking decisions requires conviction, force, confidence and complete dedication to the good of the common man. Why was the Benami property law not enforced earlier? The Benami property laws are implemented only when there is courage and a determination to do something for the country. The Jawans of our defence forces were demanding one-rank one-pension for several decades. They were not resorting to agitation because they are disciplined, but nobody was paying heed to their voices. Somebody needed to take a decision in this regard. You gave us the responsibility of taking this decision and we fulfilled it positively. My dear brothers and sisters, we are not the kind of people who work in the interest of the party. We are capable of taking tough decisions because national interest is topmost in our priority.

Fig. 16 Data set-2

As from title. What kind of visa class do I have to apply for, in order to work as an academic in Japan ?
What kind of Visa is required to work in Academia in Japan?
visajob-search
Which online resources are available for job search at the Ph.D. level in the computational chemistry field
As a computational chemist, which online resources are available for Ph.D. level jobs?
job-search
As from title. Not all journals provide the impact factor on their homepage. For those who don't
where can I find their impact factor
Where can I find the Impact Factor for a given journal?
journalsbibliometrics
8 I have seen many engineering departments want professional engineer registration. Why do they care?
8 In U.S., why do many engineering departments care about professional engineer registration?
8 job-searchengineering
What is the h-index, and how does it work
What is the h-index exactly and how does it work?
bibliometrics
8 If your institution has a subscription to Journal Citation Reports (JCR), you can check it there. Try this URL:http://isiknowledge.com/jcr
If I publish a pre-print paper on arXiv, how can I guarantee exclusive rights to the publisher afterwards? Am I unable to publish on non-open access journals after I publish a pre-prin
Does publishing a paper on arXiv prevent me to submit it to a non-open access journal?
copyrightarxiv
2 An increasing number of funding organisations require publications on the research that they fund to be open access, i.e. available to the public without having to subscribe to a jour

Fig. 17 Data set-3

Contents of directory /

Goto : [/]   [go]

Name	Type	Size	Replication	Block Size	Modification Time	Permission
Dataset1.txt	file	1.23 KB	1	128 MB	2020-08-12 21:33	rw-r--r--
Dataset2.txt	file	44.66 KB	1	128 MB	2020-08-12 21:33	rw-r--r--
Dataset3.txt	file	170.69 MB	1	128 MB	2020-08-12 21:34	rw-r--r--

Fig. 18 Data sets in HDFS

a MapReduce program [11]. As data in all the data set is separated by space we have used line offset value and string tokenizer [12] to parse the data sets. Parsing of data sets produce a bag of words (BoW) pseudocode for text parsing is defined as TEXT-PARSING(A).

TEXT-PARSING(A)

```
1        tokenizer ← tokenize(line)
2          for j ← 1 to length[tokenizer]
3            A[j] ← tokenizer[j]
```

For example assuming "My name is xyz" is a line.

Tokenize converts the line into an array of words by splitting them based on blank space. For the above example, tokenizer will create ['My', 'name', 'is', 'xyz'] for a line "My name is xyz".

Text parsing is done in mapper side and all the further steps are done in reducer side.

9.3 Text Filtering

Text filtering removes the unexpected words from the bag of words. It is done by passing the tokenized array into a stop word filter. For example, the word "an" do not contribute any morphological interpretation in the analysis. Therefore, it needs to be removed from the analysis. When a word passes through the stop word list than it is checked for its presence in the list [12]. If the word is present in the stop word list, then it is removed from the bag. Pseudocode for text filtering defined as TEXT-FILTERING(A) is explained below.

TEXT-FILTERING(A)

```
1        stopwords[]  =["a","as",  "able",  "about",  "above",  "accord-
         ing","accordingly","across","actually","after","afterwards","again",
         "against", "aint" .....]
2        for k ← 1 to length[A]
3          flag = true
4          for i ← 1 to length[stopwords]
5            if A[k] = stopwords[i]
6              flag = false
7            if flag = true
8              Insert A[k] into a new array B
```

After stop word removal three stemming algorithms such as Lovins stemmer, Porter stemmer, and proposed partitioned Porter stemmer (PSP) have been used for implementation of word stemming process. The proposed PSP stemmer takes care of the punctuation marks, special characters, etc., which are also not relevant to our analysis. Therefore, in the proposed PSP stemmer, a partitioner program has been used in this research. It removes all the special characters, punctuations into an unused partition, and only consider the words containing alphabets. Post-partitioning the terms are passed to the stop word list. Number of word stemmed are denoted by "S". The comparison of stemming results has been discussed in further sections.

9.4   Text Transformation

After the text filtering, the document is now converted to a numerical matrix form call as document matrix [19]. The symptomatic presentation of different terms has been explained below.

Term Frequency (λ) It is defined as the total occurrence of a stem word with respect to the total number of distinct words present in a document. Whereas the total occurrence of a root or stem word is defined as *Term Count*.

Term Count = Total count of existence of a stem word in a document.

Term frequency = Total count of occurrence of a word in document/Total Number of Word in document.

If N = Total number of word in document.

T = Term count then term frequency (λ) = T/N

Document Matrix It is a numerical representation of the document. After finding the term frequency [12] of each unique term in document, the document is presented in form of matrix as [Word (Term) Term frequency (F)] which constructs the document matrix. Figure 19 shows a portion of document matrix of data set-2.

Fig. 19 Document matrix of data set-"2"

CONTRIBUTING	1742.5
CONTRIBUTION	1161.6666
CONVEY	3485.0
CONVICTION	3485.0
COOKING	3485.0
CORNER	3485.0
CORRIDORS	3485.0
CORRODED	3485.0
CORRUPT	3485.0
CORRUPTION	1161.6666
COST	3485.0
COULD	1161.6666
COUNCILS	3485.0
COUNTED	3485.0
COUNTRIES	1742.5
COUNTRY	50.507248
COUNTRYMEN	112.41936
COUNTRY'S	580.8333
COUPLE	3485.0
COURAGE	1742.5
COURSE	3485.0
COURT	3485.0

Word Stem Factor (α) The percentage of total number of word stemmed with respect to the total unique word present in a document is defined as word stem factor. Algorithms providing higher percentage of stemming are known as *Dense Stemmer*.

S = Total number of word stemmed

U = Total unique word present in document

$\alpha = (S/U) * 100$

Stop Word Factor (β) It is defined as the percentage of total number of word stopped with respect to the total unique word present in a document.

X = Number of stopped words

$\beta = (X/U) * 100$

Cumulative Word Stem Factor (γ) It is defined as the percentage of total number of word stemmed and stopped with respect to the total unique word present in a document.

$\gamma = ((S + X)/U) * 100$
and $\gamma = \alpha + \beta$

9.5   Feature Selection

In this research, attributes such as "term," "frequency," "word stem factor," and "stop word factor" have been considered.

9.6   Evaluate

The values obtained from the model has been accepted to complete the evaluation process. The values obtained have been analyzed in further sections.

10   Result and Discussion

Table 2 shows a comparative study for word stemming capacity of the three stemming algorithms. From the result, it is clear that Porter stemmer with partitioner algorithm provides dense stemming than Lovins stemmer and Porter stemmer. Also, PSP is more accurate in stop word filtering. This performance improvement is applicable to documents of all sizes. Figure 20 shows the graph plotted for word stem factor with respect to the different stemming algorithms. From the graph, it is observed that with increase of data set size, PSP algorithm shows better result which resolves the big data volumetric issue [18], i.e., the model provides the better result when operate on huge data set. Similarly, a graph is plotted between the stop word factor and the stemming algorithms as shown in Fig. 21. From the figure, it is observed that a Porter stemmer algorithm when operated with a partitioner provides better stopping capability than Lovins stemmer and Porter stemmer. Another graph between cumulative word stem factor and stemming algorithm is shown in Fig. 22. The plot clearly points toward the better performance of Porter stemmer with partitioner than the other stemming algorithms according to the increased size of data sets. The accuracy of analysis depends on the stop word list and word stemmed. So, the stop word list is continuously updated for better results.

11   Conclusion and Future Work

From the above result analysis, it is clear that Porter stemmer algorithm with Hadoop MapReduce partitioner provided better result than Lovins stemmer and traditional Porter stemmer algorithm. Therefore, PSP algorithm can be used with big data to create an operation module which can be used in industrial applications, health care, social media, etc. The Porter stemmer with partitioner is capable of providing better result for huge amount of data sets than other stemming algorithms. The

Table 2 Comparison of stemming results

Result set	Data set	Algorithm	Total unique word in document (U)	Total word stemmed (S)	Total word stopped (X)	Word stem factor (α)	Stop word factor (β)	Cumulative word stem factor (γ)
1	CV	Lovins stemmer	221	17	15	7.49	6.79	14.48
2	CV	Porter stemmer	220	38	15	17.27	6.82	24.09
3	CV	Porter stemmer with partitioner	204	55	15	26.96	7.35	34.31
4	PM speech	Lovins stemmer	3484	304	254	8.72	7.29	16.02
5	PM speech	Porter Stemmer	3347	1028	254	30.71	7.59	38.3
6	PM speech	Porter stemmer with partitioner	3297	1075	277	32.6	8.4	41.01
7	Twitter	Lovins stemmer	5,209,760	28,756	10,015	0.55	0.19	0.74
8	Twitter	Porter stemmer	5,166,699	336,965	10,015	6.52	0.19	6.72
9	Twitter	Porter stemmer with partitioner	5,166,699	1,120,535	12,882	21.68	0.25	21.94

proposed methodology also has an extensible capability of reducing unnecessary words from the text mining and also has the capability to reduce the error in the following an iterative approach. The model can be used for CV filtration, online exam evaluation of subjective question answer, sentiment analysis, etc. In future, the model and algorithm will be implemented in other application domains such as health care and the obtained results will be compared. Also, the optimization techniques like particle swarm optimization (PSO) will be applied to enhance the model.

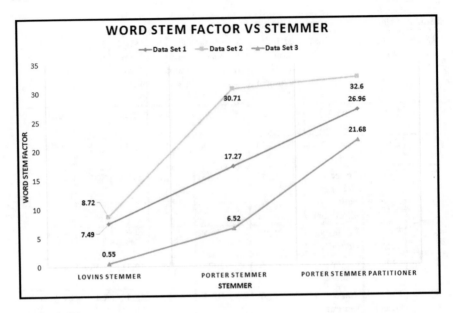

Fig. 20 WSF versus stemmer

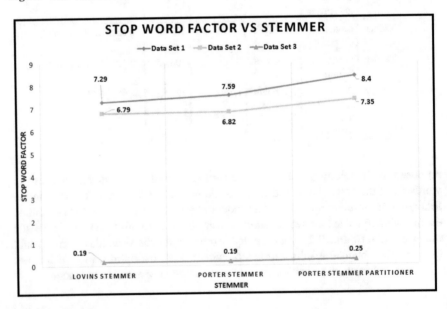

Fig. 21 SWF versus stemmer

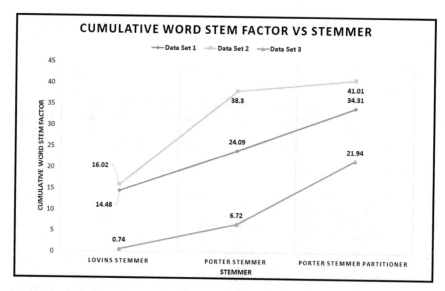

Fig. 22 CWSF versus stemmer

References

1. Sardar, T.H., and Z. Ansari. 2018. An analysis of MapReduce efficiency in document clustering using parallel K-means algorithm. *Future Computing and Informatics Journal* 3: 200–209.
2. Alexandra, A., C. Paulo, and M. Paulo. 2018. Research trends on big data in marketing: A text mining and topic modeling based literature analysis. *European Research on Management and Business Economics* 24: 1–7.
3. Sarkar, D. 2018. *Text Analytics with Python*, 109–319. New York: Apress Publication.
4. Judith, J., and J. Jayakumari. 2018. *Distributed Document Clustering Analysis Based on a Hybrid Method*, 131–142. New York: China Communications, IEEE.
5. Ramanujam, R.S., and R. Nancyamala. 2015. *Sentiment Analysis Using Big Data*.
6. Jain, A.K. 2010. Data clustering: 50 years beyond K-means. *Pattern Recognition Letters* 31: 651–666.
7. Sardar, T.H., and Z. Ansari. 2018. Partition based clustering of large data sets using MapReduce framework: An analysis of recent themes and directions. *Future Computing and Informatics Journal* 3: 247–261.
8. Lydia, E., and D. Ramya. 2018. Text mining with Lucene and Hadoop: Document clustering with updated rules of NMF non-negative matrix factorization. *International Journal of Pure and Applied Mathematics* 118: 191–198.
9. Ding, C., X. He, H.D. Simon. 2005. On the equivalence of nonnegative matrix factorization (NMF) and spectral clustering. In *International Conference on Data Mining*, 606–610. SIAM.
10. Ghazi, M.R., N.S. Raghava. 2018. MapReduce based analysis of sample applications using Hadoop. In *International Conference on Application of Computing and Communication, Technologies*, 34–44. Berlin: Springer.
11. Madasamy, K., and M. Ramaswami, 2017. Performance evaluation of word frequency count in Hadoop environment. *International Journal of Innovative Research in Science, Engineering and Technology*.
12. Gupta, G., and S. Malhotra. 2015. Text document tokenization for word frequency count using rapid miner (taking resume as an example). In *IJCA Proceedings on International Conference on Advancements in Engineering and Technology*.

13. Jiang, Y., J. Yang, L. Tang, Y. Liu, X. Zhao, and X. Hao. 2015. A distributed data mining system framework for mobile internet access log based on Hadoop. *Transactions on Edutainment* 5: 243–252.

14. Singh, A., N. Kumar, G. Sahil, and A. Mittal. 2010. Achieving magnitude order improvement in Porter stemmer algorithm over multi-core architecture. In *International Conference on Informatics and Systems*. IEEE.

15. Arianti, N.D., I. Mohamad, U. Syaripudin, D. Mariana. 2019. Porter stemmer and cosine similarity for automated essay assessment. In *International Conference on Computing Engineering and Design*. IEEE.

16. Jiang, H., K. Wang, Y. Wang, M. Gao, and Y. Zhang. 2016. *Energy Big Data: A Survey*. IEEE, 3844–3861.

17. Reinsel, D., J. Gantz, J. Rydning. 2017. *Data Age 2025—The Evolution of Data to Life-Critica*. Seagate-WP.

18. Singh, K., K. Kaur. 2014. *Hadoop—Addressing Challenges of Big Data, International Advance Computing Conference*. IEEE.

19. Kim, J., and K. Chung. 2019. Associative feature information extraction using text mining from health big data. *Wireless Personal Communications* 105: 691–707.

20. http://www.lextek.com/manuals/onix/stopwords1.html.

21. Khawlaab, T., M. Fatihaa, Z. Azeddineb, and N. Said. 2018. *A Blast Implementation in Hadoop MapReduce Using Low Cost Commodity Hardware*, 69–75. Amsterdam: Elsevier.

22. Mishra, B. 2016. Improved MapReduce K mean clustering algorithm for Hadoop architecture. *International Journal of Advanced Trends in Computer Science and Engineering*.

23. Li, H., and X. Lu. 2015. Challenges and trends of big data analytics. In *International Conference on P2P, Parallel, Grid, Cloud and Internet Computing*. IEEE.

24. Wankhede, P., and N. Paul. 2016. Secure and multi-tenant Hadoop cluster —an experience. In *International Conference on Green High Performance Computing*. IEEE.

25. https://docs.cloudera.com/documentation/enterprise/6/6.3/toics/cm_mc_service_config_ove rview.html.

26. https://docs.cloudera.com/HDPDocuments/.

27. www.freecv.com.

28. https://www.pmindia.gov.in/en/news_updates/pms-address-to-the-nation-from-the-ramparts-of-the-red-fort-on-the-72nd-independence-day/?comment=disable.

29. https://archive.org/download/stackexchange.

Business Analytics: Process and Practical Applications

Amit Kumar Gupta

Abstract Today, automation of business processes and devices like IoT for monitoring/activating services generate massive raw data, though they stand alone may not look useful but together carry domain specific signatures that are immensely useful for decision making. The problem of deducing strategic information in detecting patterns, analyzing, reasoning over it, and learning on business trends is popularly known as business analytics and uses artificial intelligence and machine intelligence techniques. This chapter while introducing basics of characteristics of business data analytics, presents types and uses of analytics, and standard processes. Further, this chapter would include an approach to design a recommendation system (with techniques such as content-based filtering, collaborative filtering, and Hybrid recommendations methods). This chapter would do a comparative analysis as well between process of business analytics, various types, and choice of recommendation systems.

Keywords Business analytics · Recommendation systems · Data analytics · Machine learning

1 Introduction

Analytics is a new buzzword, and one may have seen the same being used at many places (like HR analytics, Sports analytics, Pricing analytics, etc.). Business analytics is an area where business data is collected, and a lot of processing is done to gain insights from the data. These insights in-turn help companies to make the right decision, improve business processes, user productivity, and sales. Business analytics enables a look-forward perspective to have a possible future glimpse of the possibilities and opportunities.

A. K. Gupta (✉)
Microsoft India R&D Pvt. Ltd., Hyderabad, India
e-mail: amitgupta167@yahoo.com

© The Author(s), under exclusive license to Springer Nature Singapore Pte Ltd. 2021
S. Rautaray et al. (eds.), *Trends of Data Science and Applications*,
Studies in Computational Intelligence 954,
https://doi.org/10.1007/978-981-33-6815-6_15

In business analytics, data sets are processed in multiple ways for identifying patterns, reasoning for changes, and understanding the trends with the help of historical data (and even in absence of that), which helps us to create predictions, classify items, provide recommendations, etc. There are many statistical methods and modeling techniques that help us for predictions, classification, and recommendations, etc. Many algorithms and tools are available today for a business analyst to do all the activities required.

1.1 Definition

Business analytics involves the process of collecting business data, data processing, exploring, and generating insights to contribute to business planning. It is more prescriptive in nature (as against business intelligence) and requires involvement of business executives and solution delivery teams at larger scale.

Domain expertise plays the most critical role in overall end-to-end analytics solution (which can be patterns/models and recommendations).

1.2 Goal

Many times the problems to be addressed are not clearly defined and individual hypothesis, insights from explorations, problems being faced (by business), and market trends gives signals for a need of a specific problem that becomes the right candidate to be addressed by business analytics. The goal here is to first identify the right problem and then find the relevant metrics that would be useful to help on right decision making and improve on overall business profitability.

2 Process

The business analytics process involves a lot of data mining and below are the most popular data mining process models.

2.1 CRISP-DM (Cross-Industry Standard Process for Data Mining)

CRoss-Industry Standard Process for Data Mining, (CRISP-DM) is widely accepted as a collection of common steps for the data mining process [1]. Many organizations adopted these steps and created their own lifecycle processes as well that were suited to their needs.

CRISP-DM involves six stages (as illustrated in Fig. 1) and they are not always incremental to the next stage. The first stage is business understanding where business objective is identified, and problem formulation happens then second stage is data understanding where we try to understand the nature of data, and this exercise is popularly known as exploratory data analysis (EDA). Here, exploratory data analysis, different types of analysis such as univariate and multivariate analysis are performed. The third stage involves data preparation, where feature engineering is done. Feature engineering is one of the most critical activities in CRISP-DM where one identifies the important feature(s) that should be considered for the model building. The model that would be part of the fourth stage. The fifth stage is evaluation, where based on the nature of data, an evaluation matrix is decided, and measurement happens for that. And, the next stage is deployment where the model is deployed to be used for future sets of data, this should not be treated as final steps, because there would still be incremental changes to the model based on the changing nature of data.

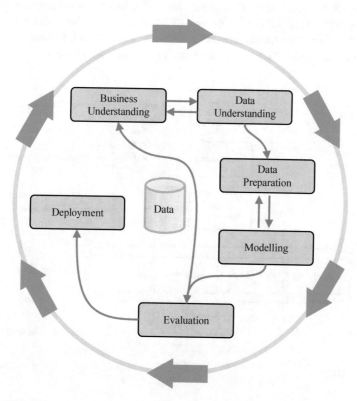

Fig. 1 CRISP-DM

2.2 SEMMA (Sample, Explore, Modify, Model, Assess)

Another method is SEMMA, which stands for Sample, Explore, Modify, Model, Assess. SEMMA was developed by SAS institute. It is more inclined toward the technical steps of mining. A graphical user interface (GUI) provides an intuitive front end to the SEMMA data mining process [2]

The first step SEMMA (stated in Fig. 2) is "Sample", where a small dataset called sample from a large dataset is selected for experimentation. A good sample should always be a true representative of the population, so appropriate sampling techniques should be applied to "Sample" steps to get the right subset of data. In the second step, "Explore", we explore the data for any anomalies, patterns, seasonality (in case of time-series data), completeness, trends, etc., for getting a better perspective of data. This is a crucial step for understanding the data and will help us to create relevant hypotheses that can be validated in current and subsequent steps. Third step is "Modify", where we prepare the data that would be suitable for the data modeling. Transformation such as grouping, ungrouping, pivoting, etc., on selected data would be in the task list of this step. Next in the "Model" step, creation of a model happens which would be used for prediction on the future data. The next step is "Assess", where we evaluate the usefulness and reliability of the model. In case the created model is not accepted, then we again go back to the first step sample.

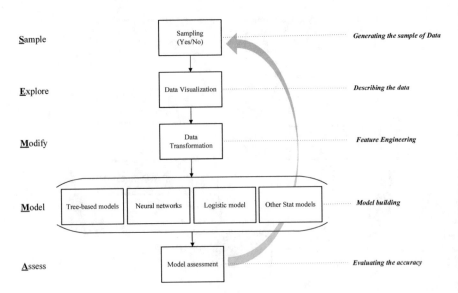

Fig. 2 SEMMA

2.3 Comparative Study

As we said that SEMMA seems to be more like a technical step in the overall process. CRISP-DM seems to cover a broader spectrum of activities. So, there are no steps that are there for business understanding and SEMMA does not include it, while second phase data understanding of CRISP-DM covers two steps (Sample and Explore) of SEMMA. The activities performed in the third phase, fourth, and fifth phase (Data Preparation, Modeling, and Evaluation) of CRISP-DM are similar to those performed in the third, fourth, and fifth steps (Modify, Model and Assessment) of SEMMA, respectively. Further in CRISP-DM, there is mention of final step Deployment, however, SEMMA does not include that assumes that it is part of post knowledge discovery in databases (KDD).

2.4 Others Approaches

Above discussed (CRISP-DM and SEMMA) are the data mining methodologies which are good practice. Additionally, industry follows and wraps a couple of more layers to these processes in order to complete the end-end process and align with engineering systems. TSDP is one of them from Microsoft, and ASUM from IBM are few examples from industry.

2.4.1 TDSP—A Broader Perspective (with Engineering)—Team Data Science Process

This is not exactly a data mining process, but a data science process introduced by Microsoft, which focuses on complete solutions from an end-to-end engineering standpoint as well. The team data science process (TDSP) is an incremental and iterative process useful for predictive analytics implementations. This contains the well-proven practices of different analytics solutions/implementations [3]. Team data science process is illustrated in Fig. 3, where flow and relationship between different stages are shown.

The process broadly goes through below 4 stages and each can be iterative (individually):

1. Business Understanding
2. Data Acquisition and Understanding
3. Modeling
4. Deployment.

In case organizations are using CRISP-DM and any other data science lifecycle, they can still use task-based TDSP. High level steps would be found common on those.

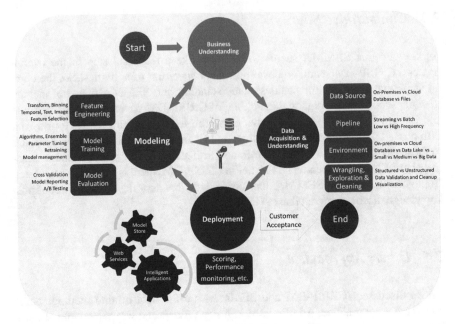

Fig. 3 Data science lifecycle [3]

2.4.2 ASUM – (Analytics Solution Unified Method)

ASUM [4] was introduced by IBM and it is also advocated as a step-by-step guideline for complete implementation lifecycle for IBM analytics solutions.

ASUM (as shown above in Fig. 4) has five phases, managed by project management streams, those work in coordination for an end-end delivery. ASUM has blended agile and traditional principles together to provide an optimal best value to users. Also, its solution project management is aligned to Project Management Institute (PMI) and Projects in Controlled Environments (PRINCE2). It contains the below phases.

1. Analyze — *Identify the solution needs*
2. Design — *Define components, dependencies, and resources*
3. Configure and build — *Configure, build, and integrate components*
4. Deploy — *Plan to deploy and maintain*
5. Operate and optimize — *Maintenance tasks and checkpoints*
6. Project management — *Processes for managing and monitoring the project*

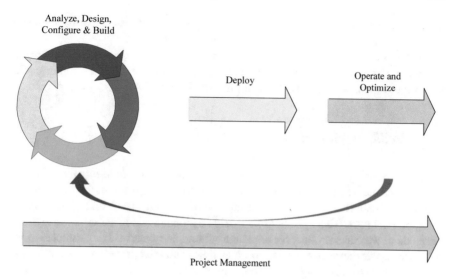

Fig. 4 ASUM

3 Types of Analytics

In business/data analytics, there is a lot of data that is processed and there are many phases based on their need/complexity on which this analysis is done. There are broadly four of them which are described below.

3.1 Descriptive Analytics

This is the analysis where we find out that *what is happening?*

Data is described or summarized here using basic statistical methods. An example can be finding a monthly sales statement which can tell profit/loss states from sales and marketing domains. Customer care churn number from past month/quarter from telecom domain (ex. 8% of the customer churn happened last month/quarter).

This is normally shared with canned and ad-hoc reporting within organizations.

3.2 Diagnostic Analytics

This is the next step of analysis where we try to find *why this is happening?*

Post descriptive analysis, based on historical data, we need to find the root cause of why a specific behavior is being exhibited by the data through diagnostics analytics. For example- once we found that there is 8% customer churn based on descriptive data, we might want to find out why this is happening. Is it because of bad/delayed service, non-pleasant experience with product or service?

This is normally shared with the audience in a well-designed dashboard. There are many tools in industry for visualizing the diagnostic analysis. Few examples could be PowerBI (from Microsoft), Tableau, Sisense, Looker, SAP, Spotfire, etc.

3.3   Predictive Analytics

In predictive analytics, the main task is to predict *what is likely to happen?*

Here, the main task is to use the all available data and apply the statistical/machine learning techniques to create a model and that model helps us to find what is likely to happen in future. These predictions would be probabilistic in nature. In case time-series data, we can also observe seasonality/trend and pattern. For example - how much customer churn is expected to happen next month/quarter from the telecom domain; what would be sales in the coming month.

3.4   Prescriptive Analytics

In predictive analytics, the main task is to predict *what is the next course of action?*

This is the last and most advanced step where once we have all the results of previous analysis, a course of action is suggested for the improvement of business function that we are targeting. There are many techniques like simulation, optimization, etc., are used for prescriptive analysis. For example—Increase the number of sales support folks to faster resolution of user queries so the churn would reduce.

3.5   Comparative Study

All these four types are basically conveying the stage and application might be in. Generally, an application goes through each one of them (i.e., descriptive, diagnostics, predictive, and prescriptive) in progressing manner one by one. An application maturity can be assessed based on which stage it is in. For example, if we are in the stage of descriptive analytics that indicates that we are just exploring the data and the various business problems that can be addressed by business analytics.

4   Domain and Applications

Business analytics is being used and making its footprints in almost every domain, be it telecommunication, healthcare, supply-chain, or HR, etc. Though many times it has been observed that each industry/sector has its own use case, but at the same time, these problem statements overlap between domains as well. For example – Fraud analytics, a very common use case in cyber security and also prevalent in banking and financial service sectors. Though a solution for fraud analytics may be different for both the domains but the concepts, based on which this would be built would be very similar. Another overlapping example could be optimal routing that has its place in supply-chain analytics, retail analytics, transportation, and logistics sectors.

Table 1 lists few of the examples (not an exhaustive list) for different domains/sectors that can be referred.

From above, it can be easily observed that almost every sector has the use cases of business analytics that many of them overlap as well.

Table 1 Analytics use-cases (across domains)

Domain/Sectors	Analytics use-case
Telecommunication	– Churn predictor – Network management and optimization – Capacity forecasting – Call details record analysis
Healthcare	– Disease prediction based on image data (like MRIs, X rays, etc.) – Helping decision support/proactive alerting – Medicine impact analysis – Diagnostics accuracy analysis and predictive medicine
Transportation	– Optimal routing: what would be optimal routing for good transport? – Dynamic fare predictor – Demand prediction
Retail analytics	– Inventory management – Store experience – Product recommendation – Behavioral analytics – Customer journey analytics – Sentiment analysis – Customer lifetime value analytics
Cyber analytics	– Fraud/anomaly detection – Threat modeling – Behavior analysis – Data exfiltration detection
Marketing analytics	– Will the customer buy? – Campaign analytics – Customer segmentation – Planning targeted campaigns
HR analytics	– Attrition risk predictor – Profile similarity mapper
Banking and finance	– Loan predictor – Credit card fraud detection

5 Recommendation System(s)—An approach

> Many times, people don't know that what they need until you show it to them—Steve Jobs.

Though there are many analytics uses-cases that are used across domain sectors, we are taking one such use-case recommendation system.

Recommendation systems provide suggestions to the user based on their interest. These interests are mined from their historical browsing history and the choices they might make. This data is then mapped with other data and based on a notion of similarity (to find the "relevant" item), the systems return a recommendation for the item. Recommendation systems are designed with few techniques and algorithms which helps to find the relevant result.

How Do You Identify It?
When visiting multiple websites, do you see this section?

Amazon → "Customers who bought this item also bought …" (*suggesting similar products that are purchased*).

Facebook → "People You May Know…" (*suggesting to who you can send the friend request and similar in LinkedIn suggesting growing your network*).

Coursera → "Recommended courses…" (*suggesting the next course based on technical skills, preferences, and interest*).

YouTube → "Recommended Videos…" (*suggestion for next video to play in YouTube*).

Netflix → "Other Movies You May Enjoy…" (*suggestion for movie tonight*).

Above sections in the above-mentioned websites are actually populated by recommendation engines. In fact, according to the McKinsey report [5], it is so popular that Amazon generates 35% of its revenue by its recommendation engine, and 75% of content being watched on Netflix comes from product recommendation.

Mathematical Model
If U = Set of users ~ $\{u_1, u_2, u_3 \ldots u_m\}$.

I = Set of Items/products ~ $\{i_1, i_2, i_3, \ldots i_n\}$.

f = objective function (to measure the relevance of item I to user U.

$f: U \times I \to R$ say R = {recommended set of items}.

---

for each user u, find the item i that
maximizes the objective function $u \in U$, $I'_u = \mathrm{argmax}_u u(u, i)$

A binary relation would be represented as below in a data store (as shown in Table 2).

Table 2 Matrix for binary representation

User/Items	i_1	i_2	i_3	...	i_n
u_1	1	0	0	...	1
u_2	0	1	1	...	0
u_3	0	0	1	...	0
...	...	...	...	...	...
u_m	1	0	1	...	0

5.1 Types of Recommendation Systems

Recommendations systems are broadly divided into 3 categories (as shown below Fig. 5) namely content-based, collaborative filtering, and hybrid methods.

Other than above, there are other categories as well such as popularity filtering models, association rule mining model, knowledge-based recommender systems.

In popularity filtering models, everyone gets the same recommendation. This is sometimes termed as over-simplification and becomes static, while still useful for most popular items (proven quality products/items). While on the other side, knowledge-based recommender systems make emphasis on the need of personalization. For knowledge-based recommender systems, user specific choices are required to be captured (by explicit survey, purchase history, keyword search, browsing page collection) which is useful to know his preference and create a model. This also helps in addressing the cold start problem [6] (discussed later in this chapter).

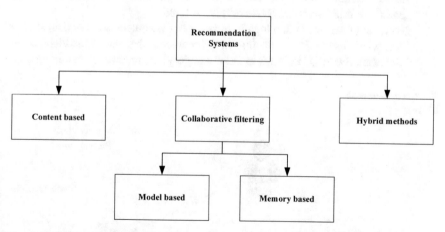

Fig. 5 Types of recommendation system

5.1.1 Content Based Recommendation System

Content-based methods depend on item description (item keywords or properties) and user's preference information. In content-based filtering methods, features are created on user and item data and that is used to recommend the relevant item to the user. Figure 6 shows the content-based recommendation system based on the example of document/article recommendation.

Example(s)

- Movie Recommendation
  - A user logs-in to a movie website, watches a "action" movie online. In the back end, the recommendation system captures the criteria that satisfies the notion of similarity (for user's preference); assume that the notion of similarity here is "Genre." So, "Genre" for the movie that the user has seen (here it is "Action") is captured.
  - Now, next time the user logs-in/browses to the movie website then similar "Action" movies (i.e., movies that are tagged/labeled as "Action" in "Genre" category) are recommended to the user.

- Document/Article Recommendation
  - A user logs-in to a content website, and reads a specific document/article. In the back end, the recommendation system captures the criteria that satisfies the notion of similarity (for user's preference); assume that the notion of similarity here is "Article Title." So, "Article Title" for the document that the user has read (here it is "Software Systems") is captured.
  - Now, next time the person logs-in/browses to the content website then similar "Software Systems" documents (i.e., documents that are tagged/labeled as "Software Systems" in "Article Title" category) are recommended to the user.

Fig. 6 Content-based filtering

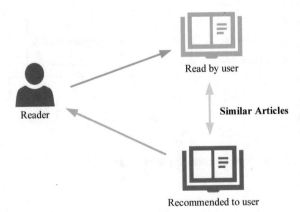

5.1.2   Collaborative Filtering Recommendation System

Collaborative filtering is based on identifying relevant items for a user by mostly from those set of users who are similar to current users (Or set of users). The user's behavior and preference are very critical in identifying similar users and further is used in recommending the new item.

There are two fundamental approaches in implementation of collaborative filtering.

1. Model-Based approach—Here, we use machine learning algorithms and probabilistic models (cluster-based, Matrix factorization-based OR deep learning-based) to establish user ratings.
2. Memory-Based approach (also called neighborhood-based)—In this, we use standard ways of finding distance/similarity between users (such as cosine similarity). These are further 2 types in memory-based approach—user-based and item-based. Emphasis is for finding—*who is the neighbor?*

Example(s)

- Movie Recommendation

  – User-1 logs-in to a movie website, watches an "Action" movie online. In the back end, the recommendation system captures the criteria that satisfies the notion of similarity (for user-1's preference); assume that the notion of similarity here is "Genre." So, "Genre" for the movie that the user-1 has seen (here it is "Action") is captured.
  – Now, another user (say User-2) logs-in to the movie website, then the above-mentioned process is again followed by user-2 as well and also similarity between both users (user-1 and user-2) is computed.
  – Finally, if user-1 and user-2 are similar, then whatever movies user-1 has seen, are also recommended to user-2.

- Document/Article Recommendation

  – User-1 logs-in to a content website, and reads a specific document/article. In the back end, the recommendation system captures the criteria that satisfies the notion of similarity (for user-1's preference); assume that the notion of similarity here is "Article Title." So, "Article Title" for the document that the user-1 has read (here it is "Software Systems") is captured.
  – Now, another user (say User-2) logs-in to the content website, then the above-mentioned process is again followed by user-2 as well and also similarity between both users (user-1 and user-2) is computed.
  – Finally, if user-1 and user-2 are similar, then whatever document user-1 has read, are also recommended to user-2.

Figure 7 illustrates the collaborative filtering.

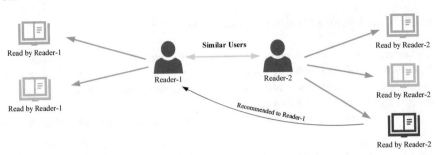

Fig. 7 Collaborative-based filtering

5.1.3 Hybrid Methods

Hybrid is amalgamation of content-based and collaborative filtering methods (as illustrated in Fig. 8). The purpose of this is to combine the powers of both the systems. Hybrid methods help in addressing the common problems (such as cold start and the sparsity problem) in recommender systems [7].

Example(s) Google news is an example of hybrid recommendation. Another example could be Netflix, where characteristic of both the methods are combined for movies recommendation.

5.2 Benefits of Recommendation System

In today's era, many companies are using one or more types of recommendation systems that are being used within or outside organization (in case of customer facing roles). There are many benefits of having a recommendation system and that is not limited to business only. Below are few benefits listed (for both customers and business).

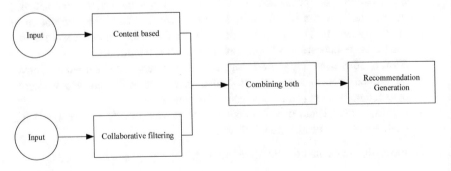

Fig. 8 Hybrid recommendation

For Customers

- Saves user's search time by recommending the similar content
- Reduction of cognitive load on the customer
- Increases the productivity
- Personalization experience and increase in satisfaction
- Have ability to compare stuff before taking a decision.

For Business

- Increases sales and revenue (by increasing average order value)
- Creates an opportunity to study user behavior
- Can know the trend for curated content
- Enables to inject new preferences
- Opportunity to cross-sell and up-sell
- Reduces customer churn
- Making a space for low popular content as well, hence an equilibrium in distribution of available options.

5.3   An Example

Let's take an example of approach for building a recommendation system.

5.3.1   Approach

We need to find and collect the relevant data for creating the recommendation system. Below is the high-level approach for the recommendation system.

First, we start collecting the relevant data and then extract the user and items data. Post that we clean the data and then the process of **scoring** starts. For scoring, we need to do feature engineering where we find the relevant attributes that would be considered for finding similarity. *Notion of similarity* is important to understand here.

Notion of Similarity—Similarity is a measurement of how much things (being compared) are similar and inversely proportional to distance. There are multiple distance/similarity formulas mentioned below.

1. **Minkowski Distance**—is a metric between two data points in multi-dimensional real space. This is the generalized distance form. The Minkowski distance [8] between two variable x and y is defined as

$$\left(\sum_{i=1}^{n} |x_i - y_i|^p \right)^{\frac{1}{p}}$$

where $p = 1$ makes it to Manhattan distance.

and $p = 2$ makes it to Euclidean distance.

and $p = \infty$ makes it to Chebychev distance.

2. **Cosine Distance**—This is used for text data. In this, the degree of angle is measured between two documents/vectors). In cases where orientation is more significant than magnitude, cosine distance can be very helpful [9].

$$\cos \theta = \frac{\vec{a} \cdot \vec{b}}{\vec{a}\vec{b}}$$

where cosine value "1" indicates vectors are same direction (means they are similar).

cosine value "0" indicates vectors are not related (but some degree of similarity found).

cosine value "−1" indicates vectors are in opposite direction (means they are not similar).

3. **Mahalanobis Distance**—is used for determining the distance between two data points in a variable space [9]. The Mahalanobis distance [10] of an data point $\vec{x} = (x_1, x_2, x_3, \ldots, x_n)^T$ from a pool of data points (with mean $\vec{\mu} = (\mu_1, u_2, u_3, \ldots, \mu_n)^T$) is

$$D_m(\vec{x}) = \sqrt{(\vec{x} - \vec{\mu})^T S^{-1}(\vec{x} - \vec{\mu})}$$

where S is the covariance matrix.

Choosing a similarity notion is a critical part of design in the recommendation system. As it would ensure the right metric for proper scoring exercise. Either we do item profiling or user profiling, and in both the cases, once we find the right features, then similarity finding would enable us to group (the items or users), which would be used in recommendations. For user profile, we calculate weighted sum of item profiles with weights being ratings user rated.

Post scoring, **re-ranking** is applied, where it is ensured that relevant information stays, and other is filtered out. For example, if many users give negative feedback for an item or it is disliked by many users then it goes lower in recommendation priority. Similarity, the most liked items find its place at the top of the recommendation list.

5.3.2 Design Flow

While designing a recommendation system, we need to ensure what kind of interface would be at the end. This interface could be a suggestion list, offers, promotions or "top-n" recommendations. Based on the chosen interface the algorithms need to be changed and deployed.

Figure 9 is the diagram showing the basic approach to design.

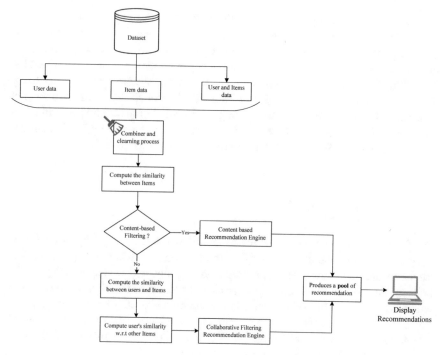

Fig. 9 Basic approach to design

5.4 Challenges of Recommendation Systems

Collaborative filtering suffers from two major problems—**data sparsity** and **cold start**. Data sparsity happens when there is a dearth of actual review. Most of the people, if satisfied with the product/service, they ignore/forget giving the review (and one of the strong reasons being the incentive around this feedback giving activity). If unsatisfied or having a strong objection to something would definitely attract a feedback and most probably a negative feedback. Cold start occurs when a new entrant (user/items) is there in the system, and lack of data about this new entrant, make it difficult for collaborative filtering to generate any recommendation initially. Guibing Guo suggested techniques for resolving these problems in his paper [11].

Further cold start could be a *Visitor cold start* problem or *Product cold start* problem. In a visitor cold start problem, the user being new to the system, the system does not have any history/preference of that user and the common solution to handle that would be showing him the most popular items (that can filtered on location-based features). While product cold start problem happens because products are recently added to the items database. This problem is often resolved by content-based filtering (initially start with feature engineering of items and then recommending based on item similarity and as user activities increases for that product then later those are used).

Other challenges are **privacy** and **shilling attacks** – feeding user preference may lead to data privacy issues, cryptographic methods can be used by providing personalized recommendation. Technique like randomized perturbation which allows users to publish their personal preferences without exposing their personal identities [12].

Shilling attacks happen when someone enters the system and tries to change the ratings (by giving false ratings) to influence the popularity of the items which recommender picks. These are more common to collaborative-based filtering. The common type of attacks such as probe, bandwagon, average attack, segment and random attack, with few algorithmic approaches are discussed in [13].

5.5 Comparative Study

To a deeper look, one can observe that content-based has the capacity to work with small amounts of data, so it is useful in the environment which has just started to use the recommendation systems. Further, while a content-based approach provides the recommendation based on a specific user, it can suffer the phenomena of over-personalization. On the other hand, in collaborative-based filtering, other user's scores are used but it needs a large amount of data. Hence, depending on the needs, current business scenario and availability of data plays a vital role in deciding the choice of approach in the recommender system.

6 Tools

There are many tools for visualizing data such as QlikView, Tableau, Microsoft PowerBI, Google Data Studio, SAS Business Objects, IBM Cognos, Sisense, Birt, Apache Zeppelin, SpagoBi. Most of these tools are designed for all segments of organizations (small/medium/large). In modern times, most of these tools also provide the software as a service (SaaS) offering. SaaS offering of tools enables users to play around the tools, evaluate them as per their needs by doing proof of concepts. Post proof of concepts, users are able to choose right business analytics visualization tools that suit their needs and budget. Many of these tools also provide student editions that can be used in academic settings for learning and experimentation purposes.

As per Gartner report 2020 [14], these tools are evaluated on completeness of vision vs. ability to execute (as illustrated in Fig. 10).

The above image is from the work summary of Gartner research.

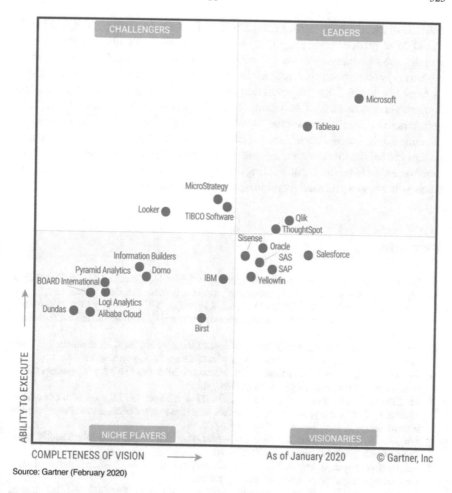

Fig. 10 Gartner quadrant for analytics and business intelligence platforms [14]

7 Conclusion

An attempt has been made to introduce business analytics in this chapter, with various processes that are in practice at industry. As mentioned, before, business analytics enables a look-forward perspective to have a possible future glimpse of the possibilities and opportunities. In this chapter, business analytics types have been introduced, and these types represent the stage at which a specific application might lie in. Also, a glance at various applications across domains (such as Telecommunication, Healthcare, Transportation, Retail Analytics, Cyber Analytics, Marketing Analytics, HR Analytics and Banking/Finance) has been listed. Post that, recommendation systems are discussed with its types (content-based and collaborative filtering) and its practical problems such as data sparsity, and cold start is elaborated. In today's world, one type of recommender system may not work for a complex application and hybrid would be an obvious choice in those cases. Though recommendation systems were initially used in many e-commerce websites (such as Amazon, Flipkart, Snapdeal, etc.), today recommendation systems have become a basic need for every other type of business domain. Recommendation system has become an important area for demand generation as well, and demand generation data is further used by the sales and marketing team to increase the overall profitability of the business. One should be planning on the business needs and define the problems to be solved before deciding on the types of recommendation system before finalizing the type to be used. Overall, the choice should be made considering the user in center and considering the benefits (such as better search/match/combine/compare capabilities) to them.

References

1. Cross-industry standard process for data mining. Wipkipedia, December 16, 2019. Available https://en.wikipedia.org/wiki/Cross-industry_standard_process_for_data_mining.
2. Introduction to SEMMA. SAS Institute Inc., August 30, 2017. Available https://documentation.sas.com/?docsetId=emref&docsetTarget=n061bzurmej4j3n1jnj8bbjjm1a2.htm&docsetVersion=14.3&locale=en.
3. What is the team data science process? Microsoft, January 10, 2020. Available:https://docs.microsoft.com/en-us/azure/machine-learning/team-data-science-process/overview
4. Analytics solutions unified methods. IBM. March 01, 2016. Available ftp://ftp.software.ibm.com/software/data/sw-library/services/ASUM.pdf.
5. MacKenzie, I., C. Meyer, and S. Noble. 2013. How retailers can keep up with consumers. McKinsey & Company. Available https://www.mckinsey.com/industries/retail/our-insights/how-retailers-can-keep-up-with-consumers.
6. Wu, J. 2019. Types of recommender systems. Medium. Available https://medium.com/@jwu2/types-of-recommender-systems-9cc216294802.
7. Recommender system. Wikipedia, July 22, 2020. Available https://en.wikipedia.org/wiki/Recommender_system#Hybrid_recommender_systems.
8. Heckert, A. 2017. Minkowski distance. National Institute of Standard and Technology. Available https://www.itl.nist.gov/div898/software/dataplot/refman2/auxillar/minkdist.htm.
9. Sharma, N. 2019. Importance of distance metrics in machine learning modelling. Towards Data Science. Available https://towardsdatascience.com/importance-of-distance-metrics-in-machine-learning-modelling-e51395ffe60d#:~:text=A%20distance%20function%20provides%20distance,formula%20used%20by%20distance%20metrics..
10. Mahalanobis distance. Wikipedia, July 31, 2020. Available https://en.wikipedia.org/wiki/Mahalanobis_distance.
11. Guo, G. 2012. Resolving data sparsity and cold start in recommender systems. In 20th international conference, UMAP 2012, Montreal, Canada.
12. Polat, H., and W. Du. 2003. Privacy-preserving collaborative filtering using randomized perturbation techniques. *Electrical Engineering and Computer Science*, 625–628. https://doi.org/10.1109/ICDM.2003.1250993.
13. Mobasher, B., R. Burke, R. Bhaumik, and J.J. Sandvig. 2007. Attacks and remedies in collaborative recommendation. *Intelligent Systems, IEEE, 22*: 56–63. https://doi.org/10.1109/MIS.2007.45
14. Richardson, J., K. Schlegel, B. Hostmann, and N. McMurchy. 2020. Gartner magic quadrant for analytics and business intelligence platforms. *Gartner Research*

Challenges and Issues of Recommender System for Big Data Applications

Chandrima Roy and Siddharth Swarup Rautaray

Abstract The knowledge overload is a big problem in today's world. In reality, the information overload implies the availability of so much data or knowledge that goes beyond the user's manageable limits and causes a great difficulty in all kinds of decision taking. The main reason we need a recommender system in modern society is that because of the proliferation of the Internet, people have so many options to choose from. A recommender system refers to a system that can predict a user's future preference for a set of items and recommends the top items. It is a knowledge retrieval application that enhances accessibility as well as efficiently and effectively suggests relevant items to users by considering the user interests and preferences. A recommendation framework tries to tackle the problem of overloading information. There are so many other instances such as these, where we have plenty of data, but we can't decide what we want. Even though volume of information has increased, a new problem has arisen as people have had difficulty selecting the items they actually want to see. Recommenders systems have the ability to change the way websites interact with users and allow businesses to optimize their Return on Investment (ROI); based on the information, they can collect on the preferences and purchases of each customer. A traditional recommendation system cannot do its work without sufficient information, and big data offers plenty of user data such as past transactions, browsing history, and reviews for recommendation systems in order to provide accurate and efficient recommendations. In short, even the most advanced recommenders without big data can't be successful. This work primarily discusses and reflects on current issues, challenges, and research gaps in the production of high-quality recommender systems. Such problems and challenges will present new paths for study, and the target can be accomplished for high-quality recommender systems. The entire research is broken into major tasks including the study of state-of-the-art approaches for recommender system and big data applications; overcoming the problem of cold start,

C. Roy (✉) · S. S. Rautaray
Kalinga Institute of Industrial Technology (KIIT) Deemed to be University, Bhubaneswar, Orissa, India
e-mail: chandrima.roy.1914@gmail.com

S. S. Rautaray
e-mail: siddharthfcs@kiit.ac.in

S. Rautaray et al. (eds.), *Trends of Data Science and Applications*,
Studies in Computational Intelligence 954,
https://doi.org/10.1007/978-981-33-6815-6_16

327

scalability, and building a proactive recommender system; this research considers the process of development for a generic intelligent recommender system that can be work on more than one domain; it also expands the basic recommender program definition.

Keywords Recommender system · Big data · Challenges · Cold start · Scalability · Proactive

1 Introduction

A recommendation system aims to tackle the problem of information overloading. In reality, the explosion of information implies the presence of a lot of data or information that goes beyond the user's manageable limits and makes it very difficult to make decisions of any sort. This challenge arises mainly when the system is unable to handle and process this large volume of knowledge in a structured manner [1]. These are the standard tools effectively implemented in many commercial surroundings such as Flipkart, Netflix, Trip Advisor, and so on. For instances, the user typically gets a lot of choices in many e-commerce web applications, with very little time to explore it all. There are also other cases like this, where we have a lot of data but we can't really determine what we want. While there has been an improvement in the amount of knowledge available, a new problem has arisen, as it has become difficult for consumers to choose the items they really want. So that's where the system of recommendations comes into play. Big data is the engine of recommendation systems. In short, without big data, even the most advanced recommenders can't be successful. This work addresses and focuses mainly on current problems, difficulties, and research gaps in delivering high-class recommendation framework. These problems and obstacles will present new research paths, and the aim for high-quality recommender systems can be reached [2]. One disadvantage of the new network of recommenders is that they are not portable. A variety of recommendation systems are currently being developed for different industries but they are only adequate to fulfill the information needs of the users on that particular application domain. The systems vary from recommending music, events, diseases, and furniture to dating profiles. User browse to listen to a song from "Spotify," but user can't decide which one. User browse "YouTube" to watch some videos, but user can't decide which one to watch. For example, the number of movies that can be put in a blockbuster store which depends on the store's size. In comparison, the Internet nowadays enables people to access plenteous online resources. For example, "Netflix" has a vast collection of movies.

In reality, the information overload means the existence of so much data or knowledge that goes beyond the user's controllable bounds and causes a great effort in all kinds of decision taking [3]. This problem arises mostly when this huge amount of information cannot be handled and processed by the system in a systematic manner.

For example, a lot of choices are usually given to the user in many e-commerce web applications, but with very little time to discover them all.

A Recommender System (RS) may be described as a collection of software program and techniques that provide users with start updates for certain things that might be of use to them. Such recommendations help users make routine decisions [4]. For example, which film to watch, which songs to listen to, which novel to buy, and what reports to read, etc. Typically, a recommendation system is pictured as a platform for information exploration because it has the ability to create new and expanded recommendations to suit its user's needs. A recommender system mainly offers its facilities to two types of companies involved in e-commerce; one is the service provider whose goals to optimize its revenues; the other is the consumer who really likes its advantages in quality decision-making with the presence of a variety of relevant data based on the knowledge or expectations of others.

1.1 Recommendation System Architecture

Customers interact with recommendation systems via a web portal or mobile user interface where a profiler extracts consumer information based on feedback from explicit and implied methods; the interest of customers in various products is predicted through ranking algorithms which provide a list of pro-positioned items based on their calculated custom relevance. The data structure of the recommendation system is basically based on a database which stores item description and customer feedback and is constantly updated (Fig. 1). Because of the clustering and filtering features, recommendation systems are commonly used within e-commerce. They direct consumers to find new and unexpected items that are not discovered by themselves [5].

Taking ebay as a case, an Ajax event is triggered every time a user visits the web and clicks on a product. This event will generate an entry database—"When User X viewed Product Y" Where do we get the User details? If he is a user registered, then we get it from HTTP Session or we extract it from the cookie scheme. If it depends on cookies, then the recommendation only works as long as the user visits the site from the same browser. If a customer adds a product to his shopping cart, then orders a product, likes a product on Facebook or shares a summary of a product, identical events are shot. The website now has data obtained for the user, and it knows what he had been viewing, how often he was watching, what brand he may be interested in, what item he really likes, and so on.

1.2 Big Data

By day-to-day activities and interaction between user and computers, data is increasing exponentially. The generation of data is not restricted to the particular

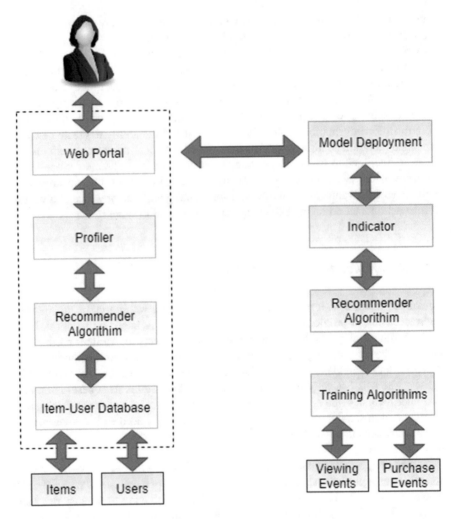

Fig. 1 General design of recommendation system [11]

domain, and any technique with which the data is collected, stored, and retrieved. Big data refers to the growing, complex sets of information that expands at rising levels [6]. This includes the amount of data, the size, or speed at which it is generated and gathered and the variety or probability of the enclosed data points. Big data often starts out from different sources and exists in various formats. The Internet of Things (IoT), the Cloud, Facebook, and Artificial Intelligence (AI) drive data sophistication through various types and sources of data. Big data, for example, comes from computers, video/audio, sensors, transactional software, networks, the online social networking sites—all of it was generated in real time and on a very huge scale [7]. The word "big data" can be defined as data that gets so large that traditional methods

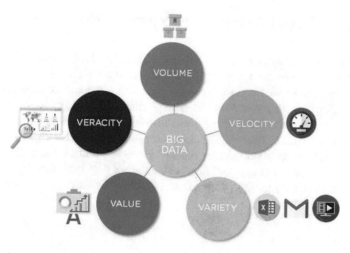

Fig. 2 5V's of big data

cannot be used to process it. The amount of data called big data is a continuously changing factor and the continuous development of new methods for handling such big data. It fully changes our planet and does not mean that it is a phenomenon that will vanish in near future at any moment. It is also broken down by five vs. (Velocity, Volumes, Value, Variety, and Veracity) to make sense of the massive amount of data (Fig. 2).

In fact, big data dictates what recommendations do. Without the continuous provision of data, recommendations can't do a thing [8]. But the role of big data goes beyond numbers. The operations require a CPU which can run for hours. It can be worked with Hadoop to do this. The role of big data can therefore be summarized in the provision of appropriate, fast-acting data, and the necessary setup for the quick processing of the data [9]. Current systems aren't designed to handle such huge data volumes quickly. But getting only big data won't be enough to send strong recommendations.

The objectives of this paper can be stated as: (1) providing the general principles and techniques associated with different types of recommender systems (2) identifying and addressing some of the prominent problems and challenges in designing and developing a high-quality recommender system, such as cold start, scalability, and proactive recommender system.

Rest of the paper is structured as Sect. 2 presents general concepts and techniques relevant to cold start problem, Sect. 3 presents problems and challenges of scalability, Sect. 4 presents several approaches and techniques to produce a proactive recommender system, and finally, Sect. 5 concludes our discussion.

2   The Cold Start Problem in Recommendation

Cold start means the system doesn't have sufficient knowledge to make recommendations for a new user or a new item [10]. Any recommendation system will face cold start problems with the introduction of new client, product, or software. A new product registry or an introduction of new products makes it impossible to introduce an item to a customer as less information is available. The collective filtering cannot efficiently make recommendations for new user and new object event [11]. The content-based recommendation method is the solution to this problem, since it does not rely on the ranking of products. Another way is to identify visitors as browsers that are only there to search items. Two basic types of cold start problem are.

2.1   New User Cold Start Problem

The term cold start means that the program does not have sufficient information to suggest a new user or a new object. With the addition of new user, an object or a program, any recommendation program will face cold start problem [12]. Limited number of information is available which makes it problematic to recommend an item to a user. The issue of cold start is linked to the sparseness of information (i.e., for users and items) available in the algorithm recommendation.

2.1.1   Approaches for Cold Start User Mitigation Technique

The model below focuses on mitigating the new user cold start problem, where a new user is asking for suggestions, and there is no data about their preferences [13]. Rating is very important, as it indicates a particular user's preferences. Additionally, there are no historical records.

The first step is responsible for identifying the new consumer into a particular category. For the classification, this model will use J48 algorithm represented in Fig. 3. The algorithm generates a decision tree based on a greedy top-down approach in which it considers each attribute that can distinguish all instances [14]. This process continues in a recursive fashion until the classification of all the results. The difference in entropy levels between attributes is used to select the best classification attribute.

During the second step, the algorithm uses a technique to locate the new user's "neighbors." This model will examine the user's important characters, and it will look for other users within the class that matches the new user best. The final result is determined in the third step. It is achieved by introducing statistical methods for predicting the new user's ratings. This model will handle the new user cold start problem.

Algorithm J48 classification
Input : an attribute-valued dataset D

```
1  :   Tree = {}
2  :   if D is "pure" OR other stopping criteria met then
3  :       terminate
4  :   end if
5  :   for all attribute a ∈ D do
6  :       Compute information-theoretic criteria if we split on a
7  :   end for
8  :   a_best = Best attribute according to above computed criteria
9  :   Tree = Create a decision node that tests a_best in the root
10 :   D_v = Induced sub-datasets from D based on a_best
11 :   for all D_v do
12 :       Tree_v = J48(D_v)
13 :       Attach Tree_v to the corresponding branch of Tree
14 :   end for
15 :   return Tree
```

Fig. 3 Pseudocode for J48 algorithm

The theory is that people with a common culture would most possibly have similar preferences [15]. A model will be implemented from the training data. Figure 4 represents the model for this process. The goal is to find a neighborhood, where the neighbors are users that belong to the same group as the group predicted by the model. After this step, similarity index will be calculated that combines similarity from the neighbors. Finally, the similarity measure and neighbors' ratings are combined to get predictions.

2.2 New Item Cold Start Problem

The issue with the item cold start will cause the new item to miss the opportunity to be recommended and stay "cold" all the time. The proposed model would like to suggest new products to potentially interested users—for which no interests have been expressed so far [16].

The method employs interrelationships mining. It can derive functionality based on a comparison of various attribute values. One of the main concepts of interrelation mining is to reflect similar characteristics based on a comparison of new attributes called interrelated attributes between values of different attributes. First, most similar item will be chosen to include a new item's neighborhood, and neighborhood rating information will be used to estimate the new item's rating value. Lastly, highest rating product will be recommended to a target customer.

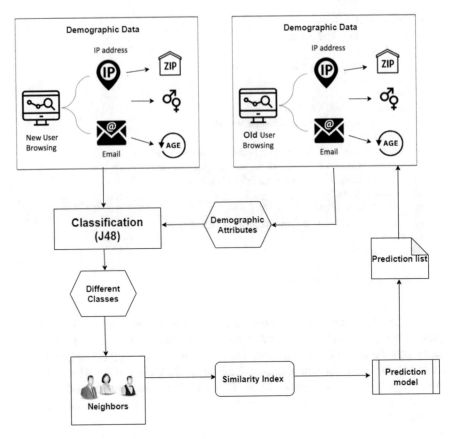

Fig. 4 Model for new user cold start problem

2.2.1 Approach for Cold Start Item Mitigation Technique

Step 1: Extracting the Relation Between Each Pair of Attributes

The purpose of this task is to extract semantic associations between pairs of co-occurring entities. This approach is mainly organized in two stages. First is defining instances of candidate relationships and then classifying instances of candidates into true and invalid ones. Classifiers are based upon training corpus as a standard classification process in which all relationships and their arguments were manually annotated [17]. The annotated relationships are viewed as positive examples of training, while the majority of the non-labeled co-occurring pairs are considered derogatory examples of training.

Step 2: Calculating Similarity Index Between the Attributes

The simplest way to determine linear dependency between two or more attributes is to use correlation. In a large set of attribute values, set of each attribute values can be treated as a vector. The scalar product of two vectors (with subtracted mean value) and normalized with both norms is a good measure of similarity [18]. When the value is 1, they are equal to a scale factor; when the value is zero, they are orthogonal.

$$R = a.b/(\|a\|\|b\|)$$

where R is correlation, and a, b are observed vectors or sets of attributes

$$a = (a_1, a_2, a_3, a_4, \ldots, a_n)b = (b_1, b_2, b_3, b_4 \ldots b_n)$$

Step 3: Selection of High Rated Item

At the last stage, the model will predict the top N high rated items which will be recommended to the user.

3 Scalability

In recent years, the rapid growth of customers and goods presents some critical challenges for recommendation systems that produce high-quality recommendations for millions of customers and goods and conducting several recommendations every second. The outcome is a rich archive of information which can be used if managed correctly to boost the user experience and rise their involvement. Both existing large-scale recommender systems in the real world have developed in scalable distributed database systems for data management and distributed and parallel recommendation finding algorithms [19].

3.1 *Scalable Neighborhood Algorithm*

Scalability is a key factor in determining the type of recommender systems that should be used. More complex systems require more people to build/maintain, potentially harder to hire, with greater hardware costs. This can be a long-term investment, and thus, business should realize the incremental business advantage versus the increased expense. Having said that here are some core elements of creating scalable systems. Offline batch processing and serving online. For a large number of users and products, one has to measure offline by lot easily obtainable recommendations.

The recommender framework for YouTube split the modeling process into two phases. During the first step, only user-item interaction data is used to pick hundreds out of millions of candidates. For more selection and ranking, it is then feasible to use more information about candidate videos in the second process. Although softmax or other functions are used for training in the output layer, the probability does not have to be measured during real-time serving time, and the closest-neighbor method can be used on the output from the last hidden layer [9].

3.1.1 Scalability in Data Management

Database Management Systems (DBMS) play a key part in social networking applications, where vast volumes of data generated and aggregated at various, geographically separate places need to be stored, extracted, and distributed in an efficient manner [20]. Centralized relational DBMSs are built to ensure uniformity but can scale-up to the needs of global social networks, because of this Twitter and Facebook have been exploring different distributed and NoSQL data storage systems like MongoDB, Megastore, Cassandra [21]. Relational database management systems are "not cloud friendly" because of the connections and interactions between stored data, even when running using cloud structure, NoSQL databases will scale out better. However, they also face many challenges which have to do with data partitioning and replication, and for this reason, middleware solutions have been developed.

3.1.2 Scalability Using Deep Learning for Big Data

The use of hybrid filtering is another way of confronting sparsity, cold start, scalability, and other challenges [22]. This approach includes the combination of various recommendation development techniques, seeking to increase prediction accuracy while highlighting shortcomings of the individual methods.

In order to learn parameters, traditional deep neural networks need vast quantities of data, which is a computationally intensive process that takes considerable time to train a model. When the data size grows exponentially and the deep learning models become more complex, more computational power and memory are required, such as high-performance computing (HPC) resources to train a model of accuracy in a timely manner. Given the ongoing efforts to train and infer deep learning models to increase competitiveness, many current deep learning algorithms are notoriously difficult to scale and parallelize due to inherent interdependencies in the computational steps as well as training data. Clustering techniques work by identifying user groups which tend to have similar tastes. By combining the views of the other users within that cluster, predictions can be made for a person once the clusters are formed. Many clustering techniques, with partial participation, represent each consumer in multiple clusters. Hence, the prediction is an average over the clusters, weighted by degree of involvement. Clustering strategies usually generate less personal recommendations than other methods, often resulting in less accuracy than neighboring algorithms [23]. Figure 5 represents the neighborhood formation after clustering.

3.2 Algorithm: Clustered Neighborhood Formation

Step 1. Using the training dataset to construct user partitions using the clustering algorithm.

Step 2. Determine a given user's neighborhood.

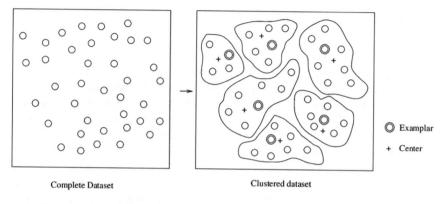

Complete Dataset   Clustered dataset

○ Examplar

+ Center

Fig. 5 Neighborhood formation after clustering

Step 3. Upon obtaining the neighborhood, the classic collaborative filtering algorithms is used to generate prediction from it.

The approach has two advantages—first, it decreases the sparsity of the data collection, and second, the prediction generation is much quicker due to the reduction of dimensionality and the use of a static precomputed neighborhood.

4 Proactive Recommender System

A proactive system would be able to deliver highly relevant information and determine at what point user need any item or information and at what moments information needs are more critical and less disruptive. Recommender systems main aim is to mitigate the effects of overloading information/choice. Normally, recommendations are prepared based on anticipating what's going to be useful or relevant for users. Therefore, it is vital that they are relevant to consumers, whether in relation to their information needs, current behavior or emotional state. This provides a detailed understanding of the users "context but also knowledge of the past of previous users" interactions within the system. A lot of existing systems need active user participation and focus. Recommendation systems manage the distribution of recommendations without taking into account the users' needs when, when, or how the recommendations are made. Proactive recommendation systems offer a more proactive approach to recommendations delivery by pre-identifying information needs and working on behalf of users with minimal effort and without interruption [24].

4.1 Proactive Recommendation

Recommended systems are a well-established and effective solution to reduce the overload of choices as customers face a range of materials, goods, applications, and services, etc. In the ideal scenario, users receive highly relevant and interesting feedback (products) while organizations (system providers) benefit from increased user interaction. Many current recommender systems need active user interaction and attention. Recommender system manages the distribution of recommendations without taking into account the expectations of users on when, where, or how the recommendations are delivered. There is still growing competition for consumer attention with an explosion of applications/services. PS aims to predict consumer needs ahead of time and then take effective action on behalf of them. Actions are typically linked to the delivery at the right time, location, etc., of the appropriate content. Therefore, if appropriate, then proactive recommenders systems should include specific suggestions, without disrupting users in any way possible [25].

Constructive systems can create highly interactive environments in which they can assist and improve user interaction and provide comfort and assistance when needed. Overall, consumers demonstrate a strong interest in implementing and following constructive suggestions. Practical implementation needs to be treated with caution, because some users feel that they will become too dependent on these systems, while others also refrain from using them because of unwanted interruptions and frustration.

4.2 Intelligent Proactive Recommender System

This strategy incorporates information relating to a user gathered from their social networking sites, along with mobile devices, to proactively suggest relevant information to the consumer. The recommendation process in Fig. 6, therefore, includes a background model that specifies when and when to suggest the relevant information to the consumer (Books, music, news, movies).

User-related information, along with low-level information, is leveraged from the user's mobile device, laptop, or tab/iPad, to infer the user's condition. The approach we are putting forward aims to suggest specific things that suit the situation of a user without waiting for the user to initiate an operation. We consider that the daily routine of the user is defined as a collection of situations and the profile of the user and organized within a knowledge database that represents a particular category of information [26].

After this knowledge, extraction can be done from the collected data about the user. Mostly, three categories of knowledge will be extracted, which are user profile: User information and associated interests; Location: User location taken from GPS coordinates; Time: numeric or intermittent labels (morning, Evening, night). To capture

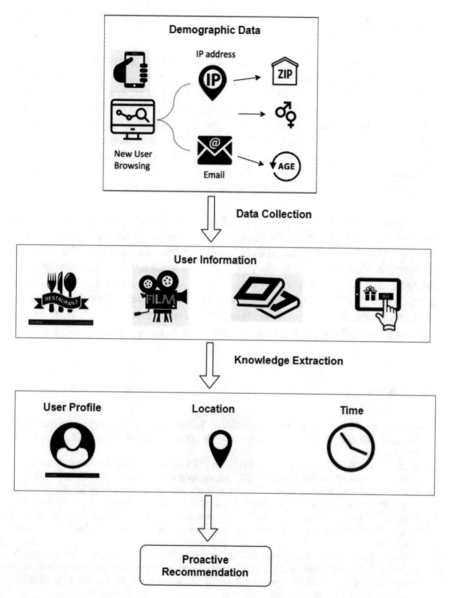

Fig. 6 A general overview of intelligent proactive recommender system

the context, these dimensions are instantiated using the sensors embedded in the mobile device of the user.

5   Conclusion

In the past few years, recommender systems have been used to alleviate knowledge and cognitive overload problems by recommending relevant and applicable items to users. Several improvements have been made in this regard to achieve a high quality, finely tuned recommendation system. This study aimed on these critical issues and challenges and proposed three models to mitigate these problems such as cold starting, scalability, and proactive recommendation system. Scaling according to the user's need and yet getting the recommendation device optimized is still an order of the day. For various applications which generate "Big Data," a new recommendation method is proposed in this research that has the ability to process heterogeneous data.

Currently, several recommendation systems are built for various sectors but they are only sufficient to satisfy the users' information needs on that specific application domain. One limitation of current recommendation system is that they are not portable. This research considers the process of development for an adaptive intelligent recommendation system that can be work on more than one domain that also extends the concept of basic recommendation system. To use dynamically auto-updated data from external websites, this recommendation framework can be designed to process complex data at the same time as active user requests arrive.

References

1. Ramzan, B., I.S. Bajwa, N. Jamil, R.U. Amin, S. Ramzan, F. Mirza, and N. Sarwar. 2019. An intelligent data analysis for recommendation systems using machine learning. *Scientific Programming*.
2. Fu, M., H. Qu, Z. Yi, L. Lu, and Y. Liu. 2018. A novel deep learning-based collaborative filtering model for recommendation system. *IEEE Transactions on Cybernetics* 49 (3): 1084–1096.
3. Zhang, Y. 2016. GroRec: A group-centric intelligent recommender system integrating social, mobile and big data technologies. *IEEE Transactions on Services Computing* 9 (5): 786–795.
4. Sabitha, S., and T. Choudhury, T. (2018). Proposed approach for book recommendation based on user k-NN. In *Advances in Computer and Computational Sciences*, 543–558. Springer: Singapore.
5. Chen, J., H. Zhang, X. He, L. Nie, W. Liu, and T.S. Chua. 2017, August. Attentive collaborative filtering: Multimedia recommendation with item-and component-level attention. In *Proceedings of the 40th International ACM SIGIR Conference on Research and Development in Information Retrieval*, 335–344.
6. Roy, C., S.S. Rautaray, and M. Pandey. 2018. Big data optimization techniques: a survey. *International Journal of Information Engineering & Electronic Business*, *10* (4)
7. Shu, J., et al. 2018. A content-based recommendation algorithm for learning resources. *Multimedia Systems* 24 (2): 163–173. https://doi.org/10.1007/s00530-017-0539-8.
8. Wang, D., Y. Liang, D. Xu, X. Feng, and R. Guan. 2018. A content-based recommender system for computer science publications. *Knowledge-Based Systems* 157: 1–9.
9. Sarwar, B.M., G. Karypis, J. Konstan, and J. Riedl. 2002, December. Recommender systems for large-scale e-commerce: Scalable neighbourhood formation using clustering. In: *Proceedings of the Fifth International Conference on Computer and Information Technology*, vol. 1, 291–324.

10. Sardianos, C., N. Tsirakis, and I. Varlamis, 2018. A survey on the scalability of recommender systems for social networks. In *Social Networks Science: Design, Implementation, Security, and Challenges*, 89–110. Cham: Springer.

11. Roy, C., M. Pandey, and S. Swarup Rautaray, 2018, April. A proposal for optimization of data node by horizontal scaling of name node using big data tools. In *2018 3rd International Conference for Convergence in Technology (I2CT)*, 1–6. IEEE.

12. Deng, H. 2019. *Recommender Systems in Practice*.https://towardsdatascience.com/recomm ender-systems-in-practice

13. Bobadilla, J., F. Ortega, A. Hernando, and A. Gutiérrez. 2013. Recommender systems survey. *Knowledge-Based Systems* 46: 109–132.

14. Maqsood, T., O. Khalid, R. Irfan, S. Madani, and S. Khan. 2016. Scalability Issues in Online Social Networks. *ACM Computing Surveys (CSUR)* 49 (2): 40.

15. Peska, L., and H. Trojanova. 2017. Towards recommender systems for Police Photo Lineup. arXiv:1707.01389.

16. Smirnova, E., and F. Vasile. 2017. Contextual sequence modeling for recommendation with recurrent neural networks. arXiv:1706.07684.

17. Roy, C., M. Pandey, and S.S. Rautaray. 2018. A proposal for optimization of horizontal scaling in big data environment. In *Advances in Data and Information Sciences*, 223–230. Singapore: Springer

18. Chatzis, S., P. Christodoulou, and A. Andreou. 2017. Recurrent latent variable networks for session-based recommendation. arXiv:1706.04026.

19. Zanotti, G., M. Horvath, L. Barbosa, V. Immedisetty, and J. Gemmell. 2016. ACM: Infusing collaborative recommenders with distributed representations. In *Conference Proceedings of the 1st Workshop on Deep Learning for Recommender Systems* 35–42.

20. Braunhofer, M., F. Ricci, B. Lamche, and W. Wörndl. 2015, August. A context-aware model for proactive recommender systems in the tourism domain. In *Proceedings of the 17th International Conference on Human-Computer Interaction with Mobile Devices and Services Adjunct*, 1070–1075.

21. Melguizo, M.C.P., L. Boves, and O.M. Ramos. 2009. A proactive recommendation system for writing: Helping without disrupting. *International Journal of Industrial Ergonomics* 39 (3): 516–523.

22. Sabic, A., and M. Zanker. 2015. Investigating user's information needs and attitudes towards proactivity in mobile tourist guides. In *Information and Communication Technologies in Tourism 2015*, 493–505. Cham: Springer.

23. Sabic, A. 2016, September. Proactive recommendation delivery. In *Proceedings of the 10th ACM Conference on Recommender Systems*, 459–462.

24. Roy C., K. Barua, S. Agarwal, M. Pandey, S.S. Rautaray. 2019. Horizontal scaling enhancement for optimized big data processing. In: A. Abraham, P. Dutta, J. Mandal, A. Bhattacharya, S. Dutta (Eds.), *Emerging Technologies in Data Mining and Information Security. Advances in Intelligent Systems and Computing*, vol 755. Singapore: Springer. https://doi.org/10.1007/978-981-13-1951-8_58

25. Kermi, I. 2017. *A hybrid model for context-aware proactive recommendation* (Doctoral dissertation). University of Toulouse, University of Toulouse III-Paul Sabatier.

26. Motadoo, S. 2018. Resolving cold start problem using user demographics and machine learning techniques for movie recommender systems.

Printed in the United States
by Baker & Taylor Publisher Services